GREEN POLYMER CHEMISTRY AND COMPOSITES

Pollution Prevention and Waste Reduction

GREEN POLYMER CHEMISTRY AND COMPOSITES

Pollution Prevention and Waste Reduction

Edited by
Neha Kanwar Rawat, PhD
Iuliana Stoica, PhD
A. K. Haghi, PhD

First edition published 2021

Apple Academic Press Inc.
1265 Goldenrod Circle, NE,
Palm Bay, FL 32905 USA
4164 Lakeshore Road, Burlington,
ON, L7L 1A4 Canada

CRC Press
6000 Broken Sound Parkway NW,
Suite 300, Boca Raton, FL 33487-2742 USA
4 Park Square, Milton Park,
Abingdon, Oxon OX14 4RN

First issued in paperback 2023

© 2021 Apple Academic Press, Inc.

Apple Academic Press exclusively co-publishes with CRC Press, an imprint of Taylor & Francis Group, LLC

Library and Archives Canada Cataloguing in Publication

Title: Green polymer chemistry and composites : pollution prevention and waste reduction / edited by Neha Kanwar Rawat, PhD, Iuliana Stoica, PhD, A.K. Haghi, PhD.
Names: Rawat, Neha Kanwar, editor. | Stoica, Iuliana, editor. | Haghi, A. K., editor.
Description: First edition. | Includes bibliographical references and index.
Identifiers: Canadiana (print) 20200387944 | Canadiana (ebook) 20200388002 | ISBN 9781771889377 (hardcover) | ISBN 9781003083917 (ebook)
Subjects: LCSH: Polymers. | LCSH: Polymerization. | LCSH: Green chemistry. | LCSH: Sustainability.
Classification: LCC QD381 .G74 2021 | DDC 668.9/20286—dc23

Library of Congress Cataloging-in-Publication Data

CIP data on file with US Library of Congress

ISBN: 978-1-77188-937-7 (hbk)
ISBN: 978-1-77463-779-1 (pbk)
ISBN: 978-1-00308-391-7 (ebk)

DOI: 10.1201/9781003083917

About the Editors

Neha Kanwar Rawat, PhD

Researcher, Materials Science Division,
CSIR-National Aerospace Laboratories, Bangalore, India

Neha Kanwar Rawat, PhD, is a recipient of a prestigious DST Young Scientist Postdoctoral Fellowship and is presently a researcher in the Materials Science Division, CSIR-National Aerospace Laboratories, Bangalore, India. She received her PhD in chemistry from Jamia Millia Islamia (a Central University), India. She has published numerous peer-reviewed research articles in journals of high repute. Her contributions have led to many chapters in international books published with the Royal Society of Chemistry, Wiley, Elsevier, Apple Academic Press, Nova U.S.A., and many others in progress. She has been working on many prestigious research and academic fellowships in her career. She is a member of many groups, including the Royal Society of Chemistry and the American Chemical Society (USA) and a life member of the Asian Polymer Association. Email: neharawatjmi@gmail.com

Iuliana Stoica, PhD

Scientific Researcher, Department of Polymer Materials Physics,
"Petru Poni" Institute of Macromolecular Chemistry, Romania

Iuliana Stoica, PhD, is currently a Scientific Researcher in physics at Romanian Academy, "Petru Poni" Institute of Macromolecular Chemistry, Department of Polymer Materials Physics. She received her PhD from the Romanian Academy, Department of Polymer Physics and Structure at the same institute. She joined a postdoctoral fellowship program at Politehnica University of Bucharest, Faculty of Applied Chemistry and Materials Science, Department of Bioresources and Polymer Science in 2014. Her area of scientific activity is focused on characterization of a wide range of polymers, copolymers, polymeric composites, and polymeric mixtures. She was a main or co-author for over 95 papers in peer-reviewed ISI journals, and she has contributed as an author to several book chapters in the field of polymer and material science. She was a member of the organizing and program committees of several scientific conferences. She was also a reviewer for a number of prestigious journals in the field of polymer science. Email: stoica_iuliana@icmpp.ro

A. K. Haghi, PhD

Editor-in-Chief, International Journal of Chemoinformatics and Chemical Engineering and Polymers Research Journal;
Member, Canadian Research and Development Center of Sciences and Cultures (CRDCSC), Montreal, Quebec, Canada

A. K. Haghi, PhD, is the author and editor of 200 books, as well as 1000 published papers in various journals and conference proceedings. Dr. Haghi has received several grants, consulted for a number of major corporations, and is a frequent speaker to national and international audiences. Since 1983, he served as professor at several universities. He served as Editor-in-Chief of the *International Journal of Chemoinformatics and Chemical Engineering* and *Polymers Research Journal* and is on the editorial boards of many International journals. He is also a member of the Canadian Research and Development Center of Sciences and Cultures (CRDCSC), Montreal, Quebec, Canada. He holds a BSc in urban and environmental engineering from the University of North Carolina (USA), an MSc in mechanical engineering from North Carolina A&T State University (USA), a DEA in applied mechanics, acoustics, and materials from the Université de Technologie de Compiègne (France), and a PhD in engineering sciences from Université de Franche-Comté (France). Email: AKHaghi@gmail.com

Contents

Contributors

Akanksha Adaval
Department of Metallurgical Engineering and Materials Science, Indian Institute of Technology, Mumbai, India

Sameer Ahmad
Department of Applied Sciences and Humanities, Faculty of Engineering and Technology, Jamia Millia Islamia, New Delhi

Mohd Danish Ansari
Laboratory of Green Synthesis, Department of Chemistry, University of Allahabad, Uttar Pradesh, India

Zubair Aslam
Material Research Laboratory, Department of Physics, Jamia Millia Islamia, New Delhi, India

Andreea Irina Barzic
Laboratory of Physical Chemistry of Polymers, "Petru Poni" Institute of Macromolecular Chemistry, 41A Grigore Ghica Voda Alley, 700487 Iasi, Romania

Shahidul Islam Bhat
Materials Research Laboratory, Department of Chemistry, Jamia Millia Islamia, New Delhi, India

Abu Darda
Department of Applied Sciences and Humanities, Faculty of Engineering and Technology, Jamia Millia Islamia, New Delhi
Materials Research Laboratory, Department of Chemistry, Jamia Millia Islamia, New Delhi, India

Adina Maria Dobos
Department of Physical Chemistry of Polymers, Petru Poni Institute of Macromolecular Chemistry, Iassy, Romania

Anca Filimon
Department of Physical Chemistry of Polymers, Petru Poni Institute of Macromolecular Chemistry, Iassy, Romania

Rimzhim Gupta
Department of Chemical Engineering, Indian Institute of Science, Bangalore, India

A. K. Haghi
Former Editor-in-Chief, International Journal of Chemoinformatics and Chemical Engineering and Polymers Research Journal; Member, Canadian Research and Development Center of Sciences and Cultures (CRDCSC), Montreal, Quebec, Canada. E-mail: akhaghi@yahoo.com

Essia Hannachi
Laboratory of Physics of Materials—Structures and Properties, Department of Physics, Faculty of Sciences of Bizerte, University of Carthage, 7021, Zarzouna, Tunisia

Athar Adil Hashmi
Department of Chemistry, Jamia Millia Islamia, New Delhi 110025, India

Sajid Iqbal
Materials Research Laboratory, Department of Chemistry, Jamia Millia Islamia, New Delhi, India

Mohd Irfan
Conservation Research Laboratory Ajanta Caves, Aurangabad, India

Shama Islam
Material Research Laboratory, Department of Physics, Jamia Millia Islamia, New Delhi, India

Hana Khan
Material Research Laboratory, Department of Physics, Jamia Millia Islamia, New Delhi, India

Halima Khatoon
Materials Research Laboratory, Department of Chemistry, Jamia Millia Islamia, New Delhi, India

Rabia Kouser
Materials Research Laboratory, Department of Chemistry, Jamia Millia Islamia, New Delhi, India

Sushant Kumar
Department of Chemical Engineering, Indian Institute of Science, Bangalore, India

Mihaela Dorina Onofrei
Department of Physical Chemistry of Polymers, Petru Poni Institute of Macromolecular Chemistry, Iassy, Romania

Sukanchan Palit
Department of Chemical Engineering, University of Petroleum and Energy Studies, Dehradun 248007, India
43, Judges Bagan, Haridevpur, Kolkata 700082, India

Ashiq Hussain Pandit
Materials Research Laboratory, Department of Chemistry, Jamia Millia Islamia, New Delhi, India

S. Preethi
Indo MIM Pvt Ltd., Bengaluru, India

Arul Maximus Rabel
Research Microscopy Solutions, Carl Zeiss India, Bengaluru, India

Neha Kanwar Rawat
Researcher, Materials Science Division, CSIR-National Aerospace Laboratories, Bangalore, India.
Email: neharawatjmi@gmail.com

S. Roopa
Department of Polymer Science and Technology, Sri Jayachamarajendra College of Engineering, JSS Science and Technology University, Mysuru 570006, India

Praveen C. Ramamurthy
Department of Materials Engineering, Indian Institute of Science, Bangalore 560012, India

Weqar Ahmad Siddiqi
Department of Applied Sciences and Humanities, Faculty of Engineering and Technology, Jamia Millia Islamia, New Delhi

Priyanka Singh
Department of Physics, Sensors & Signal Processing Laboratory, Institute of Science, Banaras Hindu University, Varanasi 221005, Uttar Pradesh, India

T. Sonamani Singh
Department of Physics, Sensors & Signal Processing Laboratory, Institute of Science, Banaras Hindu University, Varanasi 221005, Uttar Pradesh, India

Taruna Singh
Department of Chemistry, Gargi College, University of Delhi, New Delhi 110049, India

Yassine Slimani
Department of Biophysics, Institute for Research and Medical Consultations (IRMC),
Imam Abdulrahman Bin Faisal University, P.O. Box 1982, 31441, Dammam, Saudi Arabia

B. Sowmya
Department of Polymer Science and Technology, Sri Jayachamarajendra College of Engineering,
JSS Science and Technology University, Mysuru 570006, India

Iuliana Stoica
Scientific Researcher, Department of Polymer Materials Physics,
"Petru Poni" Institute of Macromolecular Chemistry, Romania

Saravanan Subbiahraj
Department of Materials Engineering, Indian Institute of Science, Bangalore 560012, India

Vidya G.
Department of Materials Engineering, Indian Institute of Science, Bangalore 560012, India

R. D. S. Yadava
Department of Physics, Sensors & Signal Processing Laboratory, Institute of Science,
Banaras Hindu University, Varanasi 221005, Uttar Pradesh, India

Mohd Zulfequar
Material Research Laboratory, Department of Physics, Jamia Millia Islamia, New Delhi, India

Abbreviations

AC	activated carbons
ACS	American Chemical Society
AgNPs	silver nanoparticles
ANN	artificial neural network
BAs	biogenic amines
CA	cellulose acetate
CCP	composite conducting polymer
CD	cyclodextrin
CMM	couple monomer methodology
CNCs	cellulose nanocrystals
CNF	cellulose nanofibers
CNTs	carbon nanotubes
CVD	chemical vapor deposition
DBs	degree of branching
DHAP	direct aryl polymerization
DM	double monomer
DMF	dimethylformamide
DSSCs	dye sensitized solar cells
EDLC	electrochemical double-layer capacitor
EPA	Environmental Protection Agency
FET	field effect transistor
FIS	fuzzy inference system
FTIR	Fourier transform infrared
GA	genetic algorithm
GChem	green chemistry
GEng	green engineering CO castor oil
HPs	hyper branched
HPUs	hyperbranched polyurethane
HSCs	hybrid solar cells
IC	inclusion complex
ICA	independent components analysis
JJO	jungle jalebi oil
LDA	linear discriminant analysis
LDH	layered double hydroxides
LO	linseed oil

MEMS	microelectromechanical system
MF	microfiltration
MIMO	multi-input/multi-output
MOFs	metal organic frameworks
MOSFET	metal-oxide-semiconductor field-effect transistor
NEMS	nanoelectromechanical system
NFs	nanofibers
NMR	Nuclear Magnetic Resonance
OFETs	organic-field effect transistors
OSCs	organic solar cells
PCE	photovoltaic efficiency
PCA	principal component analysis
PCL	polycaprolactone
PE	polyethylene
PHBV	polyhydroxybutyrate-co-valerate
PLA	polylactic acid
PLSR	partial least squared regression
PO	palm oil
PU	polyurethane
QCM	quartz crystal microbalance
RBF	radial basis function
RCF	regenerated cellulose film
REE	reduction energy efficiency
RO	reverse osmosis
ROS	resultant reactive oxygen species
RSC	The Royal Society of Chemistry
SAW	surface acoustic wave
SCVP	self-condensed vinylic polymerization
SCROP	self-condensing ring-opening polymerization
SfO	safflower oil
SFO	sunflower oil
SM	single monomer
SO	soy oil
SVM	support vector machine
UF	ultrafiltration
VO	vegetable oils
VOCs	volatile organic compounds
WB	waterborne
WPU	waterborne polyurethanes

Preface

This new book examines the latest developments in the important and growing field of producing conventional polymers from sustainable sources.

With recent advancements in synthesis technologies and the discovery of new functional monomers, research shows that green polymers with better properties can be produced from renewable resources. The book describes these advances in synthesis, processing, and technology. It provides not only state-of-the-art information but also acts to stimulate research in this direction.

This book offers an excellent resource for researchers, upper-level graduate students, brand owners, environment and sustainability managers, business development and innovation professionals, chemical engineers, plastics manufacturers, biochemists, and suppliers to the industry to debate sustainable and economic solutions for polymer synthesis.

The chemical industries play an essential role to sustain the world economies and to reinforce forthcoming technologies and scientific developments in novel products, less toxicological materials, industrial procedures with high efficiency, and renewable energy products. Green chemistry seeks for the design of innovative chemical products with higher efficiency and lowest hazardous substances for the health and the environment. Chapter 1 describes the fundamentals and metrics of green chemistry and their effects on the whole life cycle of chemicals from designs across removal. After reporting the important metrics and the latest developments in the theme within this framework, the nanotechnology case was considered. Nanotechnology offers a useful context to investigate the influences and applications of green chemistry. Interdisciplinary innovations conduct both disciplines, and both aim for transforming the nature of technology. The applications and insinuations of developing green technologies are reviewed, and forthcoming occasions for interdisciplinary associations are discussed.

Vegetable oils (VO) are superabundant, economic, nontoxic, and biodegradable resource, which are applied for surface coatings for the last two decades. Moreover, the adaptation of the ultra-modern techniques coupled with the VOs-based sustainable polymers in the field of surface coatings shows an outstanding physico-chemical, physicomechanical, electrical, and thermal properties. On account of this, Chapter 2 focuses on the classifications, properties, and modifications of the sustainable polymers. Over the past decades, the theory of environment friendly, solventless, UV-curable, waterborne (WB), and hyperbranched coatings that are developed from sustainable resources have been discussed. The chapter further highlights the modification of VO-based polymers and their nanocomposites for enduring sustainable and green future. Moreover, the future scopes of these polymers have also been discussed in detail.

In Chapter 3, it is shown that the key constraint of the applied assimilation of renewable energy technologies to assimilate the power grid is the alternative attributes of renewable energy resources. The avenue against decentralized energy systems that can aftermath an on-demand power to achieve beneath alternative cycles of power bearing requires the addition of accessories and the ability of electronics activated for conversion systems. In this manner, energy storage devices with energy technology can accommodate the temporal adaptability bare to antithesis bounded power of bearing and power of consumption. The energy storage devices ahead use of mechanical force have emerged as a capable technology for the conception of independent systems for multifunctional, maintenance-free, independent operations are called as self-powered electrochemical energy storage devices. A self-charging capacitor as an efficient solar energy storage device was fabricated driven by light. The device which achieves the name the photocapacitor works with at high quantum conversion efficiency which converts or stores the visible light energy in the form of electrical power. The structure of a photocapacitor is based on a multilayered photoelectrode with the presence of an organic electrolyte. Materials comprising dye-sensitized semiconductor nanoparticles/activated carbon particles /hole-trapping layer in contact are responsible for stored photogenerated charges at the electric double layer. Such a photocapacitor would be a more desirable device for storing energy over conventional batteries due to its rapid change to current response and also has high quantum conversion efficiency.

We are very much at the eye of a global crisis like climate change, global warming, and emission of toxic gases into our atmosphere because of the excess usage of synthetic polymers in the packaging industry. Hence the onus is on biopolymers and materials which are superior in their biodegradability and recyclable property. Biopolymers are gaining more attention by academicians, researchers, and in the industries to replace the materials which are derived from fossil fuels. These bio-based polymers provide solutions for environmental issues such as plastic waste management, depletion of fossil fuels, emission of greenhouse and toxic gases, etc. Also because of increasing involvement of the Food and Drug Administration and related organizations in terms of food safety, favorable government policies toward promoting bioplastics and the awareness of consumers about the safety and environmental effect of conventional polymers, the usage of biodegradable and food-grade plastic packaging materials is increasing. Due to these reasons, the demand of bioplastics is increasing along with the production capacity day by day. The total production capacity of bioplastic in 2018 was 2.1 million tons. Polylactic acid (PLA), thermoplastic aliphatic polyester derived from agricultural products such as potato, wheat, and corn starch is one such promising biopolymer. Due to its attractive properties such as strength, stiffness, and biodegradability, PLA is widely used in packaging and biomedical applications. According to a recent market survey, the use of PLA in the packaging industry (over 50% of the global bio-PLA market) dominated the market in 2018 and also it is expected to grow during the forecast period due to its usage in fruit and vegetable packaging, shopping or carrier bags, bread bags, bakery boxes, bottles, etc. (Source: Mordor intelligence; bio-polylactic acid (PLA) market-growth, trends, and forecast (2019–2024)). The literature survey also reveals that the interest has shifted toward the search of biodegradable/bio compostable PLA polymers, blends, and composites as an alternative to conventional polymers in packaging applications. Chapter 4 describes the production, structure–property relationship, chemical and physical modification of PLA to suit packaging applications and its characterization.

The field of "green electronics" is a cross-disciplinary research in which the recognition of naturally-occurring or nature-inspired compounds and the making of environmentally benign synthetic-organic materials that consist of its utility in environmentally safe biodegradable/or biocompatible as *green* for device making. The actual goal of this research field is to produce environmentally viable semiconducting materials and their applications in devices such as organic electronics to carry low energy intensity, low cost, negligible waste production, and also that accomplish good functionalities. Chapter 5 briefly reviews the major research work based on ideologies of green chemistry in the synthetic strategies in making of these semiconducting materials including small molecules and polymers or nature-inspired biocompatible materials and their applications in organic-green electronics.

Chapter 6 intends to show that natural oils are admirable sources of polymer as they are renewable, economical, and have various degrees of unsaturation. Direct polymerization of oils has been used in coatings for a long time. Oils can be thermally polymerized on heating at high temperatures. The products are oligomers of high viscosity. For high strength materials, the introduction of functional groups such as hydroxyls, carboxyls, amines, etc is necessary. Hydroxyls are the most useful ones as they open the whole area of polyurethanes. The triglyceride structure or fatty acid derivatives open a wide range of properties and modification of oils as polymers. Various polymerization methods like cationic, radical, condensation, and metathesis have been applied for the synthesis of polymers.

VO-based WB polymeric nanocomposite coatings emerged as important alternatives to the petro-based polymers due to increasing pressure to limit detrimental health and environmental effects. In this approach, the toxic and expensive volatile organic solvents are replaced by water as an environmentally benign solvent, resulting in minimal volatile organic contents. Waterborne polymeric nanocomposites have attracted great attention in the field of anticorrosive coating materials due to the synergistic effect of polymeric matrix and nanofillers. Chapter 7 deals with the classification of waterborne polymers (WBPS), various synthetic approaches, advantages, and shortcomings. In addition, the chapter also elaborated on the various VO-based WBPS in the field of anticorrosive nanocomposite coatings. The future perspectives in the development of WB polymers are also discussed.

The aim of Chapter 8 is to show that green catalysis is an important branch of green engineering as it is inherently derived from green chemistry and possesses the capability to bring a substantial change in this field by moving a step forward toward foreseeable sustainability. Demand of energy inputs and high operating conditions for fuel generation by direct or indirect usage of fossil fuels in any chemical process can be suppressed by reducing the energy barrier to facilitate the reaction. Catalysts are known to have the ability to suppress the activation energy of the chemical processes, accelerate the reaction rate, increase the yield and selectivity, and limit the excessive usage of stoichiometric reagents by providing surface sites without being involved in the reaction. Energy inputs are as much responsible as the direct usage of chemicals in adding toxins in the environment posing serious health problems. Therefore, nonhazardous ways to deliver energy inputs and reduction of waste from the overall process is the first and foremost requirement of moving toward the green approach. In this regard, solar irradiation, potential difference across electrodes as the energy source/driving force for the reactions operating at room temperature is a kick start to demonstrate the preventive as well as the after-treatment for pollution prevention measures. Photons absorbed via solar energy promotes the generation of charge carriers and are widely used for fuel generation by replacing the thermocatalytic processes using high temperature and high pressure as the reaction parameters. Electricity production via solar cells and photovoltaic systems, solar energy responsible for decontaminating the air and water with high efficiency are the steps forward to the greener approach. Green engineering and green chemistry provide an enormous scope to direct the research toward designing highly efficient materials, catalysts, and processes with high environmental benefits.

The world of environmental protection and nanotechnology are today in the avenues of new scientific regeneration. The scientific domain and the civil society today stand shocked with the evergrowing challenges of global climate change, global warming, and frequent environmental disasters. Also, loss of ecological biodiversity and depletion of fossil fuel resources are veritably challenging the scientific firmament. In Chapter 9, the author with insight and academic rigor delineates the recent and significant advances in the field of bionanocomposites, biofuels, and environmental protection with a clear vision toward true realization of environmental or green sustainability. The author also stresses the application of nanotechnology and composite science in environmental remediation and green sustainability. A larger investigation in the field of biofuels and application of biofuels are the other cornerstones of this paper. Mankind's immense scientific perseverance and determination and the scientific ingenuity of the field of environmental protection will surely be the torchbearers toward a new era in environmental sustainability today. The application of nanomaterials and engineered nanomaterials in energy engineering and environmental engineering stands as veritable pillars of this treatise. Material science and composite science are today in the process of re-envisioning and deep regeneration. This chapter will surely open new doors of innovation and scientific sagacity in the field of environmental engineering science. A new chapter of engineering science and technology will evolve as civilization confronts the issues of green engineering, green sustainability, and green chemistry. This chapter will validate these issues.

In Chapter 10, we focus on another green analytical methodology that makes use of chemical sensors based electronic nose (eNose) technology. The eNose technology is fairly advanced for the detection of organic vapor analytes in the air for a variety of application fields like explosive detection, food quality monitoring, industrial emission monitoring, and air pollution monitoring and it can be seen through several recent reviews. However, its potentiality as a green analytical methodology has not received much attention in green chemistry literature. By managing cross-selectivity, integration with standard microelectronics, microfluidics, and communication technologies eNoses may prove enablers for future generation green chemistry analytics. In this chapter, we shall first briefly outline the basics of eNose systems and associated sensor technologies, and then focus on their real-time chemometric aspects in an attempt to present perspectives for applications in green chemistry activities.

The progressive contamination of our planet with waste from various industries is an actual problem that demands urgent solutions. For this reason, scientists and engineers must find new ways to produce materials that generate a minimized impact on the environment. Chapter 11 focused on the presentation of the state-of-art in the field of multiphase materials based on green polymers designed for the electronic industry. The overall picture and current progress concerning the preparation routes of the green polymer composites are reviewed. The surface processing techniques of reinforced green polymer materials are also described. The latest trends regarding the application of the multiphase green polymer materials in electronics are presented. The environmental concerns are reflected in the developments in this domain, opening novel perspectives in the devices for future electronics.

As a result of the recent environmental problems, such as pollution and water contamination, in-depth studies have been carried out in order to develop systems for wastewater treatment and cleaning, some of these systems involving the use of membranes technology. There are three types of wastewater resulting from industry, agriculture, and those from households, these sectors generate the largest amount of wastewater. Chapter 12 presents a series of membranes based on eco-friendly polymers designed in order to remove the main contaminants of the water, namely: dyes, metal ions, fertilizers, feces, bacteria, or oils. The technique of membrane systems using is effective, ease of implementation, and does not involve high costs, especially when they are obtained from natural resources. The creation of the improved systems containing new and more efficient membranes with enhanced porosity, selectivity, antifouling, and antibacterial properties is and will be the subject of future research because it ensures the decontamination of the environment, improving the life quality.

CHAPTER 1

Green Chemistry and Sustainable Nanotechnological Developments: Principles, Designs, Applications, and Efficiency

YASSINE SLIMANI[1*] and ESSIA HANNACHI[2]

1Department of Biophysics, Institute for Research and Medical Consultations (IRMC), Imam Abdulrahman Bin Faisal University, P.O. Box 1982, 31441, Dammam, Saudi Arabia

2Laboratory of Physics of Materials—Structures and Properties, Department of Physics, Faculty of Sciences of Bizerte, University of Carthage, 7021, Zarzouna, Tunisia

*Corresponding author.
E-mail: yaslimani@iau.edu.sa; slimaniyassine18@gmail.com

ABSTRACT

The chemical industries play an essential role to sustain the world economies and to reinforce forthcoming technologies and scientific developments in novel products, less toxicological materials, industrial procedures with high efficiency, and renewable energy products. Green chemistry (GChem) seeks for the design of innovative chemical products with higher efficiency and lowest hazardous substances for the health and the environment. This chapter describes the fundamentals and metrics of GChem and their effects on the whole life cycle of chemicals from designs across removal. After reporting the important metrics and the latest developments in the theme within this framework, the nanotechnology case was considered. The nanotechnology offers a useful context to investigate the influences and applications of GChem. Interdisciplinary innovations conduct both disciplines, and both aim for transforming the nature of technology. The applications and insinuations of developing green technologies are reviewed, and forthcoming occasions for interdisciplinary associations are discussed.

1.1 INTRODUCTION

During the 20th century, fast tendencies and advances of the global techno-logical and economic growth pushed researchers to recognize that additional progress of human population and the execution of sociological and economic requirements of the current generations could be only feasible when the natural sources are correctly accomplished and the relation among economic evolution and the care of the environment of the current and upcoming generations is determinedly kept. In 1980s, it is clear that the majority of the global envi-ronmental defies have not been satisfactorily addressed. In 1980s, it is clear that the majority of the global environmental defies have not been satisfactorily addressed, in particular, the complications of poverty in developing and Third World countries over worldwide, the local and global environmental pollutions, and the unsustainable use of natural resources. The environmental pollution warnings varying from the pollution of air in municipalities, desertification and deforestation, public solid wastes, acid rains, the reduction of the layer of ozone, and the climate modifications were overlooked. In 1987, the idea of sustainable ecological developments was first announced.[1]

The worldwide chemical industries play a key task in the areas of science and technology related with the prospect of sustainable progress in both devel-oping and developed states. From the outset, the leaders of most chemical factories contributed in the deliberation on the activities and modifications required to realize the sustainable progress objectives and defined their part of responsibility for the sustainable progress aims. Engineers and chemists are addressing the challenges of sustainability to reduce the potential impacts of their technologies on the environment and health. In the 1990s, the American Chemical Society (ACS) endorsed green chemistry (GChem), sustainability, and green engineering (GEng) coupled with incentives to adopt sustainable technology and innovative controlled approaches.[2-3] The word GChem was invented in 1991. As a consequence, it was involving the support of various government agencies and research institutes, international scientific collabo-ration, and global actions in the area of dissemination information and educa-tion.[4-8] Recently, the European chemical industry struggles to be sustainable in its activities and play a leading role in a sustainable society by fulfilling eco-friendly technology and science, rational use of natural resources, and safe chemicals for chemical consumers.[9]

1.2 PRINCIPLES OF GREEN CHEMISTRY AND ENGINEERING

Table 1.1 illustrates the up-to-date principles of GChem and GEng that have to be tracked to attain a sustainable, greener, and eco-friendly progress of the chemical industry and design aspects to recycle products, energy proficiency, and renewable raw materials in the future.[10–14]

TABLE 1.1 Principles of GChem.

No.	Principles	Details
1	Prevention	It is better to avoid wastes than treating or cleaning up wastes after it has been produced.
2	Atoms economy	Synthetic techniques must be created to increase the integration of all products utilized in the procedure within the final product.
3	Minor harmful chemicals	Productions anywhere practicable, must be constructed to operate and create substances possessing slight or no toxicological risk to the environment and health.
4	Constructing more safe chemicals	With anticipated purpose whilst reducing their toxicity.
5	More safe solvents and auxiliaries' materials	Like separation agents, solvents, etc. for the environment and the employees.
6	Design for energy proficiency	Energy needs of chemical procedures must be identified for their economic and environmental influences and must be reduced.
7	Utilization of renewable feedstocks	Natural resources or feedstocks must be renewable rather than exhausting.
8	Decrease derivatives	Needless derivatization (impermanent change, protection/deprotection, blocking groups, etc.) must be reduced or prevented.
9	Catalysis and innovative catalytic chemicals	Like enzymes, as discerning as possible, are excellent to stoichiometric components.
10	Design products for degradation	Chemical products must be designed so that at the end of their function or use they break down into harmless degradation products and do not remain in the environment.
11	Real-time investigation to prevent pollution	Analytical methods require to be more advanced to permit for real-time, in-process checking and control prior to the creation of dangerous substances.
12	More safe natural chemistry for accident prevention	Materials and chemical method must be selected to reduce the capability for chemical accidents.

1.3 AREAS OF GREEN CHEMISTRY WITH RECENT DEVELOPED TECHNOLOGIES

During recent years, GChem and GEng have developed for a large kind of technology and research disciplines offering pioneering researches and promising applications for a broad variety of chemical products and innovated technologies. The largely crucial researches and technological disciplines of GEng and GChem involve solutions. Amongst other fields, renewable energy sources, renewable products, biopolymer technology, biocatalysis and catalysis, innovative food products, eco-friendly dyes and pigments, wastewater and waste management, phytoremediation, solvent-free reactions, electrochemical and ultrasound synthetic approaches, microwave, diminution of worldwide warming, and utilization of CO_2 as a resources for chemical production, etc. Though there exist several disciplines of innovation for GChem and GEng products, we report here some essential fields:[15–17]

- Solvents and selection of solvents in industrial production
- GChem and organic solar cells
- GChem and biodegradable polymers
- Green flow chemistry and uninterrupted procedures in chemical industries
- GChem and multicomponent reactions
- GChem and agricultural technologies benevolent to environment
- Green and renewable energy sources
- Hydrogen creation through catalytic splitting of water
- GChem and synthetic procedures in the pharmaceutical industry
- Directed evolutions and new enzymes for organic production
- Biocatalysis and biotransformations procedures for practical synthetic reactions.

In addition to these, there exist also several other technological areas of GChem and GEng, which have been developed during last decades. Nowadays, certain of these innovative inventions have been applied and improved the sustainability, reduced the environmental pollutions and released less hazardous chemical products.[18–20]

1.4 GREEN CHEMISTRY, DIRECTED EVOLUTION, AND BIOCATALYSIS

In 2016, Frances Arnold obtained the Millennium Technology Prize due to her innovations and researches on the area of directed evolutions that represents natural evolution to generate new and better proteins (enzymes for biocatalysis) in laboratories with desired properties. This innovation has resolved numerous crucial synthetic industrial troubles, frequently to replace less effective synthetic techniques and sometimes dangerous technologies. Because of directed evolutions, sustainable developments and clean technologies (biocatalysis) are suited in several areas of chemical industries.[21-22] The technique proposed by Arnold lead to generate arbitrary mutations in the DNA like that it occurs naturally.[23] The altered genes create proteins with new features from which one could select the valuable ones, repeat the procedure until the performance level required by industries is attained. Directed evolutions could create enzymes, which are utilized in industry using biotechnologies. They have been implemented in fields of renewable energies along with GChem. For instance, directed evolutions are employed to enhance enzymes that transform celluloses or other plant sugars to chemicals and biofuels.[24-26]

The very fascinating technological achievements of directed evolutions are the engineering of adapted industrial enzymes for biocatalysis. During the past few years, several innovations have been proposed and developed. For instance, an effective para-nitro-benzyl-esterase across six generations of arbitrary point mutagenesis and regrouping has been established. The greatest duplicates found after four generations of progressive arbitrary mutagenesis and two generations of random regrouping demonstrated over than 150 times the *p*-nitro-benzyl-esterase activities of wild kind. The collection of multiple mutations by directed evolutions permitted considerable enhancement of the biocatalysts for reactions on substrates.[27] These improvements have developed the time-consuming and expensive procedure of proteins adjustment. Directed evolutions are employed in several companies and laboratories around the world. Adapted proteins are utilized to change procedures in the creation of agricultural chemicals, textiles, pharmaceuticals, paper products, and fuels.[28-29]

Recently, researchers have showed that cytochrome P450 (CYPs)-derived enzymes could also catalyze valuable reactions that are before available only for synthetic chemistry. The growth and manufacturing of these enzymes delivers an outstanding case of investigation in order to encode genetically new chemistry and enlarge the reaction space of biology.[30] Directed evolutions succeeded to transform these licentious generalist enzymes to specialist ones that are efficient to mediate reactions of industries with elegant stereo- and regioselectivity. Development has been effected by adjusting the biocatalysts P450's characteristics (stability, cofactor preference, and substrate range) with prominence in industrial applications.[31] This direct evolution method has been assisted by several developments in bioinformatics, metagenomics, and DNA technologies that have allowed the utilization of biological systems (i.e., cells, metabolic pathways, and enzymes) for chemical production and fabrication of compounds from renewable sources (like sugars). There exist numerous examples of enzymes and metabolic engineering for the creation of organic molecules having superior selectivity and proficiency.[32]

1.5   HYDROGEN PRODUCTION VIA CATALYTIC SPLITTING OF WATER

The hydrogen created from solar-powered water separation has the capability to be an abundant, sustainable, and clean energy resource. Originated from natural photosynthesis, synthetic solar water separating tools are currently being constructed and examined for proficiency. While the sunlight-powered water separation is a favorable route to produce sustainable hydrogen (H_2) as fuels extensive applications are obstructed by the cost of the photoelectro-chemical and photovoltaic systems. Different catalysts and integrated tools have been utilized to produce hydrogen from water.[33-34] Numerous scientists have achieved to improve the proficiency for direct solar water separation with cycle solar cells where surfaces have been selectively adjusted with as new record of 14% proficiency.[35]

Nowadays, there exists new utilized generation of hydrogen fuel cell vehicles having null emissions. The H_2 cars run by compressing hydrogen feds within stack of fuel cells, which can generate electricity to drive the vehicles. Fuel cells could be utilized by combining them with an electric motor to power vehicles cleanly, powerfully, and quietly. Numerous efforts have been concentrated on hydrogen as potential energies and on the utilization of water-splitting technologies as renewable and clean resources to produce hydrogen by means of solar energy. Several challenges have been constructed for developing photocatalysts, which can operate not only under UV light, nevertheless also under visible light illumination to proficiently use solar energy. Some promise resources of hydrogen are thermal-biochemical, photo-biochemical, photoelectric, photothermal, electrothermal, photonic, biochemical, thermal, and electrical. Certain kinds of energies could be resulted from renewable resources and from energy recovery procedures for hydrogen production goals.[36-37]

During the past years, innovative research works announced the "bionic leaf" conception for the effective separation of water by photochemical utilization of sunlight.[38] The objective has continuously been to exploit sunlight and employ it to produce liquid fuels preferably than electricity, which should be then stored in batteries. The conventional experiments were intended to employ solar power to split oxygen atoms in water from hydrogen, which is afterward transformed within isopropyl alcohol via bacteria. However, recent attempts had utilized nickel-molybdenum-zinc (NiMoZn) catalysts and the resultant reactive oxygen species (ROS) would damage the DNA of bacteria C. Liu et al.[38] described the synthetic photosynthetic scheme (bionic leaf 2.0), which could be efficient for solar energy storage and chemical reduction of CO_2. Researchers industrialized a hybrid water separating technique with biocompatible inorganic catalysts (exchange the oldest NiMoZn with a cobalt–phosphorus alloy catalysts). The new catalysts do not generate ROS that would damage the bacteria. The separation of water generated O_2 and H_2 at lower voltages.

The hydrogen in interaction with the *Ralstoniaeutropha* bacteria was consumed to produce biomass and fuel products. The accessible technique showed a CO_2 reduction energy efficiency (REE) of about 50% once a bacteria biomass is produced and liquid alcohols are scrubbed 180g of CO_2 per kW h of electricity. Connecting the hybrid devices to existent photovoltaic tools could produce a CO_2 REE of around 10%, which exceeds that of natural photosynthetic tools.[38] Coupling this hybrid device to existing photovoltaic systems can yield a CO_2 reduction energy efficiency of ~10%, exceeding that of natural photosynthetic systems.[38] Although the investigations showed that the system could be employed to create useful fuels, its potential does not limit there. The technique could nowadays transform solar energy to biomass with 10% proficiency, which is higher than 1% as shown in the quickest growing plants.

1.6 GREEN BIOCATALYSIS FOR PHARMACEUTICAL INDUSTRIES

After an initial delay stage, the pharmaceutical industries involved GChem for economy and prestige objectives with prominence in greener artificial processes, protection of environment, and reduced solvents. During recent years, biocatalysis has been recognized as accessible and green technology that lead to produce a wide variety of pharmaceutical materials and intermediates. Biocatalysts used on broad range in pharmaceutical production provide benefits to the pharmaceutical industries in terms of costs and qualities. For this regard, there exist numerous inventions of advanced biocatalytic methods utilizing oxidases, transaminases, reductases, hydrolases, etc., which are utilized to prepare therapeutic agents. Pioneering enzymes employed for biocatalysis provided significant economic advantages to the pharmaceutical industries.[39–40] The biosynthesis in organic production in the pharmaceuticals, vitamins, flavors and fragrances, and fine chemical industries is still not completely developed;

hence, there are still further required investigations to be performed in the employment of new catalytic enzymes in organic production of drugs and other therapeutic ingredients.[41-42] Recently, innovative technological and scientific advancements have improved gene production and DNA sequencing in order to tailor biocatalysts by protein engineering and design, and the capability to rearrange enzymes within new biosynthetic routes.[43-46]

1.7   RECENT ADVANCEMENTS IN DEGRADABLE AND RECYCLED POLYMERS

Green economy endeavors to endorse sustainability and alternate approaches to reduce the demands for raw resource materials and energy, to improve industrial procedures, to decrease the emissions of greenhouse gas, to avoid environmental pollutions and hazards, to diminish wastes, and to recycle effectively wastes and expired products. In the initial industrial procedures, the fabrication of polymers used mostly oil raw materials, elevated energy inputs, and generated nondegradable plastics and huge quantities of wastes. In 1930, the initial manufacturing method to make polyethylene (PE) needed temperatures exceeding 150°C and very high pressures above 1000 bar. Catalytic olefin polymerizations were developed in 1950s at temperatures below 100°C and with lower pressures inferior than 10 bar.[47] The worldwide fabrication of plastic materials and polymers grew significantly in the last decades achieving in 2014 the incredible quantity of 315 million metric tons. The greatest producers are China with about 25% of the worldwide fabrication and the United States, Canada, and Mexico with around 20%. There exist more than 70,000 of plastics industries in Europe. Nowadays, statistics indicated 100 kg of plastics is consumed each year by one person in advanced countries.[48-49]

In 1980, the energy-effective catalytic copolymerization of ethylene for manufacturing linear low-density polyethylene (used in food packaging) became a million-ton business. During the last few years, the diversity of plastics enlarged considerably, and the first biodegradable plastics were presented. The plastic scattering difficulties grew rapidly, and the pollution of oceans with plastics became a critical and emergent matter of environmental pollution. This is the consequence of a deficiency of recycling and illegal removal of wastes in most of the developed countries. The GChem of bio-based plastics and degradable polymer-based products are being in the proper way for the area of sustainable polymers.[50]

Nowadays, the reduction of anthropogenic carbon dioxide (CO_2) emissions has become a serious climate and environmental matter. The recovery of

CO_2 could be done on a wide range in carbon capture of power plants, burn up fossil fuels, and in steam upgrading to generate hydrogen from water and coals. It is needed to couple carbon capture materials with a technique for CO_2 release in order to further improve storage and transport. Several methods have been investigated for recovering CO_2.[51] Recently, several new catalytic mechanisms have been established to produce cyclic carbonates. One of the processes utilized active bifunctional porphyrin catalysts, which show great income number for the production of cyclic carbonates from CO_2 and epoxides under solvent-free circumstances.[52] Another method provided that the cyclic amidine hydroiodides efficiently catalyzed the reactions of CO_2 and epoxides under moderate circumstances like room temperature and regular pressures, and the equivalent five-membered cyclic carbonates were attained in medium to high produces.[53] The harshness of the forthcoming problems of worldwide warming have generated worldwide attempts to diminish the amount of atmospheric CO_2. CO_2 capture and storage are being central strategies in order to meet CO_2 emissions reduction targets. Nowadays, there exist numerous researches and diverse technologies to capture, separate, transport, store, leak, monitor, and analyze the life cycle of CO_2.[54] The conversion of captured CO_2 into products like building materials, fuels, plastics, chemicals, and other products is being a vital issue around the world. This method can be particularly effective to reduce carbon emissions in regions where geologic storage of CO_2 is not applied.

1.8 ORGANIC PHOTOVOLTAIC SOLAR CELLS AND GREEN CHEMISTRY

The technology of organic photovoltaic solar cells show the aptitude to deliver cheaper electricity compared to first- and second-generation solar technologies. For the reason that numerous absorbers could be employed to generate transparent or colored organic photovoltaic devices, this technology is especially attractive to the building-integrated photovoltaic markets. Organic photovoltaic solar cells have attained proficiencies close to 12%; however, proficiency restrictions along with long-term reliability persists important obstacles. Different to the largest inorganic solar cells, organic photovoltaic cells utilize polymeric or molecular absorbers that cause a localized exciton. A graphical presentation of green solvent processable organic photovoltaics and different kinds of green solvents is presented in Figure 1.1. The absorbers are employed in aggregation with electron acceptors, like fullerene, which possesses molecular orbital energy states that aid the transport of electrons.[55]

The utilization of perovskite solar cells has progressed considerably the effectiveness of organic photovoltaic solar cells. Already, there are certified proficiencies of energy conversions for perovskite-based solar cells beyond the 20%.

Recently, scientists are discovering other serious region of solar cells, like the understanding of devices hysteresis and films evolution, the change of lead, and the improvement of tandem cell stacks. The stability of cells is another vital and serious matter of research.[56] Researchers are trying to enhance the effectiveness of polymer solar cells by utilizing tandem structures. A wider region of solar radiation spectrum is utilized and the thermalization loss of photon energy is diminished. Earlier, the deficiency of low-bandgap polymers with outstanding performances has been the main restricting factor to achieve tandem solar cells with great performances. As a consequence of this advanced technique, devices with single junction show great external quantum proficiency higher than 60% and spectral responses that spread to 900 nm with a power conversion proficiency of around 8%. The polymers enable a solution-administered tandem solar cell with qualified 10.6% power conversion proficiency underneath standard reporting circumstances.[57]

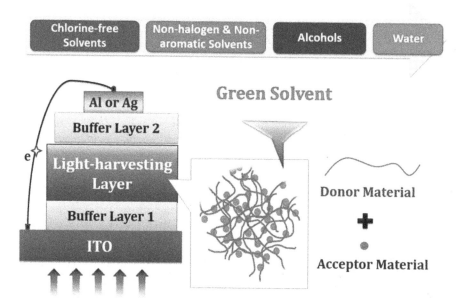

FIGURE 1.1   Graphical presentation of green solvent processable organic photovoltaics and the different kinds of green solvents.

Source: Reprinted with permission from Ref. [69]. © 2016 Elsevier.

1.9 GREEN SOLVENTS, CATALYST-FREE REACTIONS, AND IN WATER

The protection of environment and safety of workers in pharmaceutical and chemical manufactures play a progressively instrumental task in the selection of solvents. The majority of enterprises nowadays utilize greener solvents. Likewise, customers start to look on greener alternatives to traditional solvents. Figure 1.2 showed some examples of bio-based solvents. Choosing the correct solvent has always had the potential to increase competitiveness, but the environmental benefits and the safety of workers/users will be the key to product differentiation and excellent margins as the industries enter this era of enhanced environmental consciousness. At the present, green solvents show about 10% of the whole solvent markets. The consumption's increase of green solvents past 2014 was supported by the elevated costs of traditional solvents that started growing in 2004 as a consequence of highest record of feedstock prices. In the last few years, industries and laboratories shifted to catalysts-free reactions and afterward to catalysts-free reactions on water and in water. Numerous selected reactions, multicomponent reactions, and the manufacture of heterocyclic products are some typical examples.

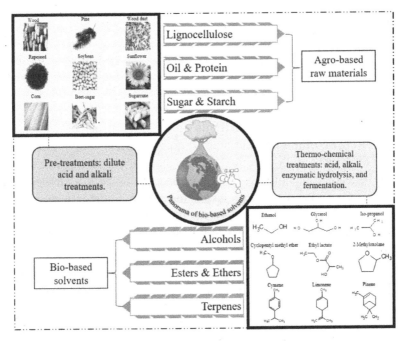

FIGURE 1.2   Examples of bio-based solvents.

Source: Reproduced from Ref. [68]. © 2019 by the authors. http://creativecommons.org/licenses/by/4.0/

During the last few years, the organic solvents with natural toxicity and elevated volatility were changed by ionic liquids, which gain immense interest from the scientific community; however, their greenness is frequently a challenge because of their inferior biodegradability. An alternate kind of solvents that represents GChem notions are deep eutectic solvents. They are identified as a combination of two or more constituents that can be liquid or solid and which at specific compositions show high melting point depression converting to liquids at ambient temperature. Deep eutectic solvents could be utilized for extraction, electrochemistry, and biocatalysis.[58]

1.10 GREEN AND RENEWABLE ENERGY SOURCES

Scientific communities and most of the countries around the world understand that the fossil fuels (natural gas, coal, and petroleum) will ultimately diminish, being extremely high-priced or excessively damaging the environment. Oppositely, several kinds of sources for renewable energy like biomass, hydroelectric, geothermal, solar, and wind energy are continuously reloaded and will never end. The alternate of oils with biomass as raw materials for fuels and chemical fabrication is an appealing opportunity to develop biorefinery complexes. Biomasses (particularly energy crops, agricultural residues, forest residues, and woody biomass) are undoubtedly the largely utilized raw materials to produce renewable energy fuels. Woody biomasses are favored materials in thermochemical procedures because of their lower residue amount and higher quality of generated bio-oils. Thermal transformation via rapid pyrolysis transforms up to 75% of the starting materials into renewable biofuels appropriate for transport. Design or combination of diverse feedstocks with chemical and/or thermal pretreatment can enable reliable, cheaper, and high-volume biomass resources to emerge biofuels industries. Carpenter et al.[59] reviewed the recent findings concerning the influence of feedstocks and pretreatments on the harvest, distribution of products, and upgradability of bio-oils.

Recently, scientists around the world have developed considerably the third-generation biofuels, particularly those that are obtained from microalgae and which are believed to be sustainable renewable energy sources. The third generation is lacking the main disadvantages related with first-generation biofuels (principally rapeseed, maize, sugar beet, sugarcane, etc.) and second-generation biofuels (resulted from agricultural and forest biomass residues). Industrial experiences in the manufacturing of biofuels indicated the significance of incorporating biofuels fabrication with the generation of high-value biomass fractions in a biorefinery notion. The sustainability of these renewable sources could be reached via the synergistic coupling of microalgae spread methods with CO_2 confiscation and bioremediation of wastewater treatments.[60]

1.11 GREEN ELECTROCHEMICAL STORAGE SYSTEMS

The polymers enable solutions-treated tandem solar cells with qualified 11% power conversion proficiency in ordinary circumstances.[61] Consequently, it was important for scientists to include plenty materials, eco-effective artificial procedures, and life-cycle examination within the design of novel electrochemical storage systems. Presently, some new current technologies consider these matters; however, ultimate technological difficulties stay to be disabled. Larcher et al.[62] summarized the present and forthcoming electrochemical storage systems further than lithium-ion batteries, their difficulty and significance of the recycle of batteries from a sustainability viewpoint.

Sodium ion batteries constructed on inter calated materials that use nonaqueous electrolytes similar to lithium-ion batteries were firstly discovered in the middle of 1980s.

During the last few years, they have suffered a revival because they provide greater energy densities compared to aqueous batteries, more sustainable and cheaper than lithium-ion batteries.[63] They are considered as green alternate for energy storage. The pioneering graphene materials are one of the most attractive issues for scientists due to their extremely interesting traits. The graphene has motivated mostly the domain of electrochemical energy-storage systems. However, in spite of extensive primary eagerness from several researchers for the task of graphene as materials for energy storage, there exists yet no developments in the domain. Recently, investigations were carried out on the uses of graphene in Li-ion batteries and supercapacitors. Moreover, graphene was utilized in developing technologies like metal–air and magnesium-ion batteries.[64] During the past years, great efforts have been done in line with the principles of GChem in order to enhance the capacity for effective transport, storage, conversion, and admittance to renewable energies. In addition, extensive research concentrate on the use of earth-plentiful and nontoxic products, thus that any advances were developed will not generate new difficulties to environment. Even though the inclusion of electrodes for lithium or sodium-ion batteries encountered primarily these necessities, researchers proceeded their investigations within diverse fields and reengineered for expansion to wider scale of applications.[65]

1.12 CONCLUSION AND CHALLENGES OF GREEN ENGINEERING AND GREEN CHEMISTRY FOR SUSTAINABILITY

GEng and GChem are getting considerable interest in the pharmaceutical and chemical factories and for innovative technological advances in an immense diversity of productions. This is a consequence of their lower cost, consumer products, less toxic chemicals, reliability of renewable raw materials, energy proficiency, and sustainable nature. Novel green products have established numerous applications in life day in appreciation of the earlier circumstances of environmental pollution by solid public waste. All these green technologies are gradually getting interest toward the worldwide degradation of environment in several countries and the population growth.

The admittance to inexpensive and consistent energy resources (gas, petroleum, coal, etc.) has been a keystone of the world's growing success and economic evolution from the initial industrial revolution. However, with the difficulties of climate change and pollution during last decades, researchers gathered that the utilization of energy in the future should be sustainable and extracted from renewable resources. Water, wave, air, and solar-based generated energies and the engineering of microbes to form biofuels are some instances of the replacements. GEng elucidations are the forthcoming prospects and paths that can conduce to successful, sustainable, and secured energies on a worldwide scale.[66]

During the past year, the significance of the managing of GChem and GEng inventions and applications has been increasing in the both industries and laboratories. Even though main engineering fields presently devote important research to green and sustainable elucidations, there exist further required achievements in green management fields.[67] The defy for forthcoming productions of environmental experts, chemical engineers, and chemists is to advance and control the technical devices and methods that will incorporate environmental purposes within the design choice in the advanced chemical industries. The current chapter gives a collection of the greatest significant pioneering advances in GChem and GEng disciplines during last year.

KEYWORDS

- **green chemistry**
- **nanotechnology**
- **toxicity**
- **environment**
- **designs**
- **sustainability**
- **efficiency**

REFERENCES

1. United Nations World Commission on Environment and Development. *Our Common Future* (Brundland Commission, *Sustainable Development. Our Common Future*). Oxford University Press, October 1987.
2. National Research Council (USA). *Our Common Journey: A Transition Toward Sustainability,* NRC; National Academy Press: Washington, D.C., 1999.
3. National Research Council. *Sustainability in the Chemical Industry,* NRC; National Academy Press: Washington, D.C., 2005.
4. Anastas, P. T.; Williamson, T. C. *Green Chemistry: DesigninGChemistry for the Environment*; American Chemical Society Publications: Washington D.C., 1996.
5. Anastas P. T.; Warner, J. C. *Green Chemistry: Theory and Practice*; Oxford University Press: New York, 1998.
6. Anastas, P. T.; Williamson, T. C., Eds. *Green Chemistry: Frontiers in Chemical Synthesis and Processes*; Oxford University Press: Oxford, 1998.
7. EPA. Green Chemistry Program. United States Environmental Protection Agency (EPA); Office of Pollution Prevention and Toxics: Washington D. C., 1999.
8. Anastas, P. T.; Zimmerman, J. B. Design through the Twelve Principles of Green Engineering. *Environ. Sci. Technol.* **2003,** *37*(5), 94A–101A.
9. European Commission. Energ-Ice. *Road Map Document for Sustainable Chemical Industry.* http://ec.europa.eu/environment/life/project/Projects/ index.cfm?fuseaction=home.show File&rep=file&fil=ENERG_ICE_Road_Map.pdf html (accessed Jan, 2013).
10. Anastas, P. T.; Kirchhoff, M. M. Origins, Current Status, and Future Challenges of Green Chemistry. *Acc. Chem. Res.* **2002,** *35*(9), 686–694.
11. Tang, S. L. Y.; Smith, R. L.; Polliakoff, M. Principles of Green Chemistry: Productively. *Green Chem.* **2005,** *7*, 761–762.
12. Anastas, P. T.; Zimmerman, J. B. Design Through the Twelve Principles of Green Engineering. *Environ. Sci. Tech.* **2003,** *37*, 5, 94A–101A.
13. Valavanidis, A.; Vlachogianni, T. H. *Green Chemistry and Green Engineering*; Scientific publications: Department of Chemistry, University of Athens, 2012.

14. Jiménez-González, C.; Constable, D. J. C. *Green Chemistry and Engineering: A Practical Design Approach*; John Wiley & Sons: New York, 2011.

15. Sheldon, R. A. Engineering for a More Sustainable World Through Catalysis and Green Chemistry. *J. Royal Soc. Interface* **2016**, *13*(116). DOI: 10.1098/rsif.2016.0087.

16. Kerton, F. M.; Marriott, R. *Alternative Solvents for Green Chemistry*, 2nd ed.; RSC Publications: Cambridge, 2013.

17. Clark, J.; Deswarte, F., Eds. *Introduction to Chemicals from Biomass*, 2nd ed.; John Wiley & Sons: Chichester, Sussex, 2015.

18. Mulvihill, M. J.; Beach, E. S.; Zimmerman, J. B.; Anastas, P. T. Green Chemistry and Green Engineering: A Framework for Sustainable Technology Development. *Ann. Rev. Environ. Resour.* **2011**, *36*, 271–293.

19. Sanghi, R.; Singh, V., Eds. *Green Chemistry for Environmental Remediation*; Wiley, Scrivener Publishing: Salem, 2012.

20. Albini, A.; Protti, S. *Paradigms in Green Chemistry and Technology*; Springer, Springer Briefs in Green Chemistry for Sustainability: Heidelberg, 2016.

21. Webb J. BBC (www.bbc.co.uk) Science and Environment. Evolutionary engineer Frances Arnold wins €1m tech prize. https://www.bbc.com/news/science-environment-36344155 (accessed May 24, 2016).

22. CALTECH05/24/2016.https://www.caltech.edu/news/frances-arnold-wins-2016-millennium-technology-prize-50784

23. Technology Academy Finland. Millennium Technology Prize. http://taf.fi/en/millennium-technology-prize/.

24. Kuchne, O.; Arnold, F. H. Directed Evolution of Enzyme Catalysts. *Trends Biotechnol.* **1997**, *15*(12), 523–530.

25. Romero, P. A.; Arnold, F. H. Exploring Protein Fitness Landscapes by Direct Evolution. *Nat. Rev. Mol. Cell Biol.* **2009**, *10*, 866–876.

26. Currin, A.; Swainston, N.; Day, P. J.; Kell, D. B. Synthetic Biology for the Directed Evolution of Protein Biocatalysts: Navigating Sequence Space Intelligently. *Chem. Soc. Rev.* **2015**, *44*, 1172–1239.

27. Arnold, F. H.; Moore, J. C. Optimizing Industrial Enzymes by Directed Evolution. *Adv. Biochem. Eng. Biotechnol.* **1997**, *58*, 1–14.

28. Otten, L. G.; Quax, W. J. Directed Evolution: Selecting Today's Biocatalysts. *Biomol. Eng.* **2005**, *22*(1–3), 1–9.

29. Turner, N. J. Directed Evolution Drives the Next Generation of Biocatalysts. *Nat. Chem. Biol.* **2009**, *5*, 567–573.

30. McIntosh, J. A.; Farwell, C. C.; Arnold, F. H. Expanding P450 Catalytic Reaction Space Through Evolution and Engineering. *Curr. Opin. Chem. Biol.* **2014**, *19*, 126–134.

31. Behrendorff, J. B.; Huang, W.; Gillam, E. M. Directed Evolution of Cytochrome P450 Enzymes for Biocatalysis: Exploiting the Catalytic Versatility of Enzymes with Relaxed Substrate Specificity. *Biochem. J.* **2015**, *467*(1), 1–15.

32. Coelho, P. S.; Arnold, F. H.; Lewis, J. C. Synthetic Biology Approaches for Organic Synthesis. In *Comprehensive Organic Synthesis II*; Elsevier: Amsterdam, 2014; pp 390–420. ISBN 978-0-08-097743-0.

33. Luo, J.; Im, J. H.; Mayer, M. T.; Schreier, M.; Mohammad Khaja Nazeeruddin, M. K.; Park, N. G.; Tilley, S. D.; Fan, H. J.; Grätzel, M. Water Photolysis at 12.3% Efficiency via Perovskite Photovoltaics and Earth-Abundant Catalysts. *Science* **2014**, *345*(6204), 1593–1596.

34. Peharz, G.; Dimroth, F. Solar Hydrogen Production by Water Splitting with a Conversion Efficiency of 18%. *Intern. J. Hydrog. Energ.* **2007**, *32*(1), 3248–3252.

35. May, M. M.; Lewerenz, H. J.; Lacner, D.; Dimroth, F.; Hannappel, T. Efficient Direct Solar-to-Hydrogen Conversion by *in situ* Interface Transformation of a Tandem Structure. *Nat. Commun.* **2015**, *6*, 8286. DOI: 10.1038/ncomms9286.

36. Dincer, I. Green Methods for Hydrogen Production. *Int. J. Hydrog. Energ.* **2012**, *37*(2), 1954–1971.

37. Ahmad, H.; Kamarudin, S. K.; Minggu, L. J.; Kassim, M. Hydrogen from Photo-Catalytic Water Splitting Process: A Review. *Renew. Sustain. Energ. Rev.* **2015**, *43*, 599–610.

38. Liu, C.; Colon, B. C.; Ziesack, M.; Silver, P. A.; Nocera, D. G. Water Splitting-Biosynthetic System with CO_2 Reduction Efficiencies Exceeding Photosynthesis. *Science* **2016**, *352* (6290), 1210–1213. DOI: 10.1126/science.aaf5039.

39. Huisman, G. W.; Collier, S. J. On the Development of New Biocatalytic Processes for Practical Pharmaceutical Synthesis. *Curr. Opin. Chem. Biol.* **2013**, *17*(2), 284–292.

40. Federsel, H. J. En Route to Full Implementation: Driving the Green Chemistry Agenda in the Pharmaceutical Industry. *Green Chem.* **2013**, *15*, 3105–3115.

41. Ciriminna, R.; Pagliaro, M. Green Chemistry in the Fine Chemicals and Pharmaceutical Industries. *Org. Progress Res. Dev.* **2013**, *17*(12), 1479–1484.

42. Jiménez-González, C.; Constable, D. J. C.; Ponder, C. S. Evaluating the Greenness of Chemical Processes and Products in the Pharmaceutical Industry—a Green Metrics Primer. *Chem. Sco. Rev.* **2012**, *41*, 1485–1498.

·43. Gupta, P.; Mahajan, A. Green Chemistry Approaches as Sustainable Alternatives to Conventional Strategies in the Pharmaceutical Industry. *Royal Soc. Chem. Adv.* **2015**, *5*, 26686–26705.

44. Meyer, H. P.; Eichhorn, E.; Hanlon, S.; Lütz, S.; Schürmann, M.; Wohlgemuthf, R.; Coppolecchiag, R. The Use of Enzymes in Organic Synthesis and the Life Sciences: Perspectives from the Swiss Industrial Biocatalysis Consortium (SIBC). *Catal. Sci. Technol.* **2013**, *3*, 29–40.

45. Bornscheuer, U. T.; Huisman, G. W.; Kazlauskas, R. J.; Lutz, S.; Moore, J. C; Robins, K. Engineering the Third Wave of Biocatalysis. *Nature* **2012**, *485*, 185–194.

46. Hoyos, P.; Pace, V.; Hermaiz, M. J.; Alcantara, A. R. Biocatalysis in the Pharmaceutical Industry. A Greener Future. *Curr. Green Chem.* **2014**, *3*(4), 155–181.

47. Chem1 General Chemistry Virtual Textbook. States of Matter: Polymers and Plastics: An Introduction. http://www.chem1.com/acad/webtext/states/polymers.html (accessed June, 2016).

48. Plastics Europe. Plastics-The Facts 2014/2015. https://www.plasticseurope.org/application/files/3715/1689/8308/2015plastics_the_facts_14122015.pdf

49. Andrady, A. L.; Neal, M. A. Applications and Societal Benefits of Plastics. *Phil. Trans. R. Soc. B.* **2009**, *364*, 1977–1984.

50. Mülhaupt, R. Green Polymer Chemistry and Bio-Based Plastics: Dreams and Reality. *Macromol. Chem. Phys.* **2013**, *214*(2), 159–174.

51. Dawson, R.; Cooper, D. I.; Admas, D. J. Chemical Functionalization Strategies for Carbon Dioxide Capture in Microporous Organic Polymers. *Polym. Int.* **2013**, *62*(3), 345–352.

52. Ema, T.; Miyazaki, Y.; Koyama, S.; Yano, Y.; Sakai, T. A Bifunctional Catalyst for Carbon Dioxide Fixation: Cooperative Double Activation of Epoxides for the Synthesis of Cyclic Carbonates. *Chem. Commun.* **2012**, *48*, 4489–4491.

53. Aoyasi, N.; Furusho, Y.; Endo, T. Effective Synthesis of Cyclic Carbonates from Carbon Dioxide and Epoxides by Phosphonium Iodides as Catalysts in Alcoholic Solvents. *Tetrahedron Lett.* **2013**, *54*(5), 7031–7034.

54. Leung, D. Y. C.; Caramanna, G.; Meroto-Valer, M. M. An Overview of Current Status of Carbon Dioxide Capture and Storage Technologies. *Renew. Sustain. Energ. Rev.* **2014**, *39*, 426–443.

55. Energy Gov. (US Dept of Energy) Office of Energy Efficiency and Renewable Energy. Organic Photovoltaic Research, 2016. http://energy.gov/eere/ sunshot/organic-photovoltaics-research.

56. Green, M. A.; Bein, T. Photovoltaics: Perovskite Cells Charge Forward. *Nat. Mater.* **2015**, *14*, 559–561.

57. You, J.; Dou, L.; Yoshimura, K.; Kato, T.; Ohya, K.; Moriarty, T.; Emery, K.; Chen, C. C.; Gao, J.; Li, G.; Yang, Y. A Polymer Tandem Solar Cell with 10.6% Power Conversion Efficiency. *Nat. Commun.* **2013**, *4*, 1446–1449.

58. Paiva, A.; Craveiro, R.; Aroso, I.; Martins, M.; Reis, R. L.; Duarte, A. R. C. Natural Deep Eutectic Solvents–Solvents for the 21st Century. *ACS Sustain. Chem. Eng.* **2014**, *2*(5), 1063–1071.

59. Carpenter, D.; Westover, T. L.; Czernik, S.; Jablonski, W. Biomass Feedstocks for Renewable Fuel Production: A Review of the Impacts of Feedstock and Pretreatment on the Yield and Product Distribution of Fast Pyrolysis Bio-Oils and Vapors. *Green Chem.* **2014**, *16*, 384–406.

60. Brennan, L.; Owende, P. Biofuels from Microalgae: Towards Meeting Advanced Fuel Standards. In *Advanced Biofuels and Bioproducts*; Lee, J. W., Ed.; Chapter 24; Springer Series + Business Media: New York, 2013; pp 553–599.

61. Cai, S.; Zhu, H. T.; Wang, J. J.; Yu, T. Y.; et al. Fertilization Impacts on Green Leafy Vegetables Supplied with Slow Release Nitrogen Fertilizers. *J. Plant Nutr.* Open Access, [Online]. DOI: 10.1080/01904167.2015.1050508 (accessed Jan 20, 2015).

62. Larcher, D.; Tarascon, J. M. Towards Greener and More Sustainable Batteries for Electrical Energy Storage. *Nat. Chem.* **2015**, *7*, 19–129.

63. Kundu, D.; Talaie, E.; Duffort, V.; Nazar, L. F. The Emerging Chemistry of Sodium Ion Batteries for Electrochemical Energy Storage. *Angewande Chemie* **2015**, *54*(11), 3431–3448.

64. Raccichini, R.; Varzi, A.; Passerini, S.; Scrosati, B. The Role of Graphene for Electrochemical Energy Storage. *Nat. Mater.* **2015**, *14*, 271–279.

65. Melot, B. C.; Tarascon, J. M. Design and Preparation of Materials for Advanced Electrochemical Storage. *Acc. Chem. Res.* **2013**, *46*(5), 1226–1238.

66. Chu, S.; Majumda, A. Opportunities and Challenges for a Sustainable Energy Future. *Nature* **2012**, *488*, 294–303.

67. Schiederig, T.; Tierze, F.; Herstatt, C. Green Innovation in Technology and Innovation Management–an Exploratory Literature Review. *R & D Manag.* **2012**, *42*(2), 180–192.

68. Chemat, F.; Vian, M. A.; Ravi, H. K.; Khadhraoui, B.; Hilali, S.; Perino, S.; Tixier, A. S. F. Review of Alternative Solvents for Green Extraction of Food and Natural Products: Panorama, Principles, Applications and Prospects. *Molecules* **2019**, *24*(16), 3007).

69. Zhang, S.; Ye, L.; Zhang, H.; Hou, J. Green-Solvent-Processable Organic Solar Cells. *Mater. Today* **2016**, *19*, 533–543). https://www.sciencedirect.com/science/article/pii/S1369702116000687

CHAPTER 2

The Next Generation Sustainable Polymers

ABU DARDA[1,2*], HALIMA KHATOON[2*], MOHD IRFAN[3],
SAMEER AHMAD[1], WEQAR AHMAD SIDDIQI[1*], SAJID IQBAL[2],
ASHIQ HUSSAIN PANDIT[2], and MOHD DANISH ANSARI[4]

[1]*Department of Applied Sciences and Humanities, Faculty of Engineering and Technology, Jamia Millia Islamia, New Delhi*

[2]*Materials Research Laboratory, Department of Chemistry, Jamia Millia Islamia, New Delhi, India*

[3]*Conservation Research Laboratory Ajanta Caves, Aurangabad, India*

[4]*Laboratory of Green Synthesis, Department of Chemistry, University of Allahabad, Uttar Pradesh, India*

Corresponding author. E-mail: abudardazilli@gmail.com; hkn.nasir02@gmail.com

ABSTRACT

Vegetable oils (VOs) are superabundant, economic, nontoxic, and biodegradable resource, which are applied for surface coatings for the last two decades. Moreover, the adaptation of the ultramodern techniques coupled with the VOs-based sustainable polymers in the field of surface coatings shows an outstanding physicochemical, physicomechanical, electrical, and thermal properties. On account of this, the present chapter focuses on the classifications, properties, and modifications of the sustainable polymers. Past decade, the theory of environment friendly, solventless, UV curable, waterborne (WB), and hyperbranched (HPs) coatings that are developed from sustainable resources have been discussed. The chapter further highlights the modification of VO-based polymers and their nanocomposites for enduring sustainable and green future. Moreover, the future scopes of these polymers have also been discussed in detail.

2.1 INTRODUCTION

Vegetable oil (VO)-based polymers can be developed via different identical or nonidentical monomeric units and processed by the evolution of greenhouse gases (CO, CO_2, N_2O) and toxic volatile organic compounds (VOCs).[1,2] The syntheses of bio-based polymers have played a vital role to overcome such type of problems by using different kind of precursors especially VOs.[3-5] VOs [sunflower oil (SFO), linseed oil (LO), soy oil (SO), jungle jalebi oil (JJO), palm oil (PO), safflower oil (SfO), castor oil (CO), etc.]-based polymers have excellent hydrophobic, enhanced hydrolytic, thermal, and UV resistive properties while releasing certain amount of VOCs.[6,7] So it is mandatory to develop a sustainable VO-based polymers such as non-solvent, UV curable, and waterborne (WB) polymers in accordance to follow the principles of green chemistry.[8]

In view of this, the first part of the present chapter deals with the introductory classification about the next generation sustainable polymers and the rest part of the chapter will concern about their modifications and the formulation of their nanocomposites. Various types of sustainable resources and their applications in different industries is shown in Figure 2.1.

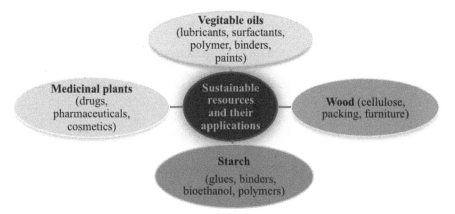

FIGURE 2.1   Various types of sustainable resources and their applications.

2.2   CLASSIFICATION OF THE SUSTAINABLE POLYMERS

2.2.1   WATERBORNE (WB) POLYMERS

The term WB refers to such polymeric materials, which contain water or blended water with green solvents in the synthesis of polymeric materials.[9,10]

The VO-based WB polymers have various applications in the field of biomedical, printing systems, sensors, additives, antibacterial, anticorrosive paints, and coatings.[11] On behalf of polymerization techniques, WB polymeric materials can be discussed as:

(a) Water-soluble polymers

These polymers when dissolved or dispersed in water to became WB by introducing the hydrophilic groups such as anionic, cationic and zwitter ions (like hydroxyl groups or polyether, amino group, NH_3CH_2COOH) incorporated in the polymer backbone (Scheme 2.1).[12]

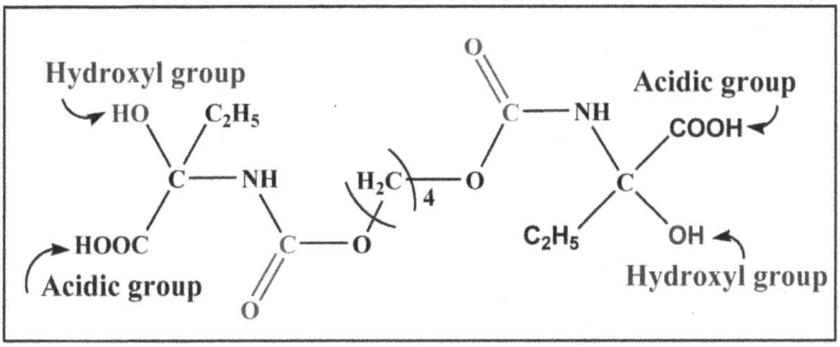

SCHEME 2.1   WB polymer by introducing ionic groups.

The hydrophilic groups are water sensitive, exhibit poor adhesion, and chemical resistance properties which can be controlled by introducing curing agents, for example, ethylene oxide.[13,14]

(b) Water-reducible polymers

Generally, copolymers are adopted in the formation of water-reducible polymers in which polymers are dissolved in water along with cosolvent (e.g., methanol and ethanol) and neutralizing agents like trimethylamine. In addition to this, the polar functionality of polymer backbone is convenient to water.[15,16] These polymers have superior clarity (transparent), high gloss, pigment wetting, and dispersion abilities.

(c) Water-dispersible polymers

The colloidal dispersion polymers or water dispersible polymers are suspended through mechanical agitation in water containing green organic solvents, and act as coalescing agents. Polar groups (acidic and basic) of these polymers impart degree of solubility of the final product, for example, styrene–butadiene copolymers, vinyl propionate copolymers, acrylate–methacrylate copolymers, and vinyl acetate copolymers, etc.[17,18]

(d) Water emulsion polymers

The term emulsion refers to as the dispersion of polymeric resins or monomer in water with the addition of an emulsifier such as alkylphenol ethoxylates,[19] alkyl diphenyloxide disulfonate, polyvinyl alcohols, hydroxyethyl celluloses, etc.[20] The general schematic representation of water emulsion polymers is shown in Figure 2.2. These polymers are dispersed and get swelled in water with the improvement of their properties without changing their viscosities after emulsification.[21] These resins show superior properties on the surface of base polymers like poly esters and alkyds with the insertion of PVA, SBR, and their derivatives.

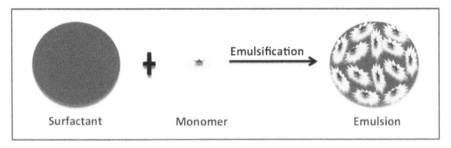

FIGURE 2.2   General schematic representations of water emulsion polymers.

2.2.2   HYPERBRANCHED POLYMERS

The hyperbranched (HPs) or dendritic polymer shows treelike morphology. The central part (core) where the branches are functionalized, is considered to have three amine groups or four ethylene diamine groups in which the linear units of polymer is associated with the core, is terminated by the multifunctional units responsible for branching.[22] HPs show poor mechanical behavior; however, the incorporation of moieties enhances the mechanical properties of HPs. For instance, insertion of polycaprolactone (PCL) into the branch of hyperbranched polyurethane (HPUs) increases the mechanical and thermal behavior. Various methodologies were adopted to develop the HPs (shown in Fig. 2.3), that is, single monomer (SM) methodology and double monomer (DM) methodology which includes the following routes:

a) AB_2 type monomeric polymerization method
b) Self-condensed vinylic polymerization (SCVP) method
c) Self-condensing ring-opening polymerization (SCROP) method
d) Proton-transfer polymerization method
e) $A_2 + B_3$ polymerization
f) Couple-monomer method

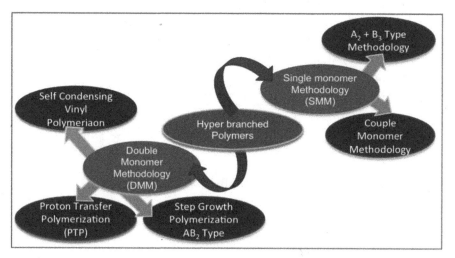

FIGURE 2.3 Synthesis routes of hyperbranched polymers.

Kricheldorf and coworkers in 1982 have reported the formation of HPs from AB_2-type of polymerization method. Here, when B groups react along with A then it results in unit D and unit L. Instead of this, when B doesn't react along with A, it results in unit T. HPs was first synthesized by Flory and groups from this methodology.[23] Another type of HPs was developed by using AB_2 type monomers treated in the light of N,N-dimethylformamide (DMF) with degree of branching (DBs) while the significances of these polymers were not reported and living chain growth mechanism results in the formation of similar HPs (Scheme 2.2).[24]

Further, SCVP method is an alternate of AB_2-type method proposed by Frechet (1995) in which AB type vinylic monomeric units were formed by AB* monomeric units, where "A" stands for vinylic monomeric units that contributes as an initiating unit to vinylic polymerization, and "B" behaves as an initiating moiety.[25] SCVP is widely used in anionic, a cationic polymerization and as a surface-initiating agent, as well as for the RAFT method. RAFT-SCVP shows superior compatibility in harsh situations with different functional groups.

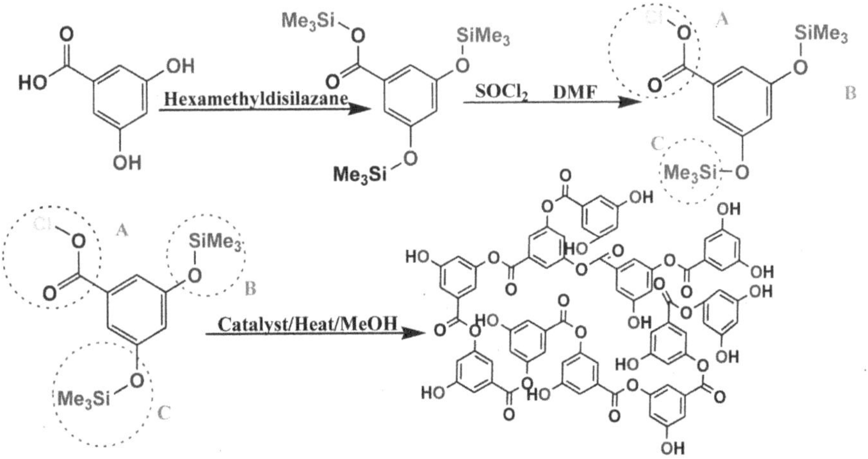

SCHEME 2.2   Schematic representation of AB$_2$-based HPs.

Frechet and Hedrick has reported the SCROP by cyclic polymerization of ester. It is much similar to the SCVP technique.[26]

In A$_2$ + B$_3$ method, two identical units of group A react with the three different identical groups of B moiety. The reaction progress of A$_2$ + B$_3$ type polymerization methods is depicted in Scheme 2.3. The postulates of A$_2$ + B$_3$ methodology are listed as follows:

  i)   Lack of cyclization in the propagation
  ii) Reaction sites A$_2$ and B$_3$ are identical in nature, and
  iii) Active sites of A$_2$ and B$_3$ are selective in nature.

These type of methodologies facilitates the feasibility of various monomers to the synthesis of HPs such as HPUs, epoxy resins, and poly(ester amide), and polyethylenes.[27]

For the first time, Yan and Gao reported the couple monomer methodology (CMM), having non-similar reactive functional groups in concerned monomers, which follow in-situ polymerization and generates an intermediate (AB$_n$) to produce HPs.[28] CMM choose first suitable monomer as a precursor and it resolves the issues of gelation and minimizes the cross-linking of monomers in the synthesis.

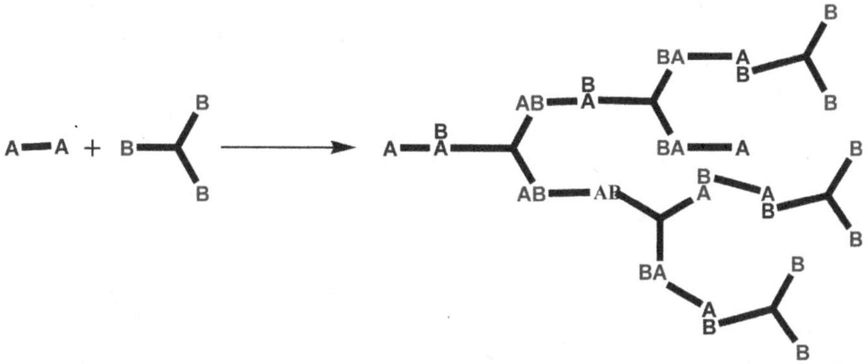

SCHEME 2.3   The reaction scheme of $A_2 + B_3$ polymerization technique.

2.2.3   UV CURABLE POLYMERS

UV curing materials are radiated and cured during polymerization from ultraviolet radiations. The graphical representation of UV curable polymers is given in Figure 2.4. These polymers are exclusively used for sealing, bonding, and surface coating. UV curable polymers have received increasing care because of their significances like as low VOC emission, lower energy consumption, low capital investment, and quick curing, while, petro-based acrylates leading to the UV curable systems. Different kind of bio-renewable materials might be adopted for the development of VOs-based UV curable materials. VOs are abundantly used because of their properties such as cost effective and easily available, nontoxic and biodegradable, and eco-friendly, etc.[29] Some of the VOs, like LO shows self-curing property through photo-oxidization mechanism with very longer drying time, when exposed to UV radiations.[30]

Allyl and vinyl groups including acrylate, methacrylates are introduced to the chemical modification of VOs to reduce the drying time.[31] Hence, acrylated epoxidized VOs are promising renewable oligomers for UV curable paint and coating industries and show superior physicomechanical and thermal properties than that of their petroleum-based counterparts. Moreover, cardinal-modified acrylates are responsible for further chemical modification of VOs and these oligomers were then formulated into UV curable paints and coating industries worldwide.[32,33]

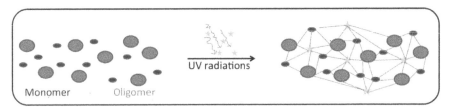

FIGURE 2.4   Graphical representation of UV curable polymers.

2.2.4   NON-SOLVENT/ZERO SOLVENT POLYMERS

Here, we are concerned about the synthesis of solventless VO-based materials. Solventless polymers such as polyesramide, polyethrimide, alkyd, polyurethanes, and epoxies might be developed from different types of VOs (LO, jatropha oil, JJO, flax oil, etc). Solventless or non-solvents means that during development of VO-based polymers, the minimal concentration of solvent is used or introduced for dilution, which results in high cross-linking, highly viscous product, and restricts the polymer chains mobility in the process of polymerization. The zero-solvent resins can be categories as follows: (1) solventless liquid coatings VOs, oligomeric binder/cross-linker combinations, moisture curable coatings, radiation-curable coatings, and thermally initiated vinyl monomer coatings; (2) Powder coatings—dry powders, powder slurries, and radiation curable powder coatings.[34] Solventless method can also be used in the fabrication of polymeric coatings, as shown in Figure 2.5.

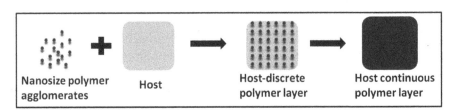

FIGURE 2.5   Graphical representation of solventless polymer coating process.

For instance, the synthesis of polyesteramide is accomplished by chemical interaction between N, N'-bis(2-hydroxyethyl) fatty amide and an anhydride (phthalic anhydride, maleic anhydride, and succcinic anhydride, etc). Solventless interaction provides ease of chemical reaction carried out at low range of reaction temperature between the reactants. The presence of small amount of metal also controls the reaction time period because of its catalytic effect.

2.3 MODIFICATION OF THE SUSTAINABLE POLYMERS AND THEIR NANOCOMPOSITES

Blending is the process from which the properties of VO-based polymers can be improved with other reinforced materials or by transforming these polymers into composites at their nanoscales improves the mechanical strength, thermal stability, hydrophilicity, and biodegradability of the product.[35] VO-based polymers are also modified by the reinforcement of nanoclays, nanofibers, carbon nanotubes (CNTs), conducting polymers, and metal oxides (Fig. 2.6) to improve their properties such as toughness, permeability, crystallinity, glass transition temperature (T_g), antimicrobial activity, as well as the for functional properties like physicomechanical characteristics, thermal stabilities, heat distortion capacities, flame retardancy, etc.[36]

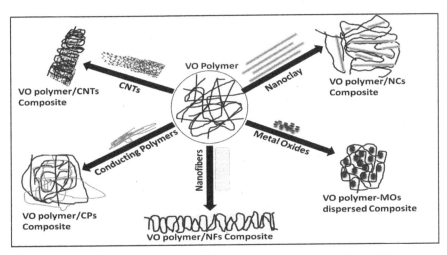

FIGURE 2.6 Modification of VO-based polymer from various types of nanofiller.

The wider range of nanofillers, cross-linkers, nanofibers, and nanoclays are being used in the development of surfaces coated materials, which act as reinforcing agents in the polymeric matrix to enhance the properties such as hardness, strength, and stiffness of the surface-coated material.[37,38] Various types of reinforced nanofillers to develop the nanocomposite anticorrosive surface-coated materials are:

I.   CNTs[39,40]
II. Natural Nanofibers[41]
III. Nanoclays[42]
IV. Conducting Polymers (CP)[43]
V.   Meta and Metal Oxide Nanoparticles (NP)[44]

2.4 SUMMARY AND FUTURE SCOPE OF THE SUSTAINABLE POLYMERS

The present chapter covered the classifications of sustainable polymers in terms of eco-friendly, non-solvent, hyperbranched, WB, and UV curable coatings. Various types of sustainable polymer nanocomposites have also been described.

Although the concept of sustainable polymers have already explored and are widely used in almost every field such as films and coatings, drug delivery, biomedical, tissue engineering, etc. However, their practical application is hampered by many problems. One of the most important problem is associated with other polymers during their blending and composite formation. During their processing the choice of appropriate solvent is prerequisite. Thus, we must have in-depth knowledge of polymer, nanofillers, and the required solvents for the development of their nanocomposites. Apart from this, new and advanced techniques should be used to modify the structure and properties of the sustainable polymer, little by little. In this regard, researchers and scientists should make use of their expertise and collaborate to develop novel materials for human prosperity and a more sustainable society.

ACKNOWLEDGMENT

One of the authors Mr. Abu Darda acknowledges to the head of Department of Chemistry for providing the research facility of its material research laboratory and also for the grant received under the UGC scheme of Non-NET fellowship.

KEYWORDS

- vegetable oils
- coatings
- sustainable
- polymers
- nanocomposites

REFERENCES

1. Patel, N. K.; Shah, S. N. *Biodiesel from Plant Oils*; Elsevier Inc.; 2015. DOI: 10.1016/B978-0-12-800211-7.00011-9.
2. Liu, B.; Wang, Y. A Novel Design for Water-Based Modified Epoxy Coating with Anti-Corrosive Application Properties. *Prog. Org. Coat.* **2014,** *77,* 219–224. DOI: 10.1016/j.porgcoat.2013.09.007.
3. Sharmin, E.; Zafar, F.; Akram, D.; Alam, M.; Ahmad, S. Recent Advances in Vegetable Oils Based Environment Friendly Coatings: A Review. *Ind. Crops Prod.* **2015,** *76,* 215–229. DOI: 10.1016/j.indcrop.2015.06.022.
4. Ghosal, A.; Ahmad, S. High Performance Anti-Corrosive Epoxy-Titania Hybrid Nanocomposite Coatings. *New J. Chem.* **2017,** *41,* 4599–4610. DOI: 10.1039/c6nj03906e.
5. Sharmin, E.; Rahman, O. U.; Zafar, F.; Akram, D.; Alam, M.; Ahmad, S. Linseed Oil Polyol/ZnO Bionanocomposite Towards Mechanically Robust, Thermally Stable, Hydrophobic Coatings: A Novel Synergistic Approach Utilising a Sustainable Resource. *RSC Adv.* **2015,** *5,* 47928–47944. DOI: 10.1039/c5ra03262h.
6. Pathan, S.; Ahmad, S. Synthesis, Characterization and the Effect of the s-triazine Ring on Physico-Mechanical and Electrochemical Corrosion Resistance Performance of Waterborne Castor Oil Alkyd. *J. Mater. Chem. A.* **2013,** *1,* 14227–14238. DOI: 10.1039/c3ta13126b.
7. Mashouf Roudsari, G.; Mohanty, A. K.; Misra, M. Green Approaches to Engineer Tough Biobased Epoxies: A Review. *ACS Sustain. Chem. Eng.* **2017,** *5,* 9528–9541. DOI: 10.1021/acssuschemeng.7b01422.
8. Pathan, S.; Ahmad, S. Green and Sustainable Anticorrosive Coating Derived from Waterborne Linseed Alkyd Using Organic-Inorganic Hybrid Cross Linker. *Prog. Org. Coat.* **2018,** *122,* 189–198. DOI: 10.1016/j.porgcoat.2018.05.026.
9. Irfan, M.; Bhat, S. I.; Ahmad, S. Reduced Graphene Oxide Reinforced Waterborne Soy Alkyd Nanocomposites: Formulation, Characterization, and Corrosion Inhibition Analysis. *ACS Sustain. Chem. Eng.* **2018,** *6,* 14820–14830. DOI: 10.1021/acssuschemeng. 8b03349.
10. Galliano, F.; Landolt, D. Evaluation of Corrosion Protection Properties of Additives for Waterborne Epoxy Coatings on Steel. *Prog. Org. Coat.* **2002,** *44,* 217–225. DOI: 10.1016/S0300-9440(02)00016-4.
11. Rahman, O. U.; Kashif, M.; Ahmad, S. Nanoferrite Dispersed Waterborne Epoxy-Acrylate: Anticorrosive Nanocomposite Coatings. *Prog. Org. Coat.* **2015,** *80,* 77–86. DOI: 10.1016/j.porgcoat.2014.11.023.

12. Panda, S. S.; Panda, B. P.; Nayak, S. K.; Mohanty, S. A Review on Waterborne Thermosetting Polyurethane Coatings Based on Castor Oil: Synthesis, Characterization, and Application. *Polym. Plast. Technol. Eng.* **2018,** *57,* 500–522. DOI: 10.1080/03602559.2016.1275681.

13. Honarkar, H.; Barmar, M.; Barikani, M. Synthesis, Characterization and Properties of Waterborne Polyurethanes Based on Two Different Ionic Centers. *Fibers Polym.* **2015,** *16,* 718–725. DOI: 10.1007/s12221-015-0718-1.

14. Jena, K. K.; Sahoo, S.; Narayan, R.; Aminabhavi, T. M.; Raju, K. Novel Hyperbranched Waterborne Polyurethane-Urea/Silica Hybrid Coatings and their Characterizations. *Polym. Int.* **2011,** *60,* 1504–1513. DOI: 10.1002/pi.3109.

15. Irfan, M.; Bhat, S. I.; Ahmad, S. Waterborne Reduced Graphene Oxide Dispersed Bio-Polyesteramide Nanocomposites: An Approach Towards Eco-Friendly Anticorrosive Coatings. *New J. Chem.* **2019,** *43*(12). DOI: 10.1039/c8nj03383h.

16. Ezidinma, T. A; Onukwuli, D. O.; Uzoh, C. F. Synthesis and Characterization of Water-Reducible Alkyd Resin from Cottonseed Oil. *Int. J. Eng. Appl. Sci.* **2015,** *2*(12), 80–84.

17. Barandiaran, M. J.; De La Cal, J. C.; Asua, J. M. Emulsion Polymerization. *Polym. React. Eng.* **2008,** 233–272. DOI: 10.1002/9780470692134.ch6.

18. Bon, S. A. F.; Ohno, K.; Haddleton, D. M. Water-Soluble and Water Dispersible Polymers by Living Radical Polymerisation. *ACS Symp. Ser.* **2001,** *780,* 148–161. DOI: 10.1021/bk-2001-0780.ch009.

19. Chen, L.; Bao, Z.; Fu, Z.; Li, W. Characterization and Particle Size Control of Acrylic Polymer Latex Prepared with Green Surfactants. *Polym. Renew. Resour.* **2015,** *6,* 65–74. DOI: 10.1177/204124791500600203.

20. Lopez, A.; Degrandi-Contraires, E.; Canetta, E.; Creton, C.; Keddie, J. L.; Asua, J. M. Waterborne Polyurethane-Acrylic Hybrid Nanoparticles by Miniemulsion Polymerization: Applications in Pressure-Sensitive Adhesives. *Langmuir* **2011,** *27,* 3878–3888. DOI: 10.1021/la104830u.

21. Berber, H. Emulsion Polymerization: Effects of Polymerization Variables on the Properties of Vinyl Acetate Based Emulsion Polymers. *Polym. Sci.* **2013.** DOI: 10.5772/51498.

22. Kricheldorf, H. R.; Zang, Q. Z.; Schwarz, G. New Polymer Syntheses: 6. Linear and Branched Poly(3-hydroxy-benzoates). *Polymer (Guildf)* **1982,** *23,* 1821–1829. DOI: 10.1016/0032-3861(82)90128-8.

23. Cook, A. B.; Barbey, R.; Burns, J. A.; Perrier, S. Hyperbranched Polymers with High Degrees of Branching and Low Dispersity Values: Pushing the Limits of Thiol-Yne Chemistry. *Macromolecules* **2016,** *49,* 1296–1304. DOI: 10.1021/acs.macromol.6b00132.

24. Liu, J.; Huang, W.; Zhou, Y.; Yan, D. Synthesis of Hyperbranched Polyphosphates by Self-Condensing Ring-Opening Polymerization of HEEP Without Catalyst. *Macromolecules* **2009,** *42,* 4394–4399. DOI: 10.1021/ma900798h.

25. Chang, H. T.; Fréchet, J. M. J. Proton-Transfer Polymerization: A New Approach to Hyperbranched Polymers. *J. Am. Chem. Soc.* **1999,** *121,* 2313–2314. DOI: 10.1021/ja983797r.

26. Barua, S.; Chattopadhyay, P.; Karak, N. S-Triazine-Based Biocompatible Hyperbranched Epoxy Adhesive with Antibacterial Attributes for Sutureless Surgical Sealing. *J. Mater. Chem. B.* **2015,** *3,* 5877–5885. DOI: 10.1039/c5tb00753d.

27. Ikladious, N. E.; Mansour, S. H.; Asaad, J. N.; Emira, H. S.; Hilt, M. Synthesis and Evaluation of New Hyperbranched Alkyds for Coatings. *Prog. Org. Coat.* **2015,** *89,* 252–259. DOI: 10.1016/j.porgcoat.2015.09.008.

28. Gao, C.; Yan, D. Synthesis of Hyperbranched Polymers from Commercially Available A2 and BB'2 Type Monomers. *Chem. Commun.* **2001,** 107–108. DOI: 10.1039/b006048h.

29. Fertier, L.; Koleilat, H.; Stemmelen, M.; Giani, O.; Joly-Duhamel, C.; Lapinte, V.; Robin, J. J. The Use of Renewable Feedstock in UV-Curable Materials-A New Age for Polymers and Green Chemistry. *Prog. Polym. Sci.* **2013**, *38*, 932–962. DOI: 10.1016/j. progpolymsci.2012.12.002.

30. Biermann, U.; Bornscheuer, U.; Meier, M. A. R.; Metzger, J. O.; Schäfer, H. J. Oils and Fats as Renewable Raw Materials in Chemistry. *Angew Chemie Int. Ed.* **2011**, *50*, 3854–3871. DOI: 10.1002/anie.201002767.

31. Yan, J.; Webster, D. C. Thermosets from Highly Functional Methacrylated Epoxidized Sucrose Soyate. *Green Mater.* **2014**, *2*, 132–143. DOI: 10.1680/gmat.14.00002.

32. Balachandran, V. S.; Jadhav, S. R.; Vemula, P. K.; John, G. Recent Advances in Cardanol Chemistry in a Nutshell: From a Nut to Nanomaterials. *Chem. Soc. Rev.* **2013**, *42*, 427–438. DOI: 10.1039/c2cs35344j.

33. Bloise, E.; Becerra-Herrera, M.; Mele, G.; Sayago, A.; Carbone, L.; D'Accolti, L; Mazzetto, S. E.; Vasapollo, G. Sustainable Preparation of Cardanol-Based Nanocarriers with Embedded Natural Phenolic Compounds. *ACS Sustain. Chem. Eng.* **2014**, *2*, 1299–1304. DOI: 10.1021/sc500123r.

34. Rahman, I. M. M.; Begum, Z. A.; Yahya, S.; Lisar, S.; Motafakkerazad, R. Cell AS-P. Complimentary Contributor Copy, 2016.

35. Ahmad, S. Preparation of Eco-Friendly Natural Hair Fiber Reinforced Polymeric Composite (FRPC) Material by Using Of Polypropylene and Fly Ash: A Review. *Int. J. Sci. Eng. Res.* **2014**, *5*, 969–972.

36. Rawat, N. K.; Khatoon, H.; Kahtun, S.; Ahmad, S. Conducting Polyborozirconia(o-toluidine) Nanostructures: Effect of Boron and Zirconia Doping on Synthesis, Characterization and their Corrosion Protective Performance. *Compos. Commun.* **2019**, *16*, 143–149. DOI: 10.1016/j.coco.2019.10.002.

37. Ahmad, S.; Ansari, A.; Siddiqi, W. A.; Akram, M. K. Nanoporous Metal-Organic-Framework. *Mater. Res.* **2019**, *58*, 107–139. DOI: 10.21741/9781644900437-6.

38. Khatoon, H.; Iqbal, S.; Ahmad, S. Influence of Medium on Structure, Morphology and Electrochemical Properties of Polydiphenylamine/Vanadium Pentoxide Composite. *SN Appl. Sci.* **2019**, *1*, 1–11. DOI: 10.1007/s42452-019-0285-y.

39. Dresselhaus, M. S. Carbon Nanotubes. **1995**, *33*. DOI: 10.1016/0008-6223(95)90067-5.

40. Khatoon, H.; Iqbal, S.; Ahmad, S. Influence of Carbon Nanodots Encapsulated Polycarbazole Hybrid on the Corrosion Inhibition Performance of Polyurethane Nanocomposite Coatings. *New J. Chem.* **2019**, *43*, 10278–10290. DOI: 10.1039/c9nj01671f.

41. Suzuki, A.; Nagata, F.; Inagaki, M.; Kato, K. Surface Modification of PLA Nanofibers for Coating with Calcium Phosphate. *Trans. Mater. Res. Soc. Japan* **2018**, *43*, 271–274. DOI: 10.14723/tmrsj.43.271.

42. Dong, J.; Zhang, J. Biomimetic Super Anti-Wetting Coatings from Natural Materials: Superamphiphobic Coatings Based on Nanoclays. *Sci. Rep.* **2018**, *8*, 1–12. DOI: 10.1038/s41598-018-30586-4.

43. Khatoon, H.; Ahmad, S. A Review on Conducting Polymer Reinforced Polyurethane Composites. *J. Ind. Eng. Chem.* **2017**, *53*, 1–22. DOI: 10.1016/j.jiec.2017.03.036.

44. Khatoon, H.; Ahmad, S. Vanadium Pentoxide-Enwrapped Polydiphenylamine/Polyurethane Nanocomposite: High-Performance Anticorrosive Coating. *ACS Appl. Mater. Interfaces* **2019**, *11*, 2374–2385. DOI: 10.1021/acsami.8b17861.

CHAPTER 3

An Efficient Self-Charging Photocapacitor for Integrated Energy Storage and Conversion

SHAMA ISLAM[1*], ZUBAIR ASLAM[1], HALIMA KHATOON[2], HANA KHAN[1], and MOHD ZULFEQUAR[1]

[1]*Material Research Laboratory, Department of Physics, Jamia Millia Islamia, New Delhi, India*

[2]*Material Research Laboratory, Department of Chemistry, Jamia Millia Islamia, New Delhi, India*

Corresponding author. E-mail: shamaphysics786@gmail.com

ABSTRACT

The key constraint of the applied assimilation of renewable energy technologies to assimilate the power grid is the alternative attributes of renewable energy resources. The avenue against decentralized energy systems that can aftermath an on-demand power to achieve beneath alternative cycles of power bearing requires the addition of accessories and the ability of electronics activated for conversion systems. In this manner, energy storage devices with energy technology can accommodate the temporal adaptability bare to antithesis bounded power of bearing and power of consumption. The energy storage devices ahead use of mechanical force, have emerged as a capable technology for the conception of independent systems for multifunctional, maintenance-free, independent operations are called self-powered electrochemical energy storage devices. A self-charging capacitor as an efficient solar energy storage device was fabricated driven by light. The device which achieves the name, the photocapacitor, works with a high quantum conversion efficiency which converts or stores the visible light energy in the form of electrical power. The structure of a photocapacitor is based on a multilayered photoelectrode with the presence of organic electrolyte. Materials comprising dye-sensitized semiconductor nanoparticles/activated carbon particles /hole-trapping layer in contact; those are responsible for stored photogenerated charges at the electric double layer. Such a photocapacitor would be more desirable device for storing energy over conventional batteries due to its rapid change to current response also has a high quantum conversion efficiency.

3.1 INTRODUCTION

The utility of sun-based quality has the viable to outfit a great answer for the determination crisis.[1,2] The challenge for straight solar energy storage is to fabricate a device incorporating both photoelectric and storing functions in an alone cell makeup. To this passing, efforts have been made to confederate a photovoltaic (PV) electrode and redox brisk materials of rechargeable batteries.[3,4] Recently, polymer nanocomposites have been extensively used in power-harvesting devices, such as perovskite hybrid solar cells (HSCs), nanogenerators, and fuel cells, due to their noble ability and lightweight, high efficiency, and uncompounded in qualification, which is promising as a renewable resolution funds for sustainable revelation of the future. At the same time, polymer nanocomposites have been used in spirit energy storage, such as supercapacitors, due to their high energy density and long life cycle.[5,6] These power harvesting and storage approaches are improved as independent technologies but on whole are taken together as a power system. Generally, the power pack is supported on a silicon solar compartment and a strong state lithium battery as two free parts, which are spacious, profound, and firm. Anyway the single crystalline silicon-based solar cells improve proficiency for business items; however, the cost of these solar cells is generally high. Undefined silicon-based solar cells get corrupted when presented to light and their proficiency diminishes by 10–20%.[8] On the other hand, HSCs also have the potential to exceed better performance while still retaining the benefits such as low-cost, flexible, and easy to produce with polymer nanocomposites. Also, an authoritative review of polymer nanocomposites-based supercapacitors covers polymer materials with carbon-based (multi-walled carbon nanotubes [CNTs] and graphene) nanomaterials show good advantages.[7] The requirement to improve the unique needs in a few fields such as hybridizing energy harvesting and capacity units as coordinated power pack upheld on same polymer nanostructured materials. This might be an agent approach to acquire a little size, lightweight, and respectable thickness vitality framework. The huge improvements in current devices boost the need for sensible improvements finished by means of coordinating energy reaping and capacities in a solitary device as a self-controlling unit. Precise incorporated devices depending on the transformation of solar-powered, mechanical, or heat energy by way of vitality accumulating parts and the direct charging of energy stockpiling parts had been created at some point previously.[1-7] As a boundless, across the board, and sustainable clean asset, sunlight-based energy has for a while been a promising choice in evaluation of normal nonrenewable electricity supply. Silicon solar-powered cells are by way of an extended shot the pleasant marketed PV devices. In any case, they are inflexible gadgets with a stressed manufacture technique, which ruins their application in adaptable and handy hardware. New-age solar-based cells, which includes color sharpened sun based cells (dye sensitized solar cells; DSSCs), natural sun orientated cells, and the developing perovskite solar oriented cells (PVSCs), have the upsides of excessive productivity, easy manufacture, minimal attempt, and flexibility.[8,9]

The continuous supply of solar energy from the sun is estimated to be 3×10^{24} J/year, which is 10^4 times more than the consumption rate of mankind.[10] This has impelled the development of PV devices, typically solar cells as prime energy production devices in conjunction with the depletion of fossil fuels and mineral energy sources.[11] An arms-race evolution and development of solar cells have arisen from the first generation of conventional single silicon solar cells to the second generation of semiconductor thin film-based solar cells ascribed to the employment of a thinner film as the working electrode as compared to the first-generation PV. Now, a third-generation PV is emerging, namely molecular absorber solar cells, which comprise polymer solar cells, DSSCs, quantum dot solar cells, and the recently developed perovskite solar cells. A DSSC is a photoelectrochemical device in which sensitizers generate electrons upon light absorption between a dye-adhered metal oxide surface compartment and hole-conducting electrolyte.[12] The third generation of solar cells are said to be compatible with any flexible substrates, with lower material and manufacturing costs, even though the first- and second-generation solar cells have attained conversion efficiencies of ~15–25%, as reported in 2010.[13] The emerging trend of using third-generation solar cells over well-established first- and second-generation solar cells is ascribed to properties such as the best performance under low-light or indoor conditions, transparency, cost effectiveness, and easy integration in buildings as solar windows. Silicon PV cells were reported to exhibit an energy conversion efficiency of 25% in 2010, outperforming the third generation of solar cells (DSSC), which have a conversion efficiency of 15%, as reported in 2014.[14] The advantages of a DSSC over a silicon solar cell are better stability, higher efficiency, low production cost, and easy fabrication.[1,7,8] While solar-driven supercapacitors have still a long way from down to earth applications, the related research is rising. This chapter explains the utilization of polymer nanocomposites to investigate the making of ptotocapacitors with expanded specific capacitances and high energy densities.[11] As of late, Bullock et al. enhanced a photogalvanic cells bolstered on lyotropic smooth pellucid nanosystems tranquilize by photoactive atoms and clarify concur for repeat as an electrochemical capacitor.[12] Among the conducting polymers, polypyrrole (PPy) has special attention which evolves as cheap and abundantly available polymers and is widely known to exhibit good absorption characteristics. PPy is deeply the most largely intent, which is due to its facility of composition and admirable redox properties. Compared to other polymers, their high conductivity also contributes in the improvement of transportation phenomenon despite of possessing regular properties of polymers. Apart from this, PPy have other properties probably mound compactness, exalted capacity appraise, and tranquility in preparation. Incorporation of CNTs with PPy preserves mechanical changes resulting to store a large amount of electric charge.[13–15] As charge is quantized into basic charges, that is, electrons, and each such basic charge needs a base space, a huge part of cathode surface is not accessible for capacity since empty spaces are not good with charge's prerequisites. With nearness of CNTs spaces might be custom fitted to estimate few too substantial or too little and therefore limit significantly expanded. Because of tunable bandgap, high absorption coefficient, and high natural charge carrier mobility CNTs can likewise be utilized as transporting material of holes. These nanocomposites do not need any binding substance that is an important practical advantage.[16]

3.2   STRUCTURE AND WORKING PRINCIPLE OF PHOTOCAPACITOR

Dates back a decade, a device that combines the function of energy conversion and energy storage was said to be photocapacitor, with practical efficiencies aspects it has only been recently that such devices have been reported. In addition to this, mutistep fabrication techniques are required to fabricate most devices that energy conversion and integrate energy storage into a single template that further strains the commercial outlook of the (typically) third-generation solar devices with nanocarbon materials and/or nanoscale metal-oxides, because the grid-scale deployment is depend to choose the raw materials costs, manufacturing costs, and combined efficiency, these are deciding factor. Figure 3.1 shows the systematic representation of a photocapacitor.

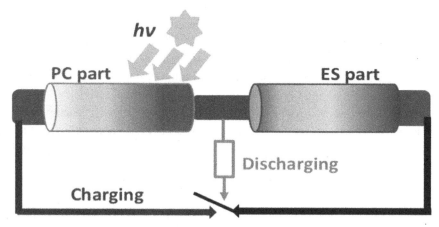

FIGURE 3.1 Schematic representation of photocapacitor with solar cell and a supercapacitor.

However, to authorize the affiliation avenue as getting meaningful, any accumulated arrangement has to accent a cost, efficiency, or accomplishment account of affiliation against the appliance of abstracted embodiments of activity accumulator and conversion. Figure 3.2 shows the charging during sunlight and discharging in the absence of sunlight of a photocapacitor.

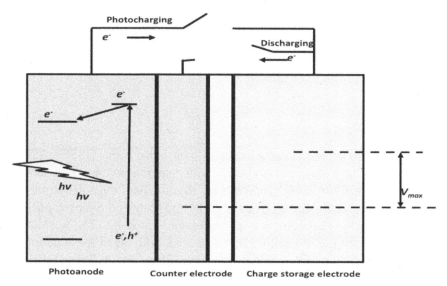

FIGURE 3.2 Illustration of charging during sunlight and discharging in the absence of sunlight of a photocapacitor.

This accurately poses a claiming for accessories acute assorted adult accomplish for actual and accessory fabrication, accessories relying on big-ticket materials, and accessories area circuitous packaging requirements are all-important due to on-board electrolytes. There are two capital configurations of the chip device: the acceptable collapsed anatomy and the new arising fiber-shaped one. The affiliation of the solar activity about-face allotment and the electrical accumulator allotment can be accomplished either in one accessory or by the administration of an accepted electrode as a connection. A completed solar capacitor should abide by a foreground electrode for the solar corpuscle and an adverse electrode for the supercapacitor.

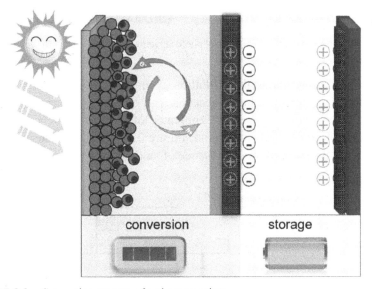

FIGURE 3.3 Conversion process of a photocapacitor.

A variety of solar cells, including silicon solar cells, DSSCs, OSCs, and PVSCs, have auspiciously chip with supercapacitors to assemble self-charging activity units. Upon a blaze illumination, the solar corpuscle converts solar activity into electrical activity and accuses the supercapacitor, which is usually declared as a photocharging process. The absolution action occurs if the capacitors are affiliated with an alien amount to accumulation power, either under light or in the dark. A light-absorbing electrode (photoelectrode) and a counter electrode are simple sandwich-type electrochemical cell consisting of a redox free liquid electrolyte, called photocapacitor. The porous layer of activated carbon particles and a heterojunction of dye adsorbed semiconductor nanoparticles bear photoelectrode; and also a porous activated carbon layer associated with counter-electrode. In principle, when the light-induced charge separation of dye molecules at the heterojunction interface, the charging reaction is initiated, that is, the mechanism of dye sensitization happens when there is an electron injection from photoexcited dye molecules to the semiconductor conduction band. After charge separation, electrons and the photogenerated holes of the dye transfer to activated carbon layers immobilized at the counter-electrode and photoelectrode, respectively. Positive and abrogating accuse are ultimately accumulated on the microporous apparent of activated carbon that holds the electric double layer in acquaintance with an organic electrolyte band-aid of top ionic concentration. The PV energy conversion part and the electrochemical energy-storage are responsible for overall efficiency of the integrated device. The overall device efficiency is depending on the electrical resistance between these parts notably. The storage efficiency of the supercapacitors multiply by the PCE (photovoltaic efficiency) of the solar cell gives electrical resistance or this can be calculated by the ratio of energy of the illuminated light and the energy stored during the photocharging process.

3.3   TYPES OF PHOTOCAPACITORS

The first ever photocapacitor reported was by Miyasaka and Murukami which was a multilayered photoelectrode. On the basis of charging process, there are two basic types of capacitors: electrochemical double-layer capacitor (EDLCs) and pseudocapaciors.

3.3.1   *ELECTROCHEMICAL DOUBLE-LAYER CAPACITOR*

EDLCs are distinctive capacitor with the properties lying in between near-middle of that of the regular capacitor and secondary (rechargeable) batteries. The construction of an EDLC involves two electrodes with a current collector, an electrolyte, and a separator (Fig. 3.4).

The two electrodes with a current collector are immersed in the electrolyte separated by a porous separator. These capacitors store the charges by forming an electric double layer formed by ions accumulated to the surface of the electrode. In EDLCs, the capacitive charging depends only on conductor–electrolyte interface electrostatic charging. The electric double layer point is where electrons-holes are electrically attracted and building an organized state between the solid electrode and the electrolyte. The conductor may be of nonmetallic material of metallic conductivity or metal itself. Activated carbon as electrode has great capabilities for an EDLC electrode material because of its porosity and has a large effective surface area that gives higher charge storage.[17] Some other potential electrode materials for EDLCs are RuO_2, RuO_2 with activated carbon composite, copper sulfide minerals, etc.[18,19] The main characteristics of EDLC are capacitance, internal resistance, and leakage current. The capacitance has a greater dependency on the surface structure like specific surface area, pore diameter, volume, etc. of the electrode material. An EDLC can be considered an equivalent circuit where a large number of miniature capacitors having internal resistances are connected in parallel. The resistance component of the electrolyte and electrodes creates internal resistance which causes a drop in effective voltage. When an EDLC is charged for an extended period of time, the charge current decreases but it does not become zero. Rather it settles at a certain constant value, which is called the leakage current. The magnitude of this current is determined by factors such as electrode material, cell construction, usage temperature, etc.

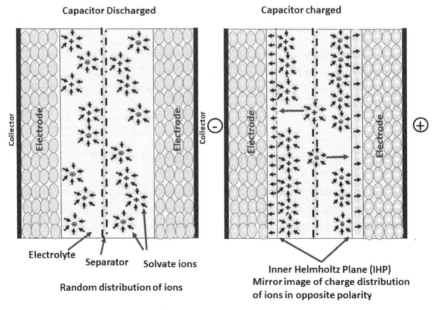

FIGURE 3.4   Mechanism of EDLCs.

3.3.2 PSEUDOCAPACITOR

A Pseudocapacitor is a cross between a battery and an EDLC capacitor. Its construction also consists of two electrodes immersed in an electrolyte and separated by a porous separator. In pseudocapacitors, the capacitive effect is because of two capacitances: double layer capacitance which is the same capacitance in EDLC capacitors and pseudocapacitance which is the result of the chemical process involving charge transfer by reversible redox (reduction–oxidation) reaction in which there is a flow of ions from the electrolyte into the electrode. This charge transfer is Faradaic in nature. A redox reaction is a type of reaction where reduction and oxidation both occur together. In a reduction mechanism, there is a gain of electrons and in the oxidation mechanism, there is a loss of electrons. The species that donate electrons are known as reducing agents while the species that accept electrons are known as oxidizing agents. This is why these capacitors are also known as redox capacitors. The two capacitances combined parallelly constitute the total capacitance for the pseudocapacitor. The fact it is called pseudocapacitor is because of the fact that the electrochemical character-istics of the pseudocapacitive materials imitate the EDLC electrochemical characteristics like cyclic voltammetry and galvanostatic discharge. Some of the pseudocapacitive electrode materials are transition metal oxides, transition metal sulfides and their composites, metal nitrides nanomaterials and composites, metal oxides, lithium/sodium metal oxide-based materials, transition metal carbides, and some conducting polymers. These materials have lower conductivity and lower surface area accessible for electrolytes in comparison with carbon materials. To overcome these shortcomings various approaches have been reported.

RuO_2 is the most studied transition metal oxide material for pseudoca-pacitive material because of their high conductivity and high capacitance in the presence of water.[20] Huang et al. reported that ultrathin MnO_2 have enhanced ion-accessible surface area and greater specific capacitance (1000 F/g) in comparison with the bulk MnO_2 (200–250 F/g).[21] Hou et al. reported the synthesis of spherical microstructural Co_3O_4 which shows high capaci-tance of 720 F/g.[22] Tao et al. reported the synthesis of amorphous CoS_x which showed the capacitance of 369 F/g with the current density of 5 mA/cm^2.[23] Zhu et al. reported that NiS hollow spheres have a specific capacitance of 583–927 F/g at 4.08–10.2 A/g.[24] Alshareef reported on Ni–Co–S mesoporous nanostructure on a sheet with high specific capacitance (1418 F/g at 5 A/g) with a great long-term cycling stability.[25] Metal nitrides have also shown a remarkable capacitive properties for pseudo capacitive material like MoN, VN, TiN, and their composites with the capacitance range lying in between 200 and 1350 F/g.[26-29] Layered double hydroxides (LDH) constitute some of the favorable materials for electrodes in a pseudocapacitor because of their high capacitance, cost-effectiveness, and facile synthesis process. Shao et al. synthesized core-shell NiAl-LDH microspheres with high specific capacitance of 735 F/g at 2 A/g, high specific surface area of 124.7 m^2/g, and a long term cycling life.[30] Conducting polymers are also one of the most favorable electrode material for pseudocapacitors because of their high conductivity and their ability to be oxidized or reduced at the surface. Most studied conducting polymer materials for pseudocapacitor electrode are polyaniline, poly (styrene sulfonate), polythiophene, polymethyl methacrylate, and poly(3,4-ehylenedioxythiophene) with their pseudocapacitances in the range between 500 and 340 F/g.[31,32] Lithium/sodium metal oxide-based materials are the materials for the pseudocapacitors with high energy density and high power density. Dong et al. reported the fabrication of self-doped $Li_4Ti_5O_{12}$ as anode and activated carbon as cathode with maximum energy density of 67 W/kg and maximum power density of 8000 W/kg.[33] Senthilkumar et al. reported the phosphate-based sodium nanomaterial $NaMPO_4$ (M = Ni, Co, Mn) which showed the highest capacitance with NaOH (1 M) as electrolyte. $NaNiPO_4$ showed the specific capacitance of 56 F/g and energy density of 20 Wh/kg.[34] For metal oxide materials various structures of TiO_2 have been studied for energy storage purposes. Liu et al. reported the synthesis of porous TiO_2–B nanosheets which showed the capacity of 332 mAh/g at 33.5 mAh/g and 202 mAh/g at 3350 mA/g.[35]

3.3.3 DYE SENSITIZED SOLAR CELLS

DSSCs are a type of photoelectrochemical solar cells. A DSSC consists of three main parts. On the top, there is a transparent conducting plate with the thin layer of metal oxides entered at the other side of the plate. This plate is then immersed in a solution of photosensitive molecular sensitizers (dye) for some time. As a result of this, dye that is covalently bonded with the surface of metal oxide will be left. There is another conducting plate with a thin layer of electrolyte on top of it. These two plates are connected and sealed together to form a DSSC solar cell. The dye is a photoactive material that catches the photons from the sunlight and uses the photon energy to excite the electrons. These excited electrons then injected into and conducted away by the metal dioxide layer. The electrons are returned back through other conducting plate and collected by the electrolyte completing the circuit. The schematic representation for DSSC solar cell is shown in Figure 3.5.

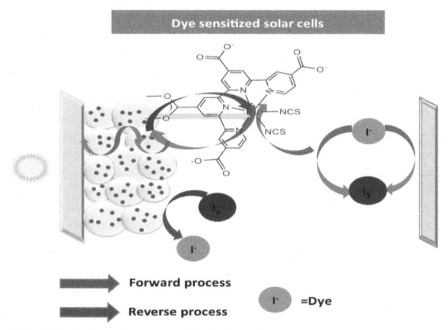

FIGURE 3.5 Schematic representation of the DSSCs.

As compared to silicon solar cells, DSSCs are cheaper, easy to manufacture, and more efficient. DSSCs have quite the advantages like it absorbs more sunlight per surface area than the silicon solar cells. These cells can function even in low-light or indirect sunlight conditions. DSSCs are mechanically robust which give higher efficiencies with different range of temperatures.

The performance of a solar cell can be evaluated by the efficiency (η) of converting the sunlight into electric power. This efficiency depends upon the open-circuit voltage, short-circuit current density, and the fill factor. To achieve maximum efficiency broadly two phenomena are considered: efficient light-harvesting and minimizing electron losses which govern the above parameters. To improve light harvesting in DSSCs, various techniques such as cosensitization, light scattering, and the use of transparent electrodes have been developed. Cosensitization is the use of a combination of two spectrum-complementary dyes to improve the range of absorption spectra. Cid et al. have used two types of dyes TT1 (60 nm) and JK2 (400–550 nm) and reported the efficiency of 7.74%, which was more than their single dye solar cells (3.52 for TT1 and 7.08% for JK2) and also showed high incident photo-to-electron conversion efficiency of 80%.[36] One more way to increase the light-harvesting is by scattering the absorbed light which can be obtained by incorporating larger particles and coating a scattering layer on the top of the active layer. 3D TiO_2 nanostructures such as nanotube arrays, aggregates, network of nanofibers have been developed to enhance the incident light scattering.[37–41] Another technique to enhance the light-harvesting is by using transparent counter electrodes for back-illuminated DSSCs. To achieve this, ultrapure CNTs are studied and resulted in achieving a transparency over 90% at 550 nm wavelength.[42] To improve the electron loss in DSSCs, development of compact layers, techniques of hydrogenation and protonation, plasma, and post-treatment techniques have been taken into account. The photoanode/ electrolyte and counter electrode/electrolyte surfaces have an impact on the rate of electron recombination. A common proposition to restrain this is to use a thin metal oxide compact layer to minimize the exposed counter electrode surface. Guan et al. reported that interfacial resistance of ZnO/FTO cell was lower (6.77 Ω) than the bare FTO substrate.[43] Another approach to improve photocatalytic and electronic property is to use hydrogenated or protonated photoanodes. Su et al. reported that H-TiO_2 nanocrystals showed improvement in short circuit current density (27%) and efficiency (28%).[44] Acid treatment introduces proton (H+) on the metal oxide surface. Wang and Zhou reported the influence of HCl treatment on TiO_2 that charge recombination is lower by a factor of five with an increase in open-circuit potential by 50 mV.[45]

3.3.4   ORGANIC SOLAR CELLS

Organic solar cell is the types of solar cells that use organic molecules and conducting polymers to absorb solar light and excite electrons to generate electricity. Based on the construction of an organic solar cell, there are various kinds of junction types: single layer junction, bilayer junction, bulk heterojunction, graded heterojunction, continuous junctions, etc. In single-layer junction type, an organic material (molecule or polymer) is sandwiched between high work function metallic conductor and low work function metallic conductor. In bilayer junction cells, the two layers of organic material with different electron affinity and ionization energy, making them electron donor and acceptor, are sandwiched between the two high and low work function metallic conductors. To overcome the hole-electron recombination in layered cells, bulk heterojunctions uses a solution containing electron donor and acceptor organic materials, dispersed between two low and high work function metallic conductors.

Organic solar cells have acceptable power conversion efficiency but this is not enough to commercialize it widely. The dissociated charge carriers are unable to reach the electrode due to several parameters hence photocurrent is unable to reach the outside circuit. It is because, mobility and diffusion length of carriers in organic solar cells are very small (<100 times) as compared to silicon solar cells. To overcome these limitations, many propositions have been addressed while designing material and devices like improving crystal structure to improve electrical conductivity, creating a large donor–acceptor interface to upgrade the dissociation of excitons and improving the collection of photogenerated carriers. Bagui and Iyer reported a 25% improvement in PCE of some organic solar cells by applying an electric field while annealing.[46] Hau et al. reported that annealing time and annealing temperatures can improve both crystallinity and PV performance of the material in organic solar cells.[47] Recently Cui et al. reported an enormous improvement in efficiency (16%) and simultaneously short-circuit current density and open-circuit voltage of organic solar cells by using chlorinated nonfullerene acceptor which shows extended optical absorption.[48]

ACKNOWLEDGMENT

This work was supported by the CSIR (file No: Pool No. 9056-A), Council of Scientific and Industrial Research, Human Resource Development Group (HRDG), Delhi, India.

KEYWORDS

- **photocapacitor**
- **device fabrication**
- **self charging**
- **energy storage**
- **faradic reactions**
- **quantum conversion efficiency**

REFERENCES

1. Meng, Q.; Cai, K.; Chen, Y.; Chen, L. Research Progress on Conducting Polymer Based Supercapacitor Electrode Materials. *Nano Energ.* **2017,** *36,* 268–285.
2. Niu, Z.; Luan, P.; Shao, Q.; Dong, H.; Li J.; Chen, J.; Zhao, D.; et al. A "Skeleton/Skin" Strategy for Preparing Ultrathin Free-standing Single-walled Carbon Nanotube/Polyaniline Films for High Performance Supercapacitor Electrodes. *Energ. Environ. Sci.* **2012,** *5* (9), 8726–8733.
3. Wang, L. P.; Leconte, Y.; Feng, Z.; Wei, C.; Zhao, Y.; Ma, Q.; Xu, W.; et al. Novel Preparation of N-Doped SnO2 Nanoparticles via Laser-Assisted Pyrolysis: Demonstration of Exceptional Lithium Storage Properties. *Adv. Mater.* **2017,** *29* (6), 1603286.
4. Feng, L.; Wang, K.; Zhang, X.; Sun, X.; Li, C.; Ge, X.; Ma, Y. Flexible Solid-State Supercapacitors with Enhanced Performance from Hierarchically Graphene Nanocomposite Electrodes and Ionic Liquid Incorporated Gel Polymer Electrolyte. *Adv. Funct. Mater.* **2018,** *28* (4), 1704463.
5. Zhang, K.; Zhang, L.; Zhao, X. S.; Wu, J. Graphene/Polyaniline Nanofiber Composites as Supercapacitor Electrodes. *Chem. Mater.* **2010,** *22* (4), 1392–1401.
6. Meng, C.; Liu, C.; Fan, S. Flexible Carbon Nanotube/Polyaniline Paper-like Films and Their Enhanced Electrochemical Properties. *Electrochem. Commun.* **2009,** *11* (1), 186–189.
7. Ke, Q.; Wang, J. Graphene-based Materials for Supercapacitor Electrodes–A Review. *J. Materiom.* **2016,** *2* (1), 37–54.
8. Qu, G., Cheng, J.; Li, X.; Yuan, D.; Chen, P.; Chen, X.; Wang, B.; Peng, H. A Fiber Supercapacitor with High Energy Density Based on Hollow Graphene/Conducting Polymer Fiber Electrode. *Adv. Mater.* **2016,** *28* (19), 3646–3652.
9. Zhang, X., Zhang, H.; Lin, Z.; Yu, M.; Lu, X.; Tong, Y. Recent Advances and Challenges of Stretchable Supercapacitors Based on Carbon Materials. *Sci. China Mater.* **2016,** *59* (6), 475–494.
10. Devaraj, S.; Munichandraiah, N. Effect of Crystallographic Structure of MnO$_2$ on Its Electrochemical Capacitance Properties. *J. Phys. Chem. C* **2008,** *112* (11), 4406–4417.
11. Kim, B. C., Ko, J. M.; Wallace, G. G. A Novel Capacitor Material Based on Nafion-doped Polypyrrole. *J. Power Sourc.* **2008,** *177* (2), 665–668.
12. Singu, B.; Srinivasan, P.; Yoon, K. R. Emulsion Polymerization Method for Polyaniline-multiwalled Carbon Nanotube Nanocomposites as Supercapacitor Materials. *J. Solid State Electrochem.* **2016,** *20* (12), 3447–3457.
13. Zhao, T. K.; Ji, X. L.; Bi, P.; Jin, W. B.; Xiong, C. Y.; Dang, A. E.; Li, H.; Li, T. H.; Shang, S. M.; Zhou, Z. F. In Situ Synthesis of Interlinked Three-Dimensional Graphene Foam/Polyaniline Nanorod Supercapacitor *Electrochim Acta.* **2017,** *230,* 342–349.
14. Chee, W. K.; Lim, H. N.; Harrison, I.; Chong, K. F.; Zainal, Z.; Ng, C. H.; Huang, N. M. Performance of Flexible and Binderless Polypyrrole/Graphene Oxide/Zinc Oxide Supercapacitor Electrode in a Symmetrical Two-electrode Configuration. *Electrochim. Acta.* **2015,** *157,* 88–94.
15. Wu, J.; Shi, X.; Song, W.; Ren, H.; Tan, C.; Tang, S.; Meng, X. Hierarchically Porous Hexagonal Microsheets Constructed by Well-interwoven MCo$_2$S$_4$ (M = Ni, Fe, Zn) Nanotube Networks via Two-step Anion-exchange for High-performance Asymmetric Supercapacitors. *Nano Energ.* **2018,** *45,* 439–447.

16. Winokur, M. J.; Wamsley, P.; Moulton, J.; Smith, P.; Heeger, A. J. Structural Evolution in Iodine-doped Poly (3-alkylthiophenes). *Macromolecules* **1991**, *24* (13), 3812–3815.

17. Qu, D.; Shi, H. Studies of Activated Carbons Used in Double-layer Capacitors. *J. Power Sourc.* **1998**, *74* (1), 99–107.

18. Panić, V. et al. The Properties of Carbon-supported Hydrous Ruthenium Oxide Obtained from RuOxHy Sol. *Electrochim. Acta* **2003**, *48* (25–26), 3805–3813.

19. Stević, Z.; Rajčić-Vujasinović, M. Chalcocite as a Potential Material for Supercapacitors. *J. Power Sourc.* **2006**, *160* (2), 1511–1517.

20. Hu, C.-C.; Chen, W.-C.; Chang, K.-H. How to Achieve Maximum Utilization of Hydrous Ruthenium Oxide for Supercapacitors. *J. Electrochem. Soc.* **2004**, *151*, A281–A290.

21. Huang, L.; et al. Nickel–Cobalt Hydroxide Nanosheets Coated on $NiCo_2O_4$ Nanowires Grown on Carbon Fiber Paper for High-performance Pseudocapacitors. *Nano Lett.* **2013**, *13* (7), 3135–3139.

22. Hou, L.; et al. Urchin-like Co_3O_4 Microspherical Hierarchical Superstructures Constructed by One-dimension Nanowires Toward Electrochemical Capacitors. *RSC Adv.* **2011**, *1* (8), 1521–1526.

23. Tao, F.; et al. Electrochemical Characterization on Cobalt Sulfide for Electrochemical Supercapacitors. *Electrochem. Commun.* **2007**, *9* (6), 1282–1287.

24. Zhu, T.; et al. Hierarchical Nickel Sulfide Hollow Spheres for High Performance Supercapacitors. *Rsc Adv.* **2011**, *1* (3), 397–400.

25. Chen, W., Xia, C.; Alshareef, H. N. One-step Electrodeposited Nickel Cobalt Sulfide Nanosheet Arrays for High-performance Asymmetric Supercapacitors. *ACS Nano* **2014**, *8* (9), 9531–9541.

26. Liu, T.-C.; et al. Behavior of Molybdenum Nitrides as Materials for Electrochemical Capacitors Comparison with Ruthenium Oxide. *J. Electrochem. Soc.* **1998**, *145* (6), 1882–1888.

27. Choi, D.; Blomgren, G. E.; Kumta, P. N. Fast and Reversible Surface Redox Reaction in Nanocrystalline Vanadium Nitride Supercapacitors. *Adv. Mater.* **2006**, *18* (9), 1178–1182.

28. Porto, R. L.; et al. Titanium and Vanadium Oxynitride Powders as Pseudo-Capacitive Materials for Electrochemical Capacitors. *Electrochim. Acta.* **2012**, *82*, 257–262.

29. Xu, Y.; et al. General Strategy to Fabricate Ternary Metal Nitride/Carbon Nanofibers for Supercapacitors. *Chem. Electro. Chem.* **2015**, *2* (12), 2020–2026.

30. Shao, M.; et al. Core–shell Layered Double Hydroxide Microspheres with Tunable Interior Architecture for Supercapacitors. *Chem. Mater.* **2012**, *24* (6), 1192–1197.

31. Hall, P. J.; et al. Energy Storage in Electrochemical Capacitors: Designing Functional Materials to Improve Performance. *Energ. Environ. Sci.* **2010**, *3* (9), 1238–1251.

32. Wang, G., Zhang, L.; Zhang, J. A Review of Electrode Materials for Electrochemical Supercapacitors. *Chem. Soc. Rev.* **2012**, *41* (2), 797–828.

33. Dong, S.; et al. Trivalent Ti Self-doped $Li_4Ti_5O_{12}$: A High Performance Anode Material for Lithium-ion Capacitors. *J. Electroanal. Chem.* **2015**, *757*, 1–7.

34. Senthilkumar, B.; et al. Synthesis and Electrochemical Performances of Maricite-$NaMPO_4$ (M = Ni, Co, Mn) Electrodes for Hybrid Supercapacitors. *RSC Adv.* **2014**, *4* (95), 53192–53200.

35. Liu, S.; et al. Nanosheet-constructed Porous TiO_2–B for Advanced Lithium Ion Batteries. *Adv. Mater.* **2012**, *24* (24), 3201–3204.

36. Cid, J.-J.; et al. Molecular Consensitization for Efficient Panchromatic Dye-sensitized Solar Cells. *Angewandte Chemie Int. Ed.* **2007**, *46* (44), 8358–8362.

37. Joshi, P.; et al. Composite of TiO$_2$ Nanofibers and Nanoparticles for Dye-sensitized Solar Cells with Significantly Improved Efficiency. *Energ. Environ. Sci.* **2010,** *3* (10), 1507–1510.

38. Kawamura, G;. et al. Ag Nanoparticle-deposited TiO$_2$ Nanotube Arrays for Electrodes of Dye-sensitized Solar Cells. *Nanoscale Res. Lett.* **2015,** *10* (1), 219.

39. Kondo, Y.; et al. Preparation, Photocatalytic Activities, and Dye-sensitized Solar-cell Performance of Submicron-scale TiO$_2$ Hollow Spheres. *Langmuir* **2008,** *24* (2), 547–550.

40. Wang, X.; et al. Rapid Construction of TiO$_2$ Aggregates Using Microwave Assisted Synthesis and Its Application for Dye-sensitized Solar Cells. *RSC Adv.* **2015,** *5* (12), 8622–8629.

41. Li, C.; et al. Mesoporous TiO$_2$ Aggregate Photoanode with High Specific Surface Area and Strong Light Scattering for Dye-sensitized Solar Cells. *J. Solid State Chem.* **2012,** 196, 504–510.

42. Trancik, J. E.; Barton, S. C.; Hone, J. Transparent and Catalytic Carbon Nanotube Films. *Nano Lett.* **2008,** *8* (4), 982–987.

43. Guan, J.; et al. Interfacial Modification of Photoelectrode in ZnO-based Dye-sensitized Solar Cells and Its Efficiency Improvement Mechanism. *RSC Adv.* **2012,** *2* (20), 7708–7713.

44. Su, T.; et al. An Insight into the Role of Oxygen Vacancy in Hydrogenated TiO$_2$ Nanocrystals in the Performance of Dye-sensitized Solar Cells. *ACS Appl. Mater. Interf.* **2015,** *7* (6), 3754–3763.

45. Wang, Z.-S.; Zhou, G. Effect of Surface Protonation of TiO$_2$ on Charge Recombination and Conduction Band Edge Movement in Dye-sensitized Solar Cells. *J. Phys. Chem. C* **2009,** *113* (34), 15417–15421.

46. Bagui, A.; Iyer, S. S. K. Effect of Solvent Annealing in the Presence of Electric Field on P3HT: PCBM Films Used in Organic Solar Cells. *IEEE T. Electron Dev.* **2011,** *58* (11), 4061–4066.

47. Hau, S. K., Yip, H.-L.; Jen, A. K.-Y. A Review on the Development of the Inverted Polymer Solar Cell Architecture. *Polym. Rev.* **2010,** *50* (4), 474–510.

48. Cui, Y.; et al. Over 16% Efficiency Organic Photovoltaic Cells Enabled by a Chlorinated Acceptor with Increased Open-circuit Voltages. *Nat. Commun.* **2019,** *10* (1), 2515.

CHAPTER 4

Polylactic Acid: An Eco-Friendly Material in the Packaging Industry, Paving the Way Toward a Greener Environment

S. ROOPA[1*], B. SOWMYA[1], S. PREETHI[2], and ARUL MAXIMUS RABEL[3]

[1]Department of Polymer Science and Technology, Sri Jayachamarajendra College of Engineering, JSS Science and Technology University, Mysuru 570006, India

[2]Indo MIM Pvt Ltd., Bengaluru, India

[3]Research Microscopy Solutions, Carl Zeiss India, Bengaluru, India

*Corresponding author. E-mail: roopasm2000@jssstuniv.in

ABSTRACT

We are very much at the eye of a global crisis like climate change, global warming, and emission of toxic gases into our atmosphere because of excess usage of synthetic polymers in the packaging industry. Hence the onus is on biopolymers and materials which are superior in their biodegradability and recyclable property. Biopolymers are gaining more attention from academicians, researchers, and the industries to replace the materials which are derived from fossil fuels. These bio-based polymers provide solutions for environmental issues such as plastic waste management, depletion of fossil fuels, emission of greenhouse and toxic gases, etc. Also because of increasing involvement of the Food and Drug Administration and related organizations in terms of food safety, favorable government policies toward promoting bio-plastics and the awareness of consumers about the safety and environmental effect of conventional polymers, the usage of biodegradable and food-grade plastic packaging materials are increasing. Due to these reasons, the demand of bioplastics is increasing along with the production capacity. The total production capacity of bioplastic in 2018 was 2.1 million tons. Polylactic acid (PLA), thermoplastic aliphatic polyester derived from agricultural products such as potato, wheat, and corn starch is one such promising biopolymer. Due to its attractive properties such as strength, stiffness, and biodegradability, PLA is widely used in packaging and biomedical applications. According to a recent market survey, the use of PLA in the packaging industry (over 50% of the global bio-PLA market) dominated the market in 2018 and also it is expected to grow during the forecast period, due to its usage in fruit and vegetable packaging, shopping or carrier bags, bread bags, bakery boxes, bottles, etc. (Source: Mordor intelligence; bio-PLA market—growth, trends, and forecast (2019–2024)). The literature survey also reveals that the interest has shifted toward the search of biodegradable/biocompostable PLA polymers, blends, and composites as an alternative to conventional polymers in packaging applications. This chapter describes the production, structure–property relationship, chemical and physical modification of PLA to suit packaging applications and its characterization.

4.1 INTRODUCTION

Urbanization, Westernized lifestyle, and online retailing have boosted a demand for packaged goods. As per a recent survey, the global packaging market has increased by 6.8% from 2013 to 2018. The four key trends that will play out across the next decade are: (1) economic and demographic growth, (2) sustainability, (3) consumer trends, and (4) brand owner trends ("Four key trends" n.d.).

Because of several advantages like lightweight, transparency, easier to handle, tunable to end-use requirements, easier sterilization, stability, and cost-effectiveness, plastics are being one of the most attractive materials for the packaging industry. As a result, waste entering into the municipal solid waste is also increasing day by day. According to the Environmental Protection Agency (EPA) survey, the United States generates almost 80 million tons of packaging waste each year and approximately half of that is used to package single-serve food items. Also, the report reveals that in 2017 roughly 8.7% of the plastic waste is recycled and incinerated to recover energy. Remaining 19.2% plastic waste entered landfilling ("Facts and Figures Fact Sheet", 2019). It is not new to us that most of the conventional plastic materials like polyethylene terephthalate (PET), polyethylene (PE), and polystyrene (PS) used in the packaging industries are nonbiodegradable, creating a menace to the environment and these plastics are derived from fossil fuels. Nowadays, consumers are more concerned about the environmental issues and are more demanding about the environmentally sound products. The number of strategies is advancing to address the issue of sustainability of the packaging materials which includes substituting to alternative materials, investing in the development of bio-based plastics (green plastics), designing packaging to make them easier to process in recycling, and improving recycling and processing of plastic waste. Investing in eco-friendly packaging is rapidly gaining momentum in many industries and hence inclining toward biodegradable green polymers derived from renewable resources. For example, The Kraft Heinz Company, a globally-trusted producer of delicious foods released a press report stating that, by 2025, the company aims to make 100% of its packaging recyclable, reusable, or compostable ("Kraft Heinz Expands Environmental", 2018). So, for a greener future, biodegradable or green polymers play an important role in the packaging industry. According to EPA, a material can be described as biodegradable when it is "capable of being decomposed by the action of biological processes." Biodegradable polymers viz, starch, cellulose, chitosan, poly(lactic acid) (PLA), polycaprolactone (PCL), and polyhydroxybutyrate (PHB) are explored in the application of the packaging field (Marija, 2017). These polymers decompose aerobically or anaerobically into CO_2, water, methane, etc., under the action of microorganisms within a short period of time after disposal.

Among biosource-based polymers emerging on the food market, PLA is one of the potential replacers of conventional plastics in the packaging industry. PLA display many advantages compared to many other packaging materials; biodegradable, biocompatible, renewable, adequate food contact properties, approved by the Food and Drug Administration, transparent, competitive cost, processability, and availability (Mattioli et al., 2013; Rhim et al., 2009).

This chapter provides an overview of the synthesis, modification, processing, testing, and characterization methods used particularly focusing on barrier properties and biodegradation of PLA along with new trends in PLA packaging to suit the present needs of consumers/society/environmental policies.

4.2 SYNTHESIS OF POLYLACTIC ACID

PLA is thermoplastic aliphatic polyester which comprises of lactic acid repeating units. Lactic acid, a natural hydroxycarboxylic acid has been used in pharmaceutical, food industries, leather tanning, personal care, and in the production of polymers. About 39% of lactic acid is consumed for the production of biodegradable/biocompatible PLA polymer (Komesu et al., 2017). The chemical structure of lactic acid is given below (Fig. 4.1):

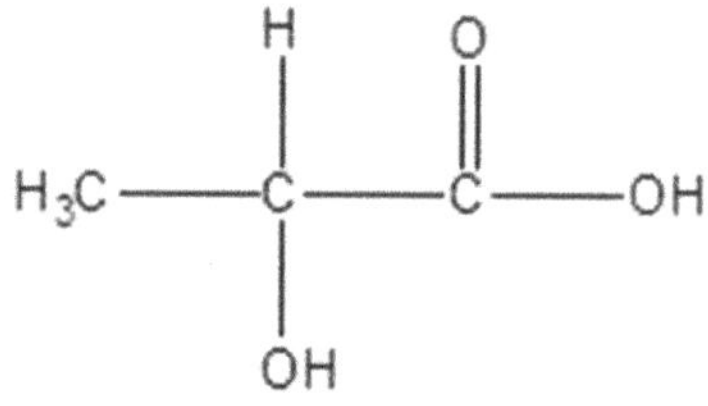

FIGURE 4.1   Chemical structure of lactic acid.

Industrially, lactic acid is produced by the chemical route and via fermentation. Fermentation process is widely used as compared to chemical synthesis of lactic acid. Starch and cellulose-based materials viz, corn, potato, rice, wheat starch, wood, and molasses are used to produce lactic acid by fermentation process along with other raw materials like bacteria, fungi, yeast, and media (Krishna, 2019). Lactic acid is available in two optically active isomeric forms as L-lactic acid and D-lactic acid or racemic form (mixture of L and D lactic acid). Depending upon the nature of lactic acid used and reaction conditions, there are different grades of PLA available in the market viz, poly (L-lactic acid) (PLLA), poly (D-lactic acid) (PDLA), and poly (DL-lactic acid) (PDLLA). The PLLA and PDLA are isotactic homopolymers whereas PDLLA is optically inactive atactic (amorphous) polymer. The properties such as mechanical, thermal, processability, and biodegradability of PLA are dependent on the ratio of stereoisomers present in the polymer backbone, the molecular weight, and the crystallinity of PLA (Mehta et al., 2005; Perego et al., 1996).

PLA is produced by direct condensation polymerization (Ajioka et al., 1995), ring opening polymerization (ROP) (Lopes et al., 2014), and azeotropic dehydrative condensation polymerization. Commercially available PLA is produced by L-lactide via ROP technique. The PLA obtained by ROP has high molecular weight and semicrystalline nature with relatively high T_m (160–180°C) and T_g (60–80°C) (Albertsson et al., 2010).

In direct condensation polymerization, the polymerization of lactic acid is carried out in bulk by distillation of condensation of water with or without the aid of a catalyst. During polymerization, the removal of water from the highly viscous PLA melt is difficult. During polymerization, if the temperature is raised to remove water, the rate of depolymerization increases when compared to the rate of polymerization. Generally from this technique, low molecular weight PLA (oligomers) is obtained. The molecular weight of the PLA lies in the range of 1000–5000 Da. This low molecular weight PLA can be converted into high molecular weight by the addition of chain coupling agents.

It was believed that the high molecular weight PLA could not be prepared until the Mitsui Chemicals Co. developed the azeotropic dehydrative condensation polymerization technique. In this technique, the water eliminated from the reaction is removed with the help of high boiling organic solvents by azeotropic distillation. This technique produces high molecular weight PLA (Ajioka et al., 1995; Ren, 2010).

Even though the direct condensation polymerization technique is an economically viable process, industrially the ROP of lactide has been a widely used method to produce PLA. ROP is carried out using cationic, anionic, coordination catalyst, or by using enzymes. Several initiators and catalyst systems were used to polymerize lactide, amongst tin (II) octanoate catalyst has gained commercial importance due to its reactivity, low toxicity, and its ability to dissolve in many organic solvents. The steps involved in the ROP of lactide are given in Figure 4.2. The detailed mechanism of the polymerization of lactic acid has been reported by several authors in earlier publications (Auras et al., 2004; Cheng et al., 2009; Chisholm et al., 2010; Ren, 2011).

4.3 TESTING AND CHARACTERIZATION

As we all know, testing and characterization play an important role in quality control, research and development, and to predict the performance and durability of the product. The performance of the PLA in packaging application is dependent on the barrier, mechanical, thermal, and optical properties. These properties and biodegradability of PLA are dependent on the molecular weight and its distribution, crystallinity, T_g, T_m, microstructure, morphology, chemical or physical modification of PLA, and processing conditions. To evaluate the characteristic properties of PLA and its derivatives, several testing and characterization tools have been used. Some of the tools used to characterize neat and modified PLA are summarized in Table 4.1.

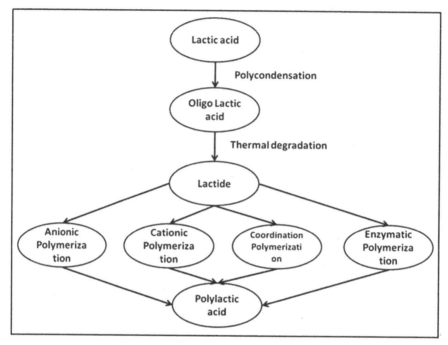

FIGURE 4.2   Process flow of ROP of lactide.

4.3.1   *BARRIER PROPERTIES*

When polymers are used for food packaging, the packaging should ensure the organoleptic quality of the food apart from providing safety to packed food. These qualities of packaging materials are highly dependent on their barrier properties. Barrier properties of packaging material encompass the permeability of small molecules like CO_2, O_2, water vapor, aroma compounds, and light. These crucial factors are essential for retaining the quality of any packaged food. The permeation of water vapor and oxygen through packaging films were studied more commonly for packaging applications. These two permeants can transfer from the internal or external environment across the thickness of the film. The interaction of water vapor or oxygen with the packed food changes the quality of the food and significantly affects the shelf life (Siracusa et al., 2013; Wihodo et al., 2013).

The permeability of oxygen and water vapor is highly influenced by the nature and composition of the PLA, internal architecture, crystallinity, molecular orientation, free volume, and T_g. Based on the nature of food and

TABLE 4.1 Testing and Characterization Tools Used for PLA Characterization

Sl. No	Characterization tools/method	Remarks	References
1	Fourier Transform Infrared (FTIR) spectrophotometer, UV spectrophotometer, Raman spectrophotometer, and NMR	• For structural analysis • To study the miscibility of the blends • Interaction between PLA matrix and reinforcement • Chain orientation	Choksi and Desai (2017), Hoidy et al. (2010), Kim et al. (2009), Notingher et al. (2002), Ramontja et al. (2009), Yang et al. (2004), and Yuniarto et al. (2016)
2	Size exclusion chromatography and gel permeation chromatography with multi-angle light scattering detector, viscometry, end group analysis	• Molecular weight and polydispersity index • Effect of thermal degradation on molecular weight	Colomines et al. (2010), Gupta et al. (1982), Liu et al. (2013), Puchalski et al. (2017), and Takizawa et al. (2008)
3	Differential scanning calorimeter, dynamic mechanical analyzer, X-ray diffractometer	• Crystallinity • Crystallization kinetics • Effect of second component on crystallization and % crystallinity • Thermal transitions (T_g, T_c, T_m) • Viscoelastic properties	Colomines et al. (2010), Siracusa et al. (2017), and Zhang et al. (2018)
4	Polarized optical microscope, scanning electron microscope, atomic force microscope, transmission electron microscope	• Spherulitic growth behavior • Morphology • Effect of degradation on the surface morphology • Filler dispersion and interaction	Jia et al. (2017), Nanthananon et al. (2015), and Zhang et al. (2018)
5	Thermo gravimetric analyzer coupled with FTIR, gas chromatography (GC) or mass spectroscopy (MS), thermal volatilization analysis	• Thermal stability • Degradation kinetics • Thermal decomposition products	McNeill et al. (1985), Mitchell et al. (2009), and Rahman and Hui (2018)

its intended application, the barrier properties of the packaging film has to be optimized. The barrier property of PLA with respect to water vapor is inferior to that of PET and PS (Colomines et al., 2010). In order to improve the barrier properties of PLA, several strategies have been employed. The addition of fillers (micro and nano) and increase in crystallinity of PLA improves the barrier properties due to increase in the tortuosity of the diffusion pathway (Arrnentano et al., 2013; Colomines et al., 2010; Drieskens et al., 2009; Guinault et al., 2012; Sawada et al., 2010). The water vapor permeability and oxygen permeation of PLA were significantly reduced by the addition of nano TiO_2, nano-silver (Ag), nano-MgO particles, hexadecylamine–montmorillonite, and dodecyltrimethyl ammonium bromide–montmorillonite (Arrieta et al., 2014; Burgos et al., 2013; Meriçer et al., 2016). The barrier property is influenced by the nature of filler and the dispersion of filler. If the filler is not distributed well in the matrix, the added nanofiller has a tendency to reduce the barrier property of the PLA film (Chang et al., 2003; Plackett et al., 2006). Fortunati et al. (2012) have prepared nanocomposite films by the addition of cellulose nanocrystals (CNCs), surfactant modified (s-CNC), and Ag nanoparticles in the PLA matrix using melt extrusion followed by a film formation process. The presence of surfactant favored the dispersion of filler in the polymer matrix. The presence of nanofillers enhances nucleation behavior and the usage of silver nanoparticles improves the antibacterial properties. Similar principle was used by Nakayama and Hayashi (2007) and TiO_2 nanoparticles surface was modified using carboxylic acid and long-chain alkyl amine. The surface-modified TiO_2 nanoparticles were uniformly dispersed into the PLA matrix without aggregation; also the photodegradation rate of nanocomposites was influenced by the content of TiO_2. Further, as reported by Rhim et al. (2009), the chemical nature of filler plays an important role in tuning the barrier properties of the PLA film. For example, the organically modified Cloisite 30B exhibited improved water vapor barrier property when compared to natural nanoclay. The barrier properties of PLA films can be improved by coating with impermeable materials like SiO_x, PVOH, aluminum, and bee-wax. The coextruded or laminated bilayer or multilayer PLA films are also more resistant to permeation of gas or vapors (Meriçer et al., 2016; Mohd Aris et al., 2019; Siracusa et al., 2012; Siracusa et al., 2017; Zhu et al., 2018).

Another strategy used to improve the PLA properties for extending its applications in packaging is the blending of PLA with other biodegradable polymers. High brittleness and cost of PLA is the major issue for their commercialization in many applications such as packaging. Therefore, to improve the properties and to lower the production cost various studies on the PLA blends were carried out. Melt blending is the common approach since this is cost-effective and readily available processing technology. The compatibility of the blends alters the physical properties of materials such as glass transition temperature, melting temperature, and crystallinity of the polymer (Ljungberg et al., 2005; Sheth et al., 1998). Without plasticizer, PLA is stiff and brittle in nature. Attempts to alter the thermal and mechanical properties have focused on blending with biocompatible polymers like PE glycol (PEG) which improves the elongation at break and flexibility as well as barrier properties of PLA (Armentano et al., 2015; Arrieta et al., 2014; Burgos et al., 2013). Sheth et al. (1997) reported that PLA/PEG blends are partially miscible, depending on the concentration. Below 50% PEG content in PLA, yielding higher ductility and lower modulus values. Above 50% PEG content the blend morphology was driven by increasing the crystallinity of PEG, resulting in increases in modulus and decrease in elongation at break. The desired thermal and mechanical properties were achieved up to the addition of 30% PEG. The addition of PEG to PLA reduces the stereoregularity and decreases the T_g from 57°C to 37°C. The equilibrium melting point of PLA decreased with increasing PEG. However, there is evidence that the blends are not stable at ambient temperature and the attractive mechanical properties are lost over time.

Bhatia et al. (2007) had performed melt blending on PLA/biodegradable polybutylene succinate (PBS). The results indicated that PLA/PBS blend is incompatible beyond 20 wt.% of PBS in the system. Tensile strength and modulus of the blends decreased with PBS content but followed approximately the mixing rule for 90/10 and 80/20 (PLA/PBS) blends. Rheological results also show compatibility between the two polymers for PBS composition less than 20% by weight. PBS reduced the brittleness of PLA, thus making it a contender to replace plastics for packaging applications. Arrieta et al. (2017) had reported that the PLA performance is substantially improved by melt blending it with high crystalline isotactic PHB. The processability and thermal stability can be improved by the plasticization and by adding microparticles and nanoparticles which lead to the improvement of small processing window of PHB. PHB is able to crystallize PLA which in turn improves the barrier properties and mechanical resistance. Moreover, PLA-PHB is a fully bio-based material that also offers composting under aerobic conditions as a sustainable end of life option. The degradation process strongly depends on the composting temperature since PLA degrades faster at 58°C, while PHB degrades faster at room temperature.

Walha et al. (2018) concluded on the improvement of melt strength of the biosourced PLA/Polyamide 11 (PA11) blends by the physical compatibilization compared to chemical modification. A new biomaterial with suitable melt strength, stiffness-to-toughness balance, and the required thermal performance for food packaging applications was proposed based on PLA, PA11, and the styrene–acrylic copolymer melt strength enhancer (BPMS-260). In BPMS, methacrylate sequence is quite miscible with PLA and may help to increase its elasticity and melt strength without any pronounced strain hardening.

4.3.2 BIODEGRADATION

In recent years, researchers and commercial sectors are looking for biodegradable polymers for food packaging applications which cause less environmental impact as compared to conventional synthetic polymers like PET, polyolefins, and PS. Biodegradable polymers undergo degradation by the action of naturally occurring microorganisms leaving behind the eco-friendly organic by-products such as carbon dioxide and water. In order to understand the biodegradability of the polymers, the international standards for the evaluation of their biodegradability are very important. The products made by these biodegradable polymers should possess biodegradability characteristics according to the international standards (Ren, 2011).

In 1993, a new working group (WG22) was established for biodegradability under subcommittee 5 (SC5) by a technical committee 61 (TC61). Since then, the ISO standards for biodegradability of plastics come under TC61/SC5/WG22. This section comprises of 23 standards, out of which nine standards are related to biodegradation testing of plastics (Funabashi et al., 2009; Ren, 2011). As with other plastics, the degradation of PLA also takes place via main chain or side chain scission through mechanical, thermal, chemical, or biological route. The rate at which PLA degrades is slow when compared with that of other biodegradable plastics (Ray et al., 2003).The degradability of PLA depends on several factors such as chemical structure, molecular weight, molecular weight distribution, T_g, T_m, crystallinity, crystal structure, modulus of elasticity, surface area, hydrophilic, hydrophobic properties, presence of additives, and environmental conditions (Nishida et al., 1992; Tokiwa et al., 2006). Several research groups have reported that amorphous PLA easily undergoes degradation when compared with the crystalline PLA (Cai et al., 1996; Iwata et al., 1998; MacDonald et al., 1996; Tsuji et al., 2001). Tanio et al. (1982) and Tokiwa et al. (1978) have studied the effect of molecular weight on the degradation of PLA. They have reported that the low molecular weight PLA degrades faster than that of high molecular weight PLA. This was attributed to the greater number of accessible chain end groups for enzymatic degradation. The recent scientific publications reported the effect of nanofillers such as nanoclays, carbon nanotubes, CNC, and hydroxyapatite on the degradation behavior of PLA (Brzeziński et al., 2014; Castro-Aguirre et al., 2018; Fukushima et al., 2009; Fukushima et al., 2010; Lee et al., 2002; Paul et al., 2005; Someya et al., 2007; Wu et al., 2006, 2007). Some authors have reported that the addition of nanoparticles increases the biodegradation due to increase in the hydrophilicity of the PLA nanocomposites (Lee et al., 2002; Fukushima et al., 2009; Ozkoc et al., 2009; Paul et al., 2005). Few other studies reported that the incorporation of nanofillers suppresses the degradation of PLA due to increase in the crystallinity (Ozkoc et al., 2009; Someya et al., 2007; Wu et al., 2006, 2007). The study on the effect of CNC and PHB by Arrieta et al. (2014) showed that the PLA composites degrade at different rates as compared to that of the blend and pristine PLA. It was noticed that due to the higher surface polarity PLA/PHB/CNC nanocomposites exhibited faster disintegration when compared to PLA/CNC films. Recently, Yang et al. (2016) have compared the effect of lignin and CNC on the degradation behavior and it was noticed that PLA/lignin nanocomposites exhibited maximum disintegration when compared to PLA/CNC composite. The addition of PEG plasticizer improves the degradation behavior of PLA due to increase in the hydrophilicity of the PLA and in turn helps in the hydrolysis of PLA (Ozkoc et al., 2009) Very recently Luo et al. (2019) have used functionalized TiO_2 nanoparticle to study the effect of surface nature of TiO_2 on the degradation property of PLA and they have noticed the faster degradation rate in PLA/TiO_2 nanocomposites compared to that of pristine PLA. This is attributed to the easy penetration of water molecules into the composite film which enhances the degradation due to hydrolysis.

The biodegradation behavior of PLA and PLA/PHB blend system was examined by Zhang et al. (2011) under soil burial conditions. It was observed that the neat PLA samples were intact at room temperature but the PLA/PHB blend system showed biodegradation. With the increase in PHB content, the biodegradability of the samples also increased which was evident from discoloration.

4.4   PROCESSING OF POLYLACTIC ACID

Well established processing methods used for commercial polymers like PS and PE is used to process PLA resin (Jamshidian, et al., 2010). Melt processing technologies such as injection molding, transfer molding, extrusion, and casting techniques are used to convert PLA resin into useful artifacts such as cutlery, packaging films, and consumer goods. During melt processing, the PLA resin is heated above its melting temperature, shaping the melt into the desired shape, and cooling the melt below its glass transition temperature in order to stabilize the dimension of the product. As PLA is hygroscopic in nature, it has to be predried for several hours before its processing (Castro-Aguirre et al., 2016).

Also, PLA is spun into fibers by melt spinning process for making products like textiles and carpets. In recent years, the electrospinning process is used to spin nanofibers. Electrospinning is a novel fabrication technology that utilizes electrostatic forces for drawing ultrathin nanofibers. Electrospinning is one of the simple and straightforward techniques used for preparing continuous nanofibers from an electrically charged jet of polymer solution. Due to very large surface area, electrospun fibers offer many advantages over conventional packaging films and sheets, like more responsive to the change in humidity and temperature, ease of encapsulation of thermally unstable bioactive agents into fibers, etc. The conventional methods like spray drying and solvent evaporation techniques are used to incorporate bioactive compounds. The working temperature of these techniques is relatively higher and affects the stability of the thermally labile bioactive compounds (Fabra et al., 2014; López-Rubio et al., 2012). The electrospinning process works in the absence of heat and hence it is easy to embed such thermally labile bioactive compounds into the nanofibers. Due to these advantages, the electrospun nanofibers are entering into biomedical and active food packaging areas.

Essential micronutrients such as vitamin A, D, E, and K possess various biochemical functions in our body. These have a tendency to degrade during processing, storage, and until the moment of their absorption in the gastrointestinal tract. To protect against degradation, they should be encapsulated e-ectively. α-Tocopherol (α-TC) is a biologically active form of Vitamin E and is one of the most studied antioxidants (Fabra et al., 2016; Pérez-Masiá et al., 2014; Wongsasulak et al., 2014). Aytac et al. (2017) encapsulated α-TC and α-TC/cyclodextrin (CD)–inclusion complex (IC) in PLA electrospun nanofiber. The antioxidant activity was examined using 2, 2-diphenyl-1-picrylhydrazyl assay method and the results revealed that

PLA/α-TC–NF and PLA/α-TC/γ-CD–IC nanofiber was adequate to avoid lipid oxidation. From the lipid oxidation analysis of thiobarbituric acid reactive substance (TBARS), it was observed that the nanofibers showed a lower TBARS content than the unpackaged meat sample. Due to the presence of IC, the electrospun PCL/α-TC/γ-CD nanofiber exhibited higher antioxidant activity compared to PCL/α-TC nanofiber.

Later in order to explore the advantages of CD-IC, researchers have made an attempt to encapsulate different bioactive compounds such as cinnamon essential oil (Wen et al., 2016), triclosan (Kayaci et al., 2013), and gallic acid, a natural phenolic antioxidant and antimicrobial agent (Aytac et al., 2016) into CD-IC. These encapsulated bioactive agents are embedded in PLA electrospun nanofibers and the antioxidant activity was evaluated using 2,2-diphenyl-1-picrylhydrazyl (DPPH) assay. This result reveals that the composite nanofibers have a higher amount of antioxidant activity and hence it was confirmed that electrospinning had no adverse effect on the the antioxidant property of bioactive agents.

Polyphenols are a class of bioactive molecules that are packed with antioxidants and have potential health benefits. polyphenols can improve or help digestion issues, weight management difficulties, diabetes, neuro-degenerative disease, and cardiovascular diseases (Ana Gotter, 2017). To make use of polyphenols in active food packaging, Liu et al. (2018) prepared PLA/tea polyphenol (PLA/TP) composite nanofibers for food packaging applications. The DPPH assay analysis reveals that PLA/TP has improved antioxidant activity as compared to neat PLA and the antioxidant activity increased with an increase in the TP content. The antimicrobial activity of composite and PLA fibers against *E. coli* and *S. aureus* was determined using the vibration method. The composite nanofibers retarded the growth of *E. coli* and *S. aureus*, whereas PLA nanofibers had no appreciable antimicrobial activity. The antimicrobial activity against *E. coli* and *S. aureus* improved with increasing TP content. It was evident that the addition of a bioactive tea polyphenol improved the oxidative stability and extended the storage period of food. These nanofibers are effective in active packaging applications.

Carvacrol is a naturally occurring monoterpenic phenol used as an anti-bactericide, antitumor agent, and food additive (Yu et al., 2012). As with other bioactive compounds, carvacrol is also volatile in nature. To improve its availability and stability, Altan et al. (2018) have encapsulated the carvacrol in zein and PLA matrix via electrospinning technique. The antioxidant activity of carvacrol loaded fibers was examined by DPPH assay and the results were compared with pristine fibers. The antioxidant capacity of carvacrol loaded zein and PLA fibers lie in the range of 62–75% and 53–65%, respectively. It was noticed that the antioxidant capacity is dependent on the carvacrol content in the fiber, the higher the carvacrol content, the higher is the antioxidant activity of the fiber. In order to study the effectiveness of the carvacrol loaded fiber in food preservation, authors used the whole wheat bread sample. The effect of carvacrol on the growth inhibition rate of aerobic mesophilic bacteria, mold, and yeast count was evaluated. The results reveal that the inhibition of both aerobic mesophilic bacteria, mold, and yeast growth in bread samples increased as the carvacrol content increased in both zein and PLA fibers.

4.5 SUMMARY

Various strategies were made to develop a sustainable and biodegradable substitute for conventional thermoplastics. PLA, a green plastic produced from a renewable resource is one of the promising materials for sustainable food packaging applications. The major economical challenge faced was to enable PLA to compete in the packaging market. Whereas, from a technical point of view, the significant limitations such as narrow process range, poor ductility, and low thermal stability were altered by understanding the properties, formulation variables, selection of apt additives (nanofillers, plasticizers), and modification of PLA polymer based on the applications. The electrospinning process is one of the emerging technologies that need to be further explored to develop the food packaging especially in the area of active food packaging. The researchers were urged to develop PLA based material to lower the greenhouse emission and to create an infinite possibility of PLA based systems to shift in contemporary food packaging systems.

KEYWORDS

- polylactic acid
- packaging
- sustainable
- biopolymer
- PLA nanofiber
- eco-friendly polymer

REFERENCES

1. Ajioka, M.; Enomoto, K.; Suzuki, K.; Yamaguchi, A. Basic Properties of Polylactic Acid Produced by the Direct Condensation Polymerization of Lactic Acid. *Bull. Chem. Soc. Jpn.* **1995**, *68* (8), 2125–2131.
2. Albertsson, A. C.; Varma, I. K.; Lochab, B.; Finne-Wistrand, A.; Kumar, K. Design and Synthesis of Different Types of Poly (Lactic Acid). *Poly (Lactic Acid) Synthesis, Structures, Properties, Processing, and Applications*, Wiley: USA; 2010, 43–58.
3. Altan, A.; Aytac, Z.; Uyar, T. Carvacrol Loaded Electrospun Fibrous Films from Zein and Poly (Lactic Acid) for Active Food Packaging. *Food Hydrocol.* **2018**, *81*, 48–59.
4. Gotter, A. Top Foods with Polyphenols, May 23, 2017. Retrieved from https://www.healthline.com/health/polyphenols-foods (accessed Jun 12, 2019).
5. Armentano, I.; Fortunati, E.; Burgos, N.; Dominici, F.; Luzi, F.; Fiori, S.; … Kenny, J. M. Processing and Characterization of Plasticized PLA/PHB Blends for Biodegradable Multiphase Systems. *Express Polym Lett.* **2015**, *9*(7), 583–596.
6. Arrieta, M. P.; Fortunati, E.; Dominici, F.; Rayón, E.; López, J.; Kenny, J. M. PLA-PHB/Cellulose Based Films: Mechanical, Barrier and Disintegration Properties. *Polym. Degrad. Stabil.* **2014**, *107*, 139–149.
7. Arrieta, M. P.; Samper, M. D.; Aldas, M.; López, J. On the use of PLA-PHB Blends for Sustainable Food Packaging Applications. *Materials* **2017**, *10* (9), 1008.
8. Arrieta, M. P.; Samper, M. D.; López, J.; Jiménez, A. Combined Effect of Poly (Hydroxybutyrate) and Plasticizers on Polylactic Acid Properties for Film Intended for Food Packaging. *J. Polym. Environ.* **2014**, *22* (4), 460–470.
9. Arrnentano, I.; Bitinis, N.; Fortunati, E.; Mattioli, S.; Rescignano, N.; Verdejo, R.; Lopez-Manchado, M. A.; Kenny, J. M. Multifunctional Nanostructured PLA Materials for Packaging and Tissue Engineering. *Prog Polym Sci.* **2013**, *38* (10–11), 1720–1747.
10. Auras, R.; Harte, B.; Selke, S. An Overview of Polylactides as Packaging Materials. *Macromol. Biosci.* **2004**, *4* (9), 835–864.
11. Aytac, Z.; Keskin, N. O. S.; Tekinay, T.; Uyar, T. Antioxidant α-tocopherol/γ-cyclodextrin–inclusion Complex Encapsulated Poly (Lactic Acid) Electrospun Nanofibrous Web for Food Packaging. *J. Appl. Polym. Sci.* **2017**, *134* (21), 1–9.
12. Aytac, Z.; Kusku, S. I.; Durgun, E.; Uyar, T. Encapsulation of Gallic Acid/Cyclodextrin Inclusion Complex in Electrospun Polylactic Acid Nanofibers: Release Behavior and Antioxidant Activity of Gallic Acid. *Mater. Sci. Eng. C* **2016**, *63*, 231–239.
13. Bhatia, A.; Gupta, R.; Bhattacharya, S.; Choi, H. Compatibility of Biodegradable Poly (Lactic Acid) (PLA) and Poly (Butylene Succinate) (PBS) Blends for Packaging Application. *Korea-Aus. Rheol. J.* **2007**, *19* (3), 125–131.
14. Brzeziński, M.; Biela, T. Polylactide Nanocomposites with Functionalized Carbon Nanotubes and Their Stereocomplexes: A Focused Review. *Mater. Lett.* **2014**, *121*, 244–250.
15. Burgos, N.; Martino, V. P.; Jiménez, A. Characterization and Ageing Study of Poly (Lactic Acid) Films Plasticized with Oligomeric Lactic Acid. *Polym. Degrad. Stabil.* **2013**, *98* (2), 651–658.
16. Cai, H.; Dave, V.; Gross, R. A.; McCarthy, S. P. Effects of Physical Aging, Crystallinity, and Orientation on the Enzymatic Degradation of Poly (Lactic Acid). *J. Polym. Sci. Part B* **1996**, *34* (16), 2701–2708.

17. Castro-Aguirre, E.; Auras, R.; Selke, S.; Rubino, M.; Marsh, T. Enhancing the Biodegradation Rate of Poly (Lactic Acid) Films and PLA Bio-nanocomposites in Simulated Composting through Bioaugmentation. *Polym. Degrad. Stabil.* **2018,** *154,* 46–54.

18. Castro-Aguirre, E.; Iniguez-Franco, F.; Samsudin, H.; Fang, X.; Auras, R. Poly (Lactic Acid)—Mass Production, Processing, Industrial Applications, and End of Life. *Adv. Drug Del. Rev.* **2016,** *107,* 333–366.

19. Chang, J. H.; An, Y. U.; Sur, G. S. Poly (lactic acid) nanocomposites with various organoclays. I. Thermomechanical Properties, Morphology, and Gas Permeability. *J. Polym. Sci. Part B* **2003,** *41* (1), 94–103.

20. Cheng, Y.; Deng, S.; Chen, P.; Ruan, R. Polylactic acid (PLA) Synthesis and Modifications: A Review. *Front. Chem. China* **2009,** *4* (3), 259–264.

21. Chisholm, M. H. Concerning the Ring-opening Polymerization of Lactide and Cyclic Esters by Coordination Metal Catalysts. *Pure Appl. Chem.* **2010,** *82* (8), 1647–1662.

22. Choksi, N.; Desai, H. Synthesis of Biodegradable Polylactic Acid Polymer by Using Lactic Acid Monomer. *Int. J. Appl. Chem.* **2017,** *13* (2), 377–384.

23. Colomines, G.; Ducruet, V.; Courgneau, C.; Guinault, A.; Domenek, S. Barrier Properties of Poly(Lactic Acid) and its Morphological Changes Induced by Aroma Compound Sorption. *Polym. Int.* **2010,** *59* (6), 818–826.

24. Drieskens, M.; Peeters, R.; Mullens, J.; Franco, D.; Lemstra, P. J.; Hristova-Bogaerds, D. G. Structure Versus Properties Relationship of Poly (Lactic Acid). I. Effect of Crystallinity on Barrier Properties. *J. Polym. Sci. Part B* **2009,** *47* (22), 2247–2258.

25. Fabra, M. J.; Jiménez, A.; Talens, P.; Chiralt, A. Influence of Homogenization Conditions on Physical Properties and Antioxidant Activity of Fully Biodegradable Pea Protein–alpha-tocopherol Films. *Food Bioproc. Technol.* **2014,** *7* (12), 3569–3578.

26. Fabra, M. J.; López-Rubio, A.; Lagaron, J. M. Use of the Electrohydrodynamic Process to Develop Active/Bioactive Bilayer Flms for Food Packaging Applications. *Food Hydrocol.* **2016,** *55,* 11–18.

27. Facts and Figures Fact Sheet. November 2019. Retrieved from https://www.epa.gov/sites/production/files/2019-11/documents/2017_facts_and_figures_fact_sheet_final.pdf. Accesed on Jul 12, 2019.

28. Fortunati, E.; Armentano, I.; Zhou, Q.; Iannoni, A.; Saino, E.; Visai, L.; … Kenny, J. M. Multifunctional Bionanocomposite Films of Poly (Lactic Acid), Cellulose Nanocrystals and Silver Nanoparticles. *Carbohydr. Polym.* **2012,** *87* (2), 1596–1605.

29. Four Key Trends That Will Shape the Future of Packaging to 2028. (n.d). Retrieved from https://www.smithers.com/resources/2019/feb/future-packaging-trends-2018-to-2028. (accessed Jul 12, 2019).

30. Fukushima, K.; Abbate, C.; Tabuani, D.; Gennari, M.; Camino, G. Biodegradation of Poly (Lactic Acid) and its Nanocomposites. *Polym. Degrad. Stabil.* **2009,** *94* (10), 1646–1655.

31. Fukushima, K.; Abbate, C.; Tabuani, D.; Gennari, M.; Rizzarelli, P.; Camino, G. Biodegradation Trend of Poly (ε-caprolactone) and Nanocomposites. *Mater. Sci. Eng. C* **2010,** *30* (4), 566–574.

32. Funabashi, M.; Ninomiya, F.; Kunioka, M. Biodegradability Evaluation of Polymers by ISO 14855-2. *Int. J. Mol. Sci.* **2009,** *10* (8), 3635–3654.

33. Guinault, A.; Sollogoub, C.; Ducruet, V.; Domenek, S. Impact of Crystallinity of Poly (Lactide) on Helium and Oxygen Barrier Properties. *Eur. Polym. J.* **2012,** *48* (4), 779–788.

34. Gupta, M. C.; Deshmukh, V. G. Thermal Oxidative Degradation of Poly-lactic Acid. *Colloid Polym. Sci.* **1982,** *260* (5), 514–517.

35. Hoidy, W. H.; Ahmad, M. B.; Al-, E. A. J.; Ibrahim, N. A. B. Preparation and Characterization of Polylactic Acid/Polycaprolactone Clay Nanocomposites. *J. Appl. Sci.* **2010,** *10* (2), 97–106.

36. Iwata, T.; Doi, Y. Morphology and Enzymatic Degradation of Poly (L-lactic Acid) Single Crystals. *Macromolecules* **1998,** *31* (8), 2461–2467.

37. Jamshidian, M.; Tehrany, E. A.; Imran, M.; Jacquot, M.; Desobry, S. Poly-Lactic Acid: Production, Applications, Nanocomposites, and Release Studies. *Compr. Rev. Food Sci. Food Saf.* **2010,** *9* (5), 552–571.

38. Jia, S.; Yu, D.; Zhu, Y.; Wang, Z.; Chen, L.; Fu, L. Morphology, Crystallization and Thermal Behaviors of PLA-Based Composites: Wonderful Effects of Hybrid GO/PEG via Dynamic Impregnating. *Polymers* **2017,** *9* (12), 528.

39. Kayaci, F.; Umu, O. C.; Tekinay, T.; Uyar, T. Antibacterial Electrospun Poly (Lactic Acid) (PLA) Nanofibrous Webs Incorporating Triclosan/Cyclodextrin Inclusion Complexes. *J. Agric. Food Chem.* **2013,** *61* (16), 3901–3908.

40. Kim, I.-H.; Lee, S. C.; Jeong, Y. G. Tensile Behavior and Structural Evolution of Poly(Lactic Acid) Monofilaments in Glass Transition Region. *Fibers Polym.* **2009,** *10* (5), 687–693.

41. Komesu, A.; Oliveira, J. A. R. D.; Martins, L. H. D. S.; Maciel, M. R. W.; Filho, R. M. Lactic Acid Production to Purification: A Review. *Bio Resourc.* **2017,** *12* (2). doi: 10.15376/biores.12.2.komesu.

42. Kraft Heinz Expands Environmental Commitments to Include Sustainable Packaging and Carbon Reduction. July 31, 2018. Retrieved from https://news.kraftheinzcompany. com/press-release/corporate/kraft-heinz-expands-environmental-commitments-include-sustainable-packaging- (accessed Jul 12, 2019).

43. Krishna, B. S.; Nikhilesh, G. S. S.; Tarun, B.; KV, N. S.; Gopinadh, R. Industrial Production of Lactic Acid and Its Applications. *Comprehensive Biotechnology*, **2019,** 208–217.

44. Lee, S. R.; Park, H. M.; Lim, H.; Kang, T.; Li, X.; Cho, W. J.; Ha, C. S. Microstructure, Tensile Properties, and Biodegradability of Aliphatic Polyester/Clay Nanocomposites. *Polymer* **2002,** *43* (8), 2495–2500.

45. Liu, C.; Jia, Y.; He, A. Preparation of Higher Molecular Weight Poly (L-lactic Acid) by Chain Extension. *Int. J. Polym. Sci.* **2013,** *2013,* 1–6.

46. Liu, Y.; Liang, X.; Wang, S.; Qin, W.; Zhang, Q. Electrospun Antimicrobial Polylactic Acid/Tea Polyphenol Nanofibers for Food-packaging Applications. *Polymers* **2018,** *10* (5), 561.

47. Ljungberg, N.; Wesslén, B. Preparation and Properties of Plasticized Poly (Lactic Acid) Films. *Biomacromolecules* **2005,** *6* (3), 1789–1796.

48. Lopes, M. S.; Jardini, A.; Maciel Filho, R. Synthesis and Characterizations of Poly (Lactic Acid) by Ring-opening Polymerization for Biomedical Applications. *Chem. Eng. Trans.* **2014,** *38,* 331–336.

49. López-Rubio, A.; Sanchez, E.; Wilkanowicz, S.; Sanz, Y.; Lagaron, J. M. Electrospinning as a Useful Technique for the Encapsulation of Living Bifidobacteria in Food Hydrocolloids. *Food Hydrocol.* **2012,** *28* (1), 159–167.

50. Luo, Y.; Lin, Z.; Guo, G. Biodegradation Assessment of Poly (Lactic Acid) Filled with Functionalized Titania Nanoparticles (PLA/TiO$_2$) Under Compost Conditions. *Nanoscale Res. Lett.* **2019,** *14* (1), 56.

51. MacDonald, R. T.; McCarthy, S. P.; Gross, R. A. Enzymatic Degradability of Poly (Lactide): Effects of Chain Stereochemistry and Material Crystallinity. *Macromolecules* **1996,** *29* (23), 7356–7361.

52. Marija, Jovic. For a Greener Future: Biodegradable Packaging Materials, April 2017. Retrieved from https://www.prescouter.com/2017/04/biodegradable-packaging-materials/ (accessed Jul 12, 2019).

53. Mattioli, S.; Peltzer, M.; Fortunati, E.; Armentano, I.; Jiménez, A.; Kenny, J. M. Structure, Gas-barrier Properties and Overall Migration of Poly (Lactic Acid) Films Coated with Hydrogenated Amorphous Carbon Layers. *Carbon* **2013,** *63,* 274–282.

54. McNeill, I. C.; Leiper, H. A. Degradation Studies of Some Polyesters and Polycarbonates—1. Polylactide: General Features of the Degradation Under Programmed Heating Conditions. *Polym. Degrad. Stabil.* **1985,** *11* (3), 267–285.

55. Mehta, R.; Kumar, V.; Bhunia, H.; Upadhyay, S. N. Synthesis of Poly (Lactic Acid): A Review. *J. Macromol. Sci., Part C* **2005,** *45* (4), 325–349.

56. Meriçer, Ç.; Minelli, M.; De Angelis, M. G.; Baschetti, M. G.; Stancampiano, A.; Laurita, R.; ... Lindström, T. Atmospheric Plasma Assisted PLA/Microfibrillated Cellulose (MFC) Multilayer Biocomposite for Sustainable Barrier Application. *Ind. Crop Prod.* **2016,** *93,* 235–243.

57. Mitchell, M. R.; Link, R. E.; Yang, M.-H.; Lin, Y.-H. Measurement and Simulation of Thermal Stability of Poly(Lactic Acid) by Thermogravimetric Analysis. *J. Test. Eval.* **2009,** *37* (4), 102271.

58. Mohd Aris, Z. F.; Bavishi, V.; Sharma, R.; Nagarajan, R. Barrier Properties and Abrasion Resistance of Biopolymer-based Coatings on Biodegradable Poly (Lactic Acid) Films. *Polym. Eng. Sci.* **2019,** *59* (9), 1874–1881.

59. Nakayama, N.; Hayashi, T. Preparation and Characterization of Poly (L-lactic Acid)/TiO2 Nanoparticle Nanocomposite Films with High Transparency and Efficient Photodegradability. *Polym. Degrad. Stabil.* **2007,** *92* (7), 1255–1264.

60. Nanthananon, P.; Seadan, M.; Pivsa-Art, S.; Suttiruengwong, S. Enhanced Crystallization of Poly (Lactic Acid) through Reactive Aliphatic Bisamide. *IOP Conf. Ser.* **2015,** *87,* 012067.

61. Nishida, H.; Tokiwa, Y. Effects of Higher-order Structure of Poly (3-hydroxybutyrate) on Its Biodegradation. I. Effects of Heat Treatment on Microbial Degradation. *J. Applied Polym. Sci.* **1992,** *46* (8), 1467–1476.

62. Notingher, I.; Boccaccini, A. R.; Jones, J.; Maquet, V.; Hench, L. L. Application of Raman Microspectroscopy to the Characterisation of Bioactive Materials. *Mater. Charact.* **2002,** *49* (3), 255–260.

63. Ozkoc, G.; Kemaloglu, S. Morphology, Biodegradability, Mechanical, and Thermal Properties of Nanocomposite Films Based on PLA and Plasticized PLA. *J. Appl. Polym. Sci.* **2009,** *114* (4), 2481–2487.

64. Paul, M. A.; Delcourt, C.; Alexandre, M.; Degée, P.; Monteverde, F.; Dubois, P. Polylactide/Montmorillonite Nanocomposites: Study of the Hydrolytic Degradation. *Polym. Degrad. Stabil.* **2005,** *87* (3), 535–542.

65. Perego, G.; Cella, G. D.; Bastioli, C. Effect of Molecular Weight and Crystallinity on Poly (Lactic Acid) Mechanical Properties. *J. Appl. Polym. Sci.* **1996,** *59* (1), 37–43.

66. Pérez-Masiá, R.; Lagaron, J. M.; López-Rubio, A. Development and Optimization of Novel Encapsulation Structures of Interest in Functional Foods through Electrospraying. *Food Bioproc. Technol.* **2014,** *7* (11), 3236–3245.

67. Plackett, D. V.; Holm, V. K.; Johansen, P.; Ndoni, S.; Nielsen, P. V.; Sipilainen-Malm, T.; ... Verstichel, S. Characterization of L-polylactide and L-polylactide–polycaprolactone Co-polymer Films for Use in Cheese-packaging Applications. *Packag. Technol. Sci.* **2006,** *19* (1), 1–24.

68. Puchalski, M.; Kwolek, S.; Szparaga, G.; Chrzanowski, M.; Krucińska, I. Investigation of the Influence of PLA Molecular Structure on the Crystalline Forms (α' and α) and Mechanical Properties of Wet Spinning Fibres. *Polymers* **2017,** *9* (12), 18.

69. Rahman, M. R.; Hui, J. L. C. Physico-mechanical and Thermal Properties of Clay/Fumed Silica Diffuse Polylactic Acid Nanocomposites. In *Silica and Clay Dispersed Polymer Nanocomposites*; Woodhead Publishing: UK, 2018, pp. 87–107.

70. Ramontja, J.; Ray, S. S.; Pillai, S. K.; Luyt, A. S. High-Performance Carbon Nanotube-Reinforced Bioplastic. *Macromol. Mater. Eng.* **2009,** *294* (12), 839–846.

71. Ray, S. S.; Okamoto, M. Biodegradable Polylactide and its Nanocomposites: Opening a New Dimension for Plastics and Composites. *Macromol. Rapid Commun.* **2003,** *24* (14), 815–840.

72. Ren, J. Synthesis and Manufacture of PLA. In *Biodegradable Poly (Lactic Acid): Synthesis, Modification, Processing and Applications* (pp. 15–37). Springer: Berlin, Heidelberg, 2010.

73. Ren, J. (Ed.). *Biodegradable Poly (Lactic Acid): Synthesis, Modification, Processing and Applications.* Springer: Science & Business Media, 2011.

74. Rhim, J. W.; Hong, S. I.; Ha, C. S. Tensile, Water Vapor Barrier and Antimicrobial Properties of PLA/Nanoclay Composite Films. *LWT-Food Sci. Technol.* **2009,** *42* (2), 612–617.

75. Sawada, H.; Takahashi, Y.; Miyata, S.; Kanehashi, S.; Sato, S.; Nagai, K. Gas Transport Properties and Crystalline Structures of Poly (Lactic Acid) Membranes. *Trans. Mater. Res. Soc. Jpn* **2010,** *35* (2), 241–246.

76. Sheth, M.; Kumar, R. A.; Davé, V.; Gross, R. A.; McCarthy, S. P. Biodegradable Polymer Blends of Poly (Lactic Acid) and Poly (Ethylene Glycol). *J. Appl. Polym. Sci.* **1997,** *66* (8), 1495–1505.

77. Siracusa, V.; Blanco, I.; Romani, S.; Tylewicz, U.; Rocculi, P.; Rosa, M. D. Poly (Lactic Acid)-Modified Films for Food Packaging Application: Physical, Mechanical, and Barrier Behavior. *J. Appl. Polym. Sci.* **2012,** *125* (S2), E390–E401.

78. Siracusa, V.; Dalla Rosa, M.; Iordanskii, A. Performance of Poly (Lactic Acid) Surface Modified Films for Food Packaging Application. *Materials* **2017,** *10* (8), 850.

79. Siracusa, V.; Rocculi, P.; Romani, S.; Dalla Rosa, M. Biodegradable Polymers for Food Packaging: A Review. *Trends Food Sci. Technol.* **2008,** *19* (12), 634–643.

80. Someya, Y.; Kondo, N.; Shibata, M. Biodegradation of Poly (butylene adipate-co-butylene terephthalate)/Layered-silicate Nanocomposites. *J. Appl. Polym. Sci.* **2007,** *106* (2), 730–736.

81. Takizawa, K.; Nulwala, H.; Hu, J.; Yoshinaga, K.; Hawker, C. J. Molecularly Defined (L)-lactic Acid Oligomers and Polymers: Synthesis and Characterization. *J. Polym. Sci. Part A: Polym. Chem.* **2008,** *46* (18), 5977–5990.

82. Tanio, T.; Fukui, T.; Shirakura, Y.; Saito, T.; Tomita, K.; Kaiho, T.; Masamune, S. An Extracellular Poly (3-hydroxybutyrate) Depolymerase from Alcaligenes Faecalis. *Eur. J. Biochem.* **1982,** *124* (1), 71–77.

83. Tokiwa, Y.; Calabia, B. P. Biodegradability and Biodegradation of Poly (Lactide). *Appl. Microbiol. Biotechnol.* **2006,** *72* (2), 244–251.

84. Tokiwa, Y.; Suzuki, T. Hydrolysis of Polyesters by Rhizopus Delemar Lipase. *Agric. Biol. Chem.* **1978,** *42* (5), 1071–1072.

85. Tsuji, H.; Miyauchi, S. Poly (L-lactide): VI Effects of Crystallinity on Enzymatic Hydrolysis of Poly (L-lactide) without Free Amorphous Region. *Polymer Degrad. Stabil.* **2001,** *71* (3), 415–424.

86. Walha, F.; Lamnawar, K.; Maazouz, A.; Jaziri, M. Biosourced Blends Based on Poly (Lactic Acid) and Polyamide 11: Structure–properties Relationships and Enhancement of Film Blowing Processability. *Adv. Polym. Technol.* **2018,** *37* (6), 2061–2074.

87. Wen, P.; Zhu, D. H.; Feng, K.; Liu, F. J.; Lou, W. Y.; Li, N.; … Wu, H. Fabrication of Electrospun Polylactic Acid Nanofilm Incorporating Cinnamon Essential Oil/β-cyclodextrin Inclusion Complex for Antimicrobial Packaging. *Food Chem.* **2016,** *196,* 996–1004.

88. Wihodo, M.; Moraru, C. I. Physical and Chemical Methods Used to Enhance the Structure and Mechanical Properties of Protein Films: A Review. *J. Food Eng.* **2013,** *114* (3), 292–302.

89. Wongsasulak, S.; Pathumban, S.; Yoovidhya, T. E-ect of Entrapped α-tocopherol on Mucoadhesivity and Evaluation of the Release, Degradation, and Swelling Characteristics of Zein–chitosan Composite Electrospun Fbers. *J. Food Eng.* **2014,** *120,* 110–117.

90. Wu, K. J.; Wu, C. S.; Chang, J. S. Biodegradability and Mechanical Properties of Polycaprolactone Composites Encapsulating Phosphate-solubilizing *Bacterium Bacillus* sp. PG01. *Process Biochem.* **2007,** *42* (4), 669–675.

91. Wu, T. M.; Wu, C. Y. Biodegradable Poly (Lactic Acid)/Chitosan-modified Montmorillonite Nanocomposites: Preparation and Characterization. *Polym. Degrad. Stabil.* **2006,** *91* (9), 2198–2204.

92. Yang, W.; Fortunati, E.; Dominici, F.; Giovanale, G.; Mazzaglia, A.; Balestra, G. M.; … Puglia, D. Effect of Cellulose and Lignin on Disintegration, Antimicrobial and Antioxidant Properties of PLA Active Films. *Int. J. Biol. Macromol.* **2016,** *89,* 360–368.

93. Yang, X.; Kang, S.; Yang, Y.; Aou, K.; Hsu, S. L. Raman Spectroscopic Study of Conformational Changes in the Amorphous Phase of Poly (Lactic Acid) During Deformation. *Polymer* **2004,** *45* (12), 4241–4248.

94. Yu, H.; Zhang, Z. L.; Chen, J.; Pei, A.; Hua, F.; Qian, X.; … Xu, X. Carvacrol, a Food-additive, Provides Neuroprotection on Focal Cerebral Ischemia/Reperfusion Injury in Mice. *PloS One* **2012,** *7* (3), e33584.

95. Yuniarto, K.; Purwanto, Y. A.; Purwanto, S.; Welt, B. A.; Purwadaria, H. K.; Sunarti, T. C. *Infrared and Raman Studies on Polylactide Acid and Polyethylene Glycol-400 Blend,* AIP Conference Proceedings, 2016, 1725.

96. Zhang, C.; Lan, Q.; Zhai, T.; Nie, S.; Luo, J.; Yan, W. Melt Crystallization Behavior and Crystalline Morphology of Polylactide/Poly(ε-caprolactone) Blends Compatibilized by Lactide-Caprolactone Copolymer. *Polymers* **2018,** *10* (11), 1181.

97. Zhang, M.; Thomas, N. L. Blending Polylactic Acid with Polyhydroxybutyrate: The Effect on Thermal, Mechanical, and Biodegradation Properties. *Adv. Polym. Technol.* **2011,** *30* (2), 67–79.

98. Zhu, J. Y.; Tang, C. H.; Yin, S. W.; Yang, X. Q. Development and Characterization of Novel Antimicrobial Bilayer Films Based on Polylactic Acid (PLA)/Pickering Emulsions. *Carbohydr. Polym.* **2018,** *181,* 727–735.

CHAPTER 5

Organic Materials for Green Electronics

VIDYA G.*, SARAVANAN SUBBIAHRAJ, and PRAVEEN C. RAMAMURTHY

Department of Materials Engineering, Indian Institute of Science, Bangalore 560012, India

Corresponding author. E-mail: vidyagopi5@gmail.com

ABSTRACT

The field of "green electronics" is a cross-disciplinary research area which recognises naturally occurring or nature-inspired compounds and use them for the design and synthesis of eco-friendly synthetic-organic materials that are environmentally safe biodegradable/or biocompatible, for making of "Green-Devices." The actual goal of this research field is to produce environmentally viable semiconducting materials and their applications in devices such as organic electronics to carry low energy intensity, low cost, negligible waste production, and also accomplish good functionalities. This chapter briefly reviews major research work based on ideologies of green chemistry in the synthetic strategies in making of semiconducting materials including small molecules and polymers or nature-inspired biocompatible materials and their applications in organic-green electronics.

5.1 INTRODUCTION

In the modern digital era, we are witnessing different evolutionary changes in making electronic technologies crossing the lines between chemical, physical, digital, and biological domains. The results of continuing electronic miniaturization in device making from inorganic based materials such as silicon, germanium, indium phosphide and gallium arsenide, the fundamental technological achievements after fifty years of research allowed the invention of materials in extremely ordered fashion starting from atomic scale to nanoscale with a precise functionality. Therefore, numerous electronic materials appear every day and slowly become part of daily life. The increasing demand for electronics is leading to a series of undesirable problems such as a large amount of e-waste, quick exhaustion of natural elements (gallium, Indium, etc.), and causing some serious environmental problems.[1] Hence, the necessity to control the electronics waste disposal and also reduce the resource misuse will be mandatory because of minimalizing the adverse effect of current and upcoming generations on the environment and to make an admissible future. According to the United Nations World Commission on Environment and Development, sustainable development is recognized when people confirm its existing requirements without conceding the capability of upcoming generations to meet their personal requirements.[2] The organic electronics was established in the mid-1970s holding the advantages based on energy-efficient materials and also delivering as low-cost devices.[3-5] But, the intense effort of research on earlier times, the stability and device performance of organic semiconductors endure at present, main horseraces in their progress as solid opponents of their inorganic counterpart. Hence, the massive replacement of inorganic electronic components such as integrated circuits, processors, and different solar cell modules, with organic counterpart is not instantly applicable.[6,7] Yet, the soft and flexible-nature organic material has substantial benefit over their inorganic counterpart which leads to permitting the fabrication flexible, conformable, and exceptionally thin to electronic devices.[8]

"Green materials" and "Green technologies" are the best way for realizing the sustainability in the field of organic electronics.[9,10] This is achieved in several ways, and these are as follows:

1. Organic synthesis of chemical compounds by maximizing the slow-down of waste
2. The use of harmless and environmentally safe chemicals
3. Atom economy
4. Renewable raw materials
5. Proper design for disposal and degradation of chemical substances after use
6. Pollution prevention actions and sustainable progress
7. Cost-effective fabrication routes in practical devices
8. Electronic devices that feature degradability in ordinary atmosphere conditions after their end of maximum lifetime

Fulfilling all the above-mentioned standards remains really challenging at this time. Because "Green" materials and technologies are now going through their initial stage of emerging concepts and for giving the correct description of "Greenness" is not clear. There is an exciting multidisciplinary research field now developing on "organic bioelectronics" where biocompatible materials are introduced in several electronic devices connected with living tissues.[11] The organic bioelectronics research field might be appropriate for introducing natural and nature-inspired organic materials and it is well suitable for realizing the determined intention of "green organics" and sustainable organic electronics future. There is a different way for estimating the "Greenness" of organic synthetic sequence.[12] The atom economy or atom efficiency was first announced by Trost in the year 1991. It is described as the ratio of the mass of the product by the mass of all the reagents involved.[13] This theoretical approach undertakes accurate amounts of the substances that were taken. And other important factors were the solvents used, exact yield, and catalyst.[14] The concept E-factor was established by Sheldon; it is explained as the ratio of the total mass of the waste to the total mass of the wanted product. The exact mass of the unwanted waste product is not significant because it was environmentally hazardous per unit mass.[15] However, the concept directed to the origin of the environmental quotient (EQ) factor having molecular weight of each by-product measured by using a Q factor and it was denoting a random number that measures the undesirability of the waste substance.[15] This EQ factor helps the quantitative monitoring of the "Greenness" of the chemical reactions. Another additional parameter is the production energy of chemical substances and EQ factor is also mainly used in the life-cycle analysis of manufactured chemicals.[14,15] The application based on EQ factor mostly related to up-front translation of production energy to carbon dioxide releases.[16] And this production energy also relates to cost. But the main disadvantage of production energy as the measurement of "Greenness" is dealing with one factor such as carbon dioxide releases and disregarding the other important issues based on the formation and disposal of toxic chemical waste.

The aim of this chapter mainly highlights brief insight of synthetic aspects of Green-Organic materials and their applicability in LEDs, organic solar cells (OSCs), organic field-effect transistors (OFETs).

5.2 SCOPE OF GREEN-ORGANIC SYNTHESIS

The purpose of this chapter is not to provide an in-depth example of such chemical synthesis and their applications on green chemistry, but to recognize a few developments in this area that look particularly favorable to large scale organic synthesis. The conventional tactics to synthesize low band-gap organic-electronic polymers are exemplified in Figure 5.1.[18] In this particular chapter, we are focussing on the synthesis and applications of organic semiconducting materials in a greener approach. For any multistep-synthesis, materials, energy, and time are the foremost characters for the purification and work-up procedures. Failure to purify an organic compound can seriously affect the optoelectronic properties of the fabricated devices. Specifically, the remaining residual palladium or nickel catalyst have an adverse effect on the electronic properties of the materials.[17] Finding an environmentally benign solvent is another significant problem for Green-Organic synthesis.

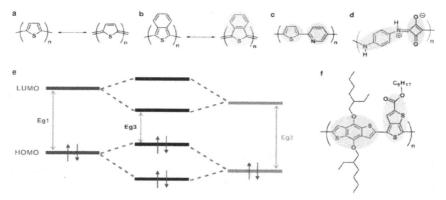

FIGURE 5.1 Approaches to synthesize low band-gap organic electronic polymers. (a) Aromatic structure and quinoid resonance structures of polythiophene. (b) Stabilization of the quinoid resonance structure of poly(isathianapthene) molecule (Green color). (c) Donor (blue part)–Acceptor (red part) copolymer synthesized by Yamamoto. (d) Donor–acceptor copolymer produced by Havinga. (e) Shortened mechanism of lowering of band-gap by donor–acceptor interaction. (f) Modern synthetic method that associates both methods.

Source: Reprinted with permission from Ref. [18]. © 2015 The Royal Society of Chemistry.

5.3 GREEN-ORGANIC REACTION METHODOLOGY FOR ELECTRONIC APPLICATIONS

5.3.1 *SUZUKI COUPLING REACTIONS*

Suzuki reaction is classified as a cross-coupling reaction having the coupling partners are boronic acid and an organohalide catalyzed with a palladium (0) complex.[19] This reaction was first published by Akira Suzuki, Richard F. Heck, and Ei-ichi Negichi in 1979 for this invention they were shared

Nobel prize in 2010.[19] The main advantages of Suzuki coupling reaction are, it should be scalable and cost-effective for the making of intermediates for pharma and other industries. This reaction in some aspects is more environmentally desirable than other coupling reactions. But in the presence of phase transfer catalyst and water, the base leads to substantial difficulty to the reaction medium, due to this sometimes-multi-phase reactions become slower and very difficult to mix.[20] Figure 5.2 displays a typical Suzuki coupling reaction by Kim et al. with carbazole, as the electron-rich unit and tetrafluoro phenylene as the electron-deficient unit.[21]

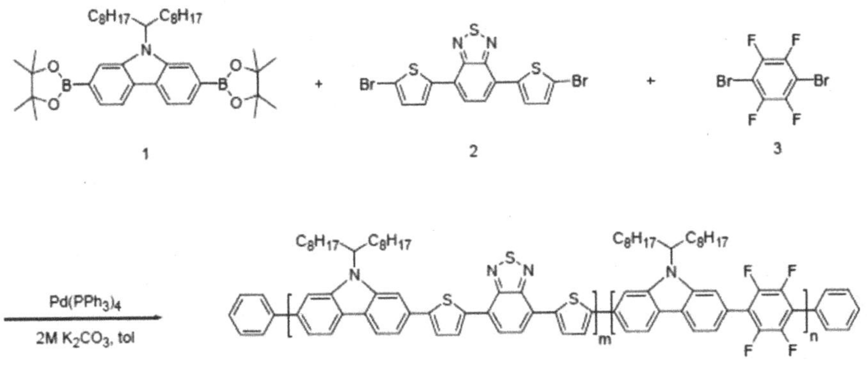

FIGURE 5.2 Suzuki coupling polymerization.

Suzuki coupling reaction Greened in different approaches such as reutilizing the palladium catalyst thus dropping the heavy metal waste produced, by using nonvolatile liquid such as water or ionic liquids and by using ligand-free environments.[22–24] Nancy E. Costa et al. developed a greenest and cost-effective Suzuki coupling method to prepare ethyl (4-phenylphenyl) acetate, it was a precursor to the drug felbinac and this was employed as an anti-inflammatory drug for arthritis treatment and the preparation of ethyl (4-phenylphenyl) acetate through Green–Suzuki coupling is depicted in Figure 5.3.[25]

5.3.2 STILLE COUPLING REACTIONS

The most well-known synthesis methodology for making donor-acceptor based low bandgap polymers/small molecules is Stille based condensation reaction between bromide terminated and trimethyl stannyl-terminated compounds.[26,27] Gopi et al. synthesized highly soluble, thermally stable, and semiconducting thienylene–biphenylene vinylene hybrid polymers via Stille coupling reaction and explore its suitability for polymer light emitting diode (PLED) applications and the reaction scheme is depicted in Figure 5.4.[28] Stille coupling reaction has prominent in the field of conducting polymers due to its dependability, but the reaction contains some drawbacks also. These are issues with its preparation and purification of stannane monomers and the use of aromatic halogenated solvents. And thus, both are environmentally harmful and need substantial input of energy to recycle.[29]

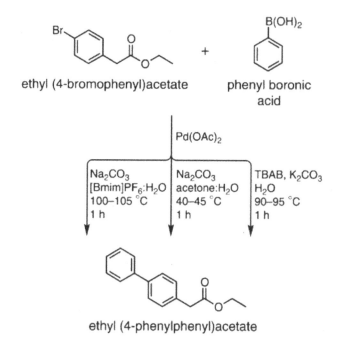

ethyl (4-bromophenyl)acetate   phenyl boronic acid

Pd(OAc)$_2$

| Na$_2$CO$_3$ [Bmim]PF$_6$:H$_2$O 100–105 °C 1 h | Na$_2$CO$_3$ acetone:H$_2$O 40–45 °C 1 h | TBAB, K$_2$CO$_3$ H$_2$O 90–95 °C 1 h |

ethyl (4-phenylphenyl)acetate

FIGURE 5.3   Synthesis of ethyl (4-phenylphenyl) acetate by using Green–Suzuki coupling reaction method.

Source: Reprinted with permission from Ref. [25]. © 2012, American Chemical Society.

Guo-ping Lu et al. reported a non-ionic amphiphile, TPGS-750-M, allows well-organized Stille couplings between varieties of substrates to be conducted in water at room temperature and the scheme is presented in Figure 5.5.[30] This reaction is a typical example of Greener approach toward the Stille coupling reaction because in this case limited amounts of water are used and work-up carried out based on in-flask extraction with minimal and recoverable solvents are used.

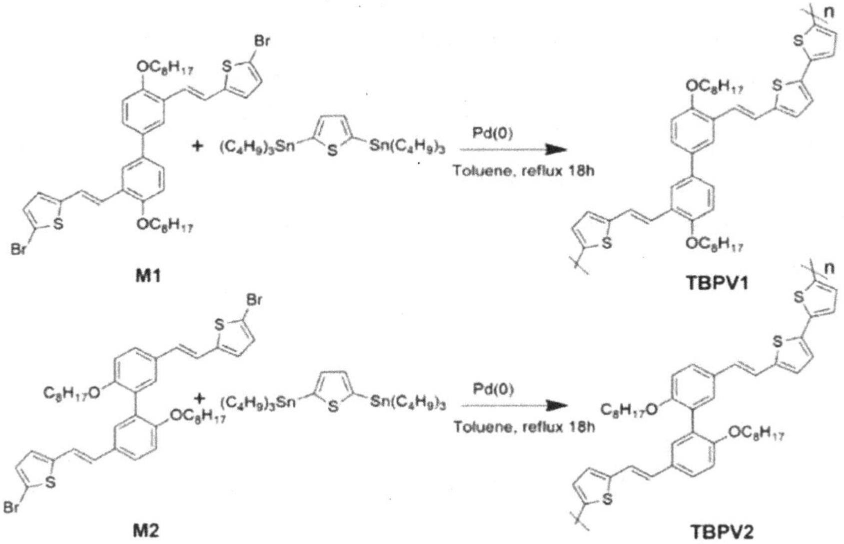

FIGURE 5.4 Synthesis thienylene–biphenylene vinylene hybrid polymers via Stille coupling polymerization reaction.

Source: Reprinted with permission from Ref. [28]. © 2016 Elsevier.

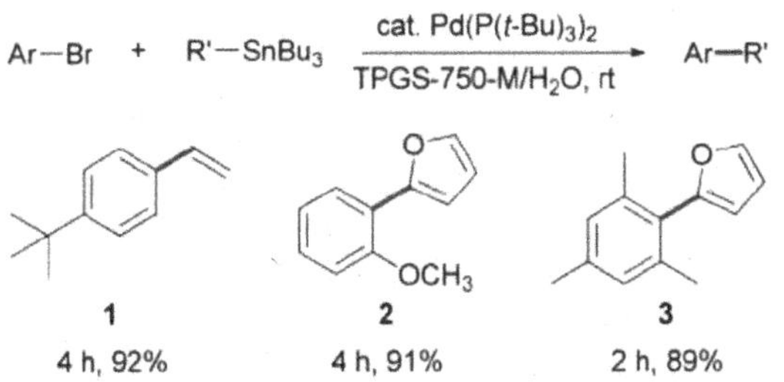

FIGURE 5.5 Stille coupling with aryl bromides in water medium at room temperature.

Source: Reprinted from Ref. [30]. Open access.

5.3.3 DIRECT ARYLATION REACTIONS

Nowadays, a developing synthetic method called direct arylation technique and are widely considered as simplistic, atom efficient, and environmentally approachable route for the synthesis of semiconducting polymers and small molecules].[31,32] Figure 5.6 demonstrates a typical direct arylation reaction between alkyl–diketopyrrolopyrrole (DPP) (Th)$_2$ and appropriate brominated end-capping units.[33] Direct arylation protocol has some greener advantages

(Fig. 5.7) over other heavy-metal catalyzed reactions such as, it may hold more possible for reducing the amount of more toxic by-products, more facile and the use of diprotic monomer in place of a distannane also helps complete purification and also gaining polymers with high molecular weight.[34] Direct aryl polymerization (DHAP) mainly applicable for the synthesis of high performing polymers for various organic-electronic applications because it should be more facile, less economic, and eco-friendly than other synthesis protocols.[35]

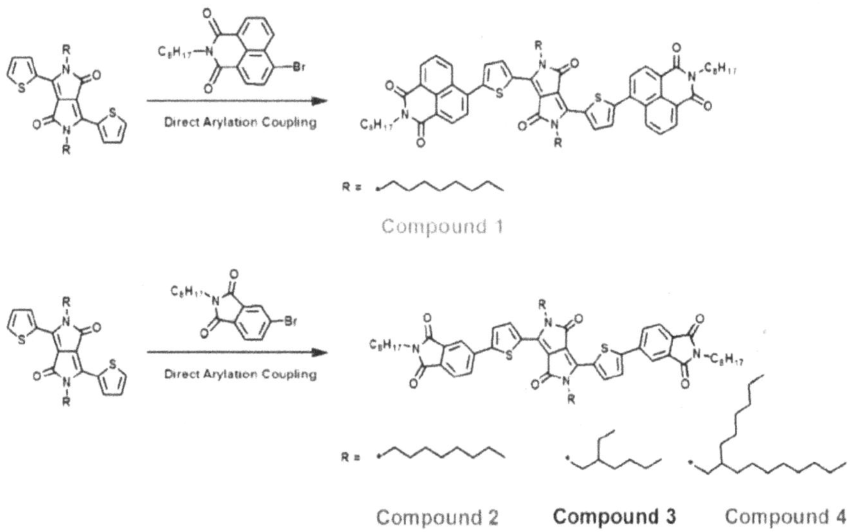
Direct Arylation Coupling
Compound 1
Direct Arylation Coupling
Compound 2
Compound 3
Compound 4

FIGURE 5.6 General synthetic route toward DPP molecules through direct arylation reaction and the conditions: Pd $(OAc)_2$, Pivalic Acid, K_2CO_3, dimethylacetamide (DMA), Microwave irradiation 170°C, 17 min.

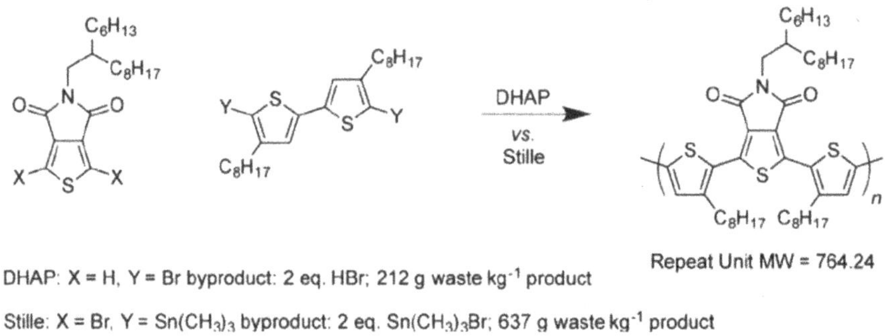
DHAP
vs.
Stille
Repeat Unit MW = 764.24
DHAP: X = H, Y = Br byproduct: 2 eq. HBr; 212 g waste kg-1 product
Stille: X = Br, Y = Sn(CH3)3 byproduct: 2 eq. Sn(CH3)3Br; 637 g waste kg-1 product

FIGURE 5.7 Direct arylation produces very little waste by mass than Stille coupling to obtain the same one.[34]

Source: Reprinted with permission from Ref. [34]. © The Royal Society of Chemistry.

5.3.4 ORGANIC MATERIALS FROM BIOLOGICALLY DERIVED MATERIALS

Improved performance, biodegradable electronic materials from biodegradable substances considering intensively with the inspiration of diminishing the usage of petroleum and reducing environmental pollution. In the case of Bio-electronic materials, biodegradability and bio compactivity are more crucial.[36,37] Cellulose is the most versatile biopolymer on earth because due its sustainability, biodegradability, bio-compactivity, mechanical strength, abundance, optical properties, etc..[38,39] Paper electronics attracted more attention due to the versatile advantages such as eco-friendly, cost-effective, large scale production, roll to roll process, and easily disposable.[40] And also, regular paper is opaque, therefore it can be used as a barrier for the organic light emitting diode (OLED) device. Other materials from naturally occurring compounds such as indigo, aurin, beta carotene, has already used in bio-electronics.[41] But these are found in very low abundance in nature, therefore the mass production of these dyes is impossible. The most dependable way for producing these bio-based dyes from chemical modification of derivatives from natural products or are synthetic counterparts prepared from ample, cheap petrochemicals and the examples are Vat yellow, Vat orange dyes, perylene diimides, etc.[42]

5.4 GREEN-ORGANIC MATERIALS FOR ELECTRONIC APPLICATIONS

The progress and execution of green electronics combining with naturally occurring materials are vital for giving a more sustainable future. The recycling of waste products should be a possible route to save the materials, energy, and money. Organic electronics device consists of several important components such as electrodes, carrier substrates, dielectrics, adhesives, organic semiconductors, etc..[43] Organic semiconductors are more sensitive toward water, moisture and also show increasing defect sites upon using. Therefore, organic layers of each device could be shielded with an encapsulant material and an adhesive.[44]

The layer structures of OLED and OSC or (OPVs) are comparable. In the case of substrate materials along with mechanical and barrier properties, the substrates are highly affecting the device lifetime. Conventional plastic substrates are polyethylene terephthalate (PET), polycarbonate (PC), etc. but the decomposition of these materials should be difficult. Therefore, nowadays novel, bio-based—biodegradable, flexible, transparent, and clear substrates have garnered substantial consideration.[44] Some examples of recognized biodegradable polymers as carrier substrates are: starch from potatoes and corn contain amylopectin and linear amylose, cellulose, poly-lactic acid (PLA), polyhydroxybutyrate-co-valerate (PHBV), etc..[44] Yang and co-workers reported a water-based adhesive for electrical interconnects and electrical circuits.[45] Bettinger et al. published biodegradable adhesives for organic electronics and it could hold good crystallinity, hydrophobic nature, and easy to process.[46] Different biodegradable adhesives are available now but their barrier properties and applicability in bio-electronics are not explored. Some natural dyes are considered as organic semiconductors therefore the scientists were focusing the naturally occurring dyes and their different derivatives, some natural dyes are represented in Figure 5.8.

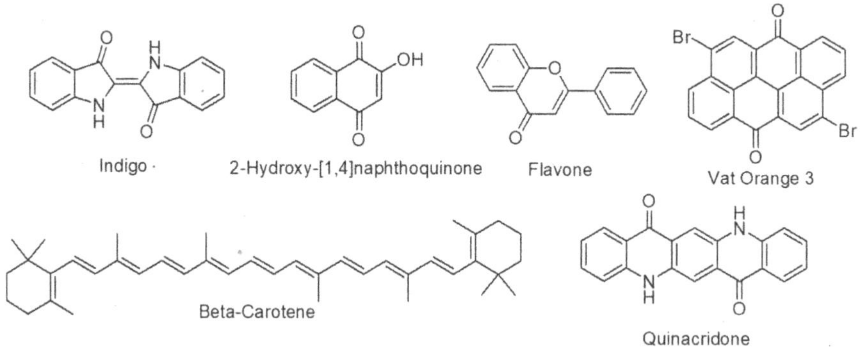

FIGURE 5.8   Structures of biodegradable and biocompatible organic dyes.

Głowacki et al. in 2011 informed that anthraquinones, carotenoids, and indigoids are giving charge carrier mobilities in the range of ~10−4 cm^2 V^{-1}·s^{-1} when applied in OFETs.[47] Ambipolar indigoids show good semiconductor characteristics because of its long-range order, crystallinity, and very good dielectric features. The conducting polymers especially polythiophene, polypyrrole, polyaniline is used as organic electrodes but their biodegradability is not fully explored till now. But biocompatible metals such as Fe, Mg, and Zn are using in biomedical applications because the metals degrade in a particular biological environment.[48] Instead of shellac, natural dielectric materials classified as two categories, first category consist of sugar-related small molecules such as fructose or glucose and the second category contain nucleobases such as adenine, guanine, cytosine, and thymine. These are using in OFETs because of their low dielectric losses, low leakage current, and also high breakdown strength.[45,46] For the functionalization of DNA with a cationic surfactant leads to show solubility in polar solvents and also permits to form wet coated thin films. Therefore, it can be easily introduced as electron blocking layers in OLEDs and gate dielectrics in OFETs.[49] Hongli Zhu et al. reported a cellulose-related transparent, biodegradable substrate joining either nano paper or a regenerated cellulose film (RCF) for highly flexible OLED applications and the performance should be stable in both the flat and bend direction.[50] In 2013, Mihai Irimia-Vladu et al. reported as natural resin shellac applied in OFETs because of its biocompatibility, simple solution processing from ethanol solution, good dielectric breakdown fields, and low temperature crosslinking at 70 ^{0}C.[51] In a photovoltaic cell, paper seems to be an ideal supporting material having its prize less than that of plastic and glass. The performance reported based on paper photovoltaic cell consists of open-circuit voltage having 50 V and a current of ~10 μA. Furthermore, the photovoltaic arrays should be twistingly bent and folded without loss of function.[44] Quinacridone is a less costly natural derivative of acridone and it displays exciting photophysics properties such as highly sensitive toward light, already invented external quantum efficiency should be 10% in the single-layer OPV configuration.[44] Organic-bioelectronics need more attention for the further development and industrialization because this field is now in its early stage and the main factors which are required to concentrate such as (1) examine the semiconducting properties of whole natural or nature-inspired materials, (2) understand their processability and film-forming properties, (3) examine electrochemical, optical and charge transport properties, (4) perform biodegradability studies and (5) use all the congregated knowledge to fabricate novel devices, etc.

5.5 SUMMARY

In this chapter, we have presented the Green synthesis methodology and applications of naturally occurring semiconducting biomaterials in organic electronics. Synthesis strategies presented here are based on small energy intensity, with the negligible release of unwanted or toxic waste, and at little cost. Overall studies demonstrate that by the introduction of green chemistry to the synthesis of organic semiconductors as the most applicable way to diminish the environmental hazardous associated on the manufacturing of OLEDs, OSCs, OFETs, etc. Green electronics or it is called as sustainable electronics consist of nature-inspired or biocompatible substrates, electrodes, dielectrics, organic semiconducting materials are now available, and these are useful for making optoelectronic devices. This interdisciplinary field suggests that there is plenty of scope between the synthetic chemists and device scientists involved in the process development support to produce energy using novel devices and are designed based on Green practices.

ACKNOWLEDGMENTS

One of the authors Vidya G., wishes to thank to University Grant Commission, New Delhi for the award of Dr. D.S. Kothari Postdoctoral Fellowship, NO.F.4-2/2006(BSR)/CH/15-16/0205.

KEYWORDS

- **green chemistry**
- **green electronics**
- **OLEDs**
- **OSCs**
- **OFETs**
- **biodegradability**
- **bio-inspired semiconductors**

REFERENCES

1. Zoeteman, B. C. J.; Krikke, H. R.; Venselaar, J. Handling WEEE Waste Flows: On the Effectiveness of Producer Responsibility in a Globalizing World. *Int. J. Adv. Manuf. Technol.* **2010,** *47*, 415–436.
2. *United Nations, Report of the World Commission on Environment and Development "Our Common Future"* Oxford University Press: Oxford, 1987; pp 37.
3. Ma. H.; Acton. O.; Hutchins. D. O.; Cernetic. N.; Jen, A. K. Y. Multifunctional Phosphonic Acid Self-assembled Monolayers on Metal Oxides as Dielectrics, Interface Modification Layers and Semiconductors for Low-voltage High-performance Organic Field-effect Transistors. *Phys. Chem. Chem. Phys.* **2012,** *14*, 14110–14126.
4. Mei, J.; Diao, Y.; Appleton, A. L.; Fang L.; Bao, Z. Integrated Materials Design of Organic Semiconductors for Field-Effect Transistors. *J. Am. Chem. Soc.* **2013,** *135*, 6724–6746.

5. Jenssen R. A.; Nelsson, J. Factors Limiting Device Efficiency in Organic Photovoltaics. *Adv. Mater.* **2013,** *25,* 1847–1858.

6. Jorgensen, M.; Norrman, K.; Gevorgyan, S. A.; Tromholt, T.; Andreasen, B.; Krebs, F. C. Stability of Polymer Solar Cells. *Adv. Mater.* **2012,** *24,* 580–612.

7. F. Ante. F.; Kälblein, D.; Zaki, T.; Zschieschang, U.; Takimiya, K.; Ikeda, M.; Sekitani, T.; Someya, T.; Burghartz, J. N.; Kern, K.; and Klauk, H. Contact Resistance and Megahertz Operation of Aggressively Scaled Organic Transistors. *Small.* **2012,** *8,* 73–79.

8. Kaltenbrunner, M.; White, M. S.; Głowacki, E. D.; Sekitani, T.; Someya, T.; Sariciftci, N. S.; Bauer, S. Ultrathin and Lightweight Organic Solar Cells with High Flexibility. *Nat. Commun.* **2012,** *3,* 770.

9. Zhu, H.; Luo, W.; Ciesielski, P. N.; Fang, Z.; Zhu, J. Y.; Henriksson, G.; Michael E. Himmel, M. E.; Hu, L. Wood-Derived Materials for Green Electronics, Biological Devices, and Energy Applications. *Chem. Rev.* **2016,** *116,* 9305–9374.

10. Liao, X.; Zhang, Z.; Liao, Q.; Liang, Q.; Ou, Y.; Xu, M.; Lia, M.; Zhanga U.; Zhang Y. Flexible and Printable Paper-based Strain Sensors for Wearable and Large-Area Green Electronics. *Nanoscale.* **2016,** *8,* 13025.

11. Owens. R. M.; Malliaras, G. G. Organic Electronics at the Interface with Biology. *Mater. Today.* **2010,** *35,* 449–456.

12. Sekitani. T.; Someya, T. Human-friendly Organic Integrated Circuits. *Mater. Today.* **2011,** *14,* 398–407.

13. Trost, B. M. *Science.* **1991,** *254,* 1471–1477.

14. Sheldon, R. A. *Green Chem.* **2007,** *9,* 1273–1283.

15. Anctil, A.; Babbitt, C. W.; Raffaelle R. P.; Landi, B. J. *Environ. Sci. Technol.* **2011,** *45,* 2353–2359.

16. U. E. P. Agency, Clean Energy Calculations and References, accessed 24 January 2013, 2013.

17. Farina, V.; Krishnamurthy V.; Scott, W. J.; *Inorganic Reactions*, John Wiley & Sons, Inc., 2004.

18. Dou, L.; Liu, Y.; Liu, L.; Hong, Z.; Gang Li, G.; Yang, Y. Low-Bandgap Near-IR Conjugated Polymers/Molecules for Organic Electronics. *Chem. Rev.* **2015,** *115,* 12633–12665.

19. Suzuki, A. *J. Organomet. Chem.* **1999,** *576,* 147.

20. Molander G. A.; N. Ellis, N. *Acc. Chem. Res.* **2007,** *40,* 275–286.

21. Kim. J.; Kim N. H.; Song, S.; Park, S. Y.; Chae, S.; Bae. E.; Kim, I.; Kim H. J.; Kim. J. Y.; Suh, H. Syntheses of PCDTBT Containing Tetra Fluorobenzene as Electron-Withdrawing Group with Deep HOMO Energy Level and Applications for Photovoltaics. *Polymer.* **2016,** *102,* 84–89.

22. Zhang, Y.; Liu, L.; Wang, Y. *Synlett.* **2005,** *20,* 3083–3086.

23. Miyaura, N.; Suzuki, A. *Chem. Commun.* **1979,** 866–867.

24. Hidehiro Sakurai, H.; Tsukuda, T.; Hirao., T. *J. Org. Chem.* **2002,** *67,* 2721–2722.

25. Costa, N. E.; Pelotte, A. L.; Simard, J. M.; Syvinski, C. A.; Deveau, A. M. Discovering Green, Aqueous Suzuki Coupling Reactions: Synthesis of Ethyl (4-Phenylphenyl) acetate, a Biaryl with Anti-Arthritic Potential. *J. Chem. Educ.* **2012,** *89,* 1064–1067.

26. Beaujuge P. M.; Fréchet, J. M. J. *J. Am. Chem. Soc.* **2011,** *133,* 20009–20029.

27. Dennler, G.; Scharber M.C.; Brabec, C.J. *Adv. Mater.* **2009,** *21,* 1323–1338.

28. Gopi V.; Varma S. J.; Kumar M. V.; Prathapan, S.; Jayalekshmi, S.; Joseph, R. Semiconducting Thienylene-Biphenylene Vinylene Hybrid Polymers: Synthesis, Characterization and Application Prospects in Polymer LEDs. *Dyes and Pigments.* **2016,** *126,* 303-312.

29. Babudri, F.; Cardone, A.; Chiavarone, L.; Ciccarella, G.; Farinola, G, M.; Naso F.; G. Scamarcio. *Chem. Commun.* **2001**, 1940–1941.
30. Lu, G.; Caib C.; Lipshutz, B. H. Stille Couplings in Water at Room Temperature. *Green Chem.* **2013**, *15*, 105.
31. Lyons T.W.; Sanford, M. S. *Chem. Rev.* **2010**, *110*, 1147–1169.
32. Alberico, D.; Scott M.E.; Lautens, M. *Chem. Rev.* **2007**, *107*, 174–238.
33. Hendsbee, A. D.; Sun, J. P.; Rutledge, L. R.; Hillb, I. G.; Welch G. C. Electron Deficient Diketopyrrolopyrrole Dyes for Organic Electronics: Synthesis by Direct Arylation, Optoelectronic Characterization, and Charge Carrier Mobility. *J. Mater. Chem. A.* **2014**, *2*, 4198.
34. Burke D. J.; Lipomi D. J. Green Chemistry for Organic Solar Cells. *Energy Environ. Sci.* **2013**, *6*, 2053–2066.
35. Bohra H.; Wang M. Direct C–H Arylation: a "Greener" Approach Towards Facile Synthesis of Organic Semiconducting Molecules and Polymers. *J. Mater. Chem. A.* **2017**, *5*, 11550–11571.
36. Berggren M.; Dahlfors, A. R. Organic Bioelectronics. *Adv. Mater.* **2007**, *19*, 3201–3213.
37. Jin, J.; Lee, D.; Gyun Im, H.; Han, Y. C.; Jeong, E. G.; Rolandi, M. R.; Choi, K. C.; Bae B. S. Chitin Nanofi ber Transparent Paper for Flexible Green Electronics. *Adv. Mater.* **2016**, *28*, 5169–5175.
38. Zhou, Y.; Khan, T.m.; Liu, J. C.; Hernandez, C. F.; Shim, J. W.; Najafabadi E.; Youngblood J. P.; Moon, R. J.; Kippelen, B. Efficient Recyclable Organic Solar Cells on Cellulose Nanocrystal Substrates with a Conducting Polymer Top Electrode Deposited by Film-Transfer Lamination. *Org. Electron.* **2014**, *15*, 661–666.
39. Seol, Y. R.; Kim, J. W.; Hoon, S.; Kim, J.; Chung, J. H.; Lim, K. T. Cellulose-based Nanocrystals: Sources and Applications via Agricultural Byproducts. *J. Biosyst. Eng.* **2018**, *43*, 59-71.
40. Liao, X.; Zhang, Z.; Liao, Q.; Liang, Q.; Ou, Y.; Xu, M.; Li, M.; Zhanga, G.; Yue Zhang, Y. Flexible and Printable Paper-based Strain Sensors for Wearable and Large-Area Green Electronics. *Nanoscale.* **2016**, *8*, 13025–13032.
41. Tobjork, D.; Osterbacka, R. Paper electronics. *Adv. Mater.* **2011**, *23*, 1935–1961.
42. Hwang, S. W.; Tao, H.; Kim, D.H.; Cheng, H. Y.; Song, J. K.; Rill, E.; Brenckle, M. A.; Panilaitis, B.; Won, S. M.; Kim, Y. S.; Song, Y. M.; Yu, K. J.; Ameen, A.; Li, R.; Su, Y. W.; Yang, M. M.; Kaplan, D. L.; Zakin, M. R.; Slepian, M. J.; Huang, Y. G.; Omenetto, F. G.; Rogers, J. A. *Science.* **2012**, *337*, 1640–1644.
43. Tress, T.; Leo, K.; Riede, M. Influence of Hole-Transport Layers and Donor Materials on Open-Circuit Voltage and Shape of I–V Curves of Organic Solar Cells. *Adv. Funct. Mater.* **2011**, *21*, 2140–2149.
44. Mühl, S.; Beyer, B. Bio-Organic Electronics—Overview and Prospects for the Future. *Electronics.* **2014**, *3*, 444-461.
45. Yang, C.; Lin, W.; Li, Z.; Zhang, R.; Wen, H.; Gao, B.; Chen, G.; Gao, P.; Yuen, M.M.F.; Wong, C.P. Water-Based Isotropically Conductive Adhesives: Towards Green and Low-Cost Flexible Electronics. *Adv. Funct. Mater.* **2011**, *21*, 4582–4588.
46. Bettinger, C.J.; Bao, Z. Biomaterials-Based Organic Electronic Devices. *Polym. Int.* **2010**, *59*, 563–567.
47. Głowacki, E.D.; Leonat, L.; Voss, G.; Bodea, M.; Bozkurt, Z.; Irimia-Vladu, M.; Bauer, S.; Sariciftci, N.S. Natural and Nature-Inspired Semiconductors for Organic Electronics. *Proc. SPIE.* **2011**, *8118*, 81180M:1–81180M:10.

48. Hermawan, H. *Biodegradable Metals: From Concept to Application.* Springer: Heidelberg, Germany. 2012, pp. 13–22.

49. Vladu, I.; Sariciftcib, M.; Bauer, N. S. Exotic Materials for Bio-Organic Electronics. *J. Mater. Chem.* **2011**, *21*, 1350–1361.

50. Zhu, H.; Xiao, Z.; Liu, D.; Li, Y.; Weadock, N. J.; Fang, Z.; Huang, J.; Hu, L.; Biodegradable Transparent Substrates for Flexible Organic-Light-Emitting Diodes. *Energy Environ. Sci.* **2013**, *6*, 2105–2111.

51. Vladu, M. I.; Głowacki, E. D.; Schwabegger, G.; Leonat, L.; Akpinar, H. Z.; Sitter, H.; Bauer, S.; Sariciftcia N.S. Natural Resin Shellac as a Substrate and a Dielectric Layer for Organic Field-Effect Transistors. *Green Chem.* **2013**, *15*, 1473–1476.

CHAPTER 6

Natural Oil-Based Polymer: A Sustainable Approach Toward Green Chemistry

TARUNA SINGH[1*] and ATHAR ADIL HASHMI[2]

[1]*Department of Chemistry, Gargi College, University of Delhi, New Delhi 110049, India*

[2]*Department of Chemistry, Jamia Millia Islamia, New Delhi 110025, India*

Corresponding author. E-mail: tarunachemistry@gmail.com

ABSTRACT

Natural oils are admirable sources of polymer as they are renewable, economical, and have various degrees of unsaturation. Direct polymerization of oils has been used in coatings for a long time. Oils can be thermally polymerized on heating at high temperature. The products are oligomers of high viscosity. For high-strength materials introduction of functional groups such as hydroxyls, carboxyls, amines, etc. are necessary. Hydroxyls are the most useful ones, as they open the whole area of polyurethanes. The triglyceride structure or fatty acid derivatives opens a wide range of properties and modification of oils as polymers. Various polymerization methods like cationic, radical, condensation, and metathesis have been applied for synthesis of polymers.

6.1 INTRODUCTION

Polymer is a large molecule composed of repeating structural units known as monomers whose subunits are connected by covalent bonds. It involves a huge number of natural and synthetic compounds with an inclusive range of properties. Owing to the remarkable range of properties polymeric materials play an essential and ubiquitous role in daily life. The vital role of polymer ranges from synthetic plastics to natural biopolymers such as proteins and nucleic acids which are essential for life.

Natural polymers are the substances which obtain from natural sources, that is, horns of animals, tortoise shell, shellac, amber, and natural rubber. Cellulose a natural polymer is the structural component of plants and is the chief component of wood and paper. Various chemicals have been prepared from these biomaterials. By the pyrolysis of wood and agricultural wastes, bio-oils and syngas are produced. Bio-oil can be upgraded for applications such as transportation fuels, while methanol can be obtained from syngas.

Polymers are formed by the process of addition polymerization or by condensation polymerization. Most of the natural polymers are obtained from condensation polymerization. Bakelite, rubber, neoprene, polyethylene, nylon, PVC, polystyrene, silicone, etc. are different types of synthetic polymers. Carbon-containing plastic is obtained by continuously linked polymer backbone like the repeating unit of polyethylene is ethylene monomer. While in polymers like polyethylene glycol (PEG), polysaccharides, and DNA both carbon and oxygen are present as polymer backbone.

In polymerization, monomers combine together to form a covalently bonded chain or network with the removal of small chemical compounds. Monomers of polyethylene terephthalate polyester are terephthalic acid ($HOOC-C_6H_4-COOH$) and ethylene glycol ($HO-CH_2-CH_2-OH$) and its repeating unit is $-OC-C_6H_4-COO-CH_2-CH_2-O-$. It corresponds to the combination of these two monomers with the removal of two water molecules. Step-growth and chain-growth polymerization are two laboratory synthesis methods of polymers. Chains of the monomers combine with one another directly in step-growth polymerization while monomers are added to the chain one at a time only in chain-growth polymerization.[1–3]

Biopolymers like polysaccharides, polypeptides, and polynucleotides are essential for living organisms. Synthesis of protein involves enzyme-mediated process to transcribe genetic information from DNA to RNA and subsequently translate that information to synthesize the protein from amino acids. Various commercial polymers are prepared by chemical amendment of naturally occurring polymers. Nitrocellulose is produced by the reaction of nitric acid and cellulose. Heating natural rubber in presence of sulphur results in the synthesis of vulcanized rubber. Polymers can be modified through various ways which include oxidation, cross-linking, and end-capping.

Polymers are divided into numerous classes based on the various physical and chemical properties. The monomeric unit of a polymer plays a major role in determining its properties. Microstructure of polymer describes the arrangement of monomers. Chain interaction through various forces at the nanoscale describe the chemical properties of the polymer. Interaction of polymer with chemicals and solvents describes its properties on the macroscale.

During 21st century sustainable development is regarded as the reformed and stable economic development for maintaining social parity and environmental eminence. The concept of sustainable development came in 1987 and is defined as "the improvement that meets the needs of the present without negotiating the ability of future generations. It was defined as the social and economic advancement to assure human beings a healthy and productive life, but one that does not compromise the ability of future generations to meet their own needs. Base chemicals are produced in large quantities, their resource-saving production is specifically significant as they directly affect the chemical products, although availability of energy was admitted to be key for future development. As fossil oil reserves are depleting with time, the utilization of renewable raw materials is an essential step toward sustainable development. Use of renewable sources provides a constant source of material and also minimizes the emission of greenhouse gases. Products obtained from renewable raw materials have lower toxicity.[4-6] Owing to its comprehensive use polymers results in generation of a large amount of solid waste. Vegetable oils as renewable sources are opening a promising route due to their ready availability, biodegradability, and low toxicity. Vegetable oil is mixtures of triglycerides and can be extracted from seed or any other part of fruit. Rapeseed oil, soybean oil, and cocoa butter are extracted from seeds while palm oil, olive oil, and rice bran oil are extracted from other parts of fruits of a plant. Animal, vegetable, algae, and fish oil are excellent raw materials for polymers. Most oils are triglycerides, that is, esters of glycerin and fatty acids, but some oils like cashew nut shell oil are not. In order to make oil combative for polymer synthesis introduction of functional groups are necessary. Industrial uses consumed 15% of all soybean oil from 2001 to 2005. Vegetable oils have been used in paints and coatings as these unsaturated oils can oligomerize or polymerize on exposure to air. Various oil-based polymeric systems have been developed since the last decade, biofuels can also be used as an alternative to engine fuels. By thermal or cationic polymerization methods unmodified vegetable oils have been synthesized. Modified vegetable oils exhibit higher reactivities and undergo free radical polymerization to form polymers with respectable mechanical and thermal properties.[7-9] Vegetable oil-based polymers are used in various fields as they are biodegradable, flexible, biocompatible, non-flammable, thermally stable, adhesive to metallic substances, resistive to chemicals, conductive, etc. Structure of monomers directly affects the properties of polymer during their synthesis. Degradable polyesters are used as disappearing surgical sutures. By heat treatment or air blowing, triglycerides yield bodied and blown oils which can be used for coating purposes. Introduction of triglycerides improve end-product properties of synthesized polymer.[10]

6.2 STRUCTURE OF VEGETABLE OILS

Vegetable oils are triglycerides that consist of esters of glycerol and various fatty acids such as palmitic, stearic, linoleic and linolenic acids, etc. (Fig. 6.1). Fatty acids are long-chain compounds with an even number of carbon atoms. Double bonds are also present in unsaturated fatty acids along with carbon. The nature and distribution of fatty acids determine the physical state of vegetable oils. The maximum weight of triglyceride molecule of oil is attributed to fatty acids.

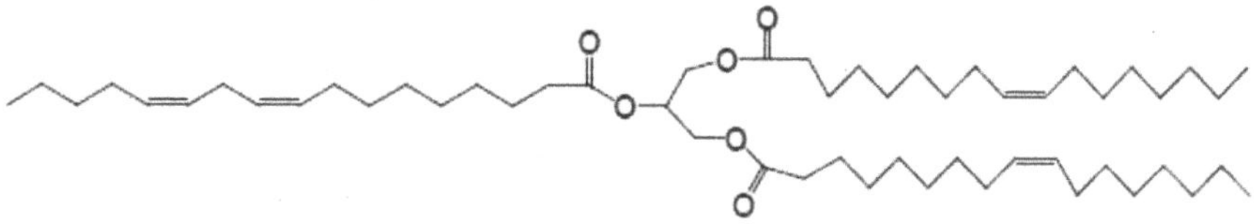

FIGURE 6.1   Structure of triglyceride molecule.

The commonly used oils for synthesis of polymers are linseed, sunflower, castor, soybean, palm, oiticia, tall and rapseed oil. Each oil has specific fatty acid distribution which depends on climate, growth conditions and purification methods. Linseed oil consists mainly of linoleic and linolenic acids while castor oil contains ricinoleic acid as one of the major fatty acids. Coconut oil contains lauric acid while soybean and safflower oil contain linoleic as one of the major fatty acids. Olive and naharseed oil consist mainly of oleic acid. Distribution of fatty acids plays a ubiquitous role in determining properties of the polymer. Physical and chemical properties of the oil are determined by the distribution of fatty acids in the oil (Tables 6.1 and 6.2).

Degree of unsaturation of an oil which is measured by calculating the iodine value is the most important factor affecting properties of the oil. Iodine value/ iodine number/ iodine adsorption/iodine index value is the mass of iodine in grams which is consumed by 100 g of an oil or fat. It is used to determine degree of unsaturation in fatty acids. Higher the iodine number, the greater will be the degree of unsaturation in the fat or oil. Depending on their iodine values triglycerides are divided into three categories, that is drying, semi-drying, and nondrying oils. For drying oil, iodine value is higher than 130, for semi-drying oils iodine value range from 90 to 130 while for non-drying oil iodine value is less than 90. Drying power of oil is directly related to the unsaturated nature of oil which on reaction with atmospheric oxygen forms a network. Choice of triglyceride plays an important role in defining the properties of polymer. Coconut oil is very saturated and is used for making soaps while linseed oil is highly unsaturated and is commonly used in the preparation of paints, coatings, inks, and resins.[11]

TABLE 6.1   Common Fatty Acids Present in Natural Oils, Their Structure and Formulae.

Name	Formula	Structure
Caprylic acid	$C_8H_{16}O_2$	$CH_3(CH_2)_6COOH$
Capric acid	$C_{10}H_{20}O_2$	$CH_3(CH_2)_8COOH$
Lauric acid	$C_{12}H_{24}O_2$	$CH_3(CH_2)_{10}COOH$
Myristic acid	$C_{14}H_{28}O_2$	$CH_3(CH_2)_{12}COOH$
Palmitic acid	$C_{16}H_{32}O_2$	$CH_3(CH_2)_{14}COOH$
Palmitoleic acid	$C_{16}H_{30}O_2$	$CH_3(CH_2)_5CH=CH(CH_2)_7COOH$
Stearic acid	$C_{18}H_{36}O_2$	$CH_3(CH_2)_{16}COOH$
Oleic acid	$C_{18}H_{34}O_2$	$CH_3(CH_2)_7CH=CH(CH_2)_7COOH$
Linoleic acid	$C_{18}H_{32}O_2$	$CH_3(CH_2)_4CH=CH-CH_2-CH=CH(CH_2)_7COOH$
Linolenic acid	$C_{18}H_{30}O_2$	$CH_3CH_2CH=CHCH_2CH=CHCH_2CH=CH(CH_2)_7COOH$
α Eleostearic acid	$C_{18}H_{30}O_2$	$CH_3(CH_2)_3CH=CHCH=CHCH=CH(CH_2)_7COOH$
Ricinoleic acid	$C_{18}H_{34}O_3$	$CH_3(CH_2)_5CH(OH)CH_2CH=CH(CH_2)_7COOH$
Arachidic acid	$C_{20}H_{40}O_2$	$CH_3(CH_2)_{18}COOH$
Erucic acid	$C_{22}H_{42}O_2$	$CH_3(CH_2)_7CH=CH(CH_2)_{11}COOH$

Triglycerides contain double bonds which polymerized through free radical or cationic mechanism. Thermal polymerization of vegetable oils at high temperature under nitrogen atmosphere gives low molecular weight polymers. Chemical modification of triglycerides results in the formation of monomers that can be polymerized to obtain linear, hyperbranched, or cross-linked polymers.

The commonly used method to characterize triglycerides is Fourier Transform Infrared (FTIR) spectroscopy that can be used for structural analysis

TABLE 6.2 Composition of Various Fatty Acids in Vegetable Oils.

Oil	Caprylic	Capric	Lauric	Myristic	Palmitic	Stearic	Arachidic	Oleic	Erucic	Linoleic	Linolenic	Ricinoleic
Palm	-	–	-	1.2	41.8	3.4	-	41.9	-	11.0	-	-
Soybean	-	-	-	-	14.0	4.0	-	23.3	-	52.2	5.6	-
Coconut	6.2	6.2	51.0	18.9	8.6	1.9	-	5.8	-	1.3	-	-
Sunflower	-	-	-	-	6.5	2.0	-	45.4	-	46.0	0.1	-
Rapeseed	-	-	-	-	4.0	2.0	-	56.0	-	26.0	10.0	-
Castor	-	-	-	-	1.5	0.5	-	5.0	-	4.0	0.5	87.5
Linseed	-	-	-	-	5.0	4.0	-	22.0	-	17.0	52.0	-
Naharseed	-	-	-	-	15.9	9.5	-	52.3	-	22.3	-	-
Corn	-	-	-	-	10.0	4.0	-	34.0	-	48.0	-	-
Olive	-	-	-	-	6.0	4.0	-	83.0	-	7.0	-	-
Sesame	-	-	-	0.1	8.2	3.6	-	42.1	-	43.4	-	-
Safflower	-	-	-	0.1	6.8	2.3	0.3	12.0	-	77.7	0.4	-
Refined Tall	-	-	-	-	4.0	3.0	-	46.0	-	35.0	12.0	-
Oticia	-	-	-	-	6.0	4.0	-	8.0	-	8.0	74.0	-

of oil (Table 6.3). Nuclear Magnetic Resonance (NMR) spectroscopy is used for calculating the fatty acid content of triglycerides (Table 6.4). Gas Chromatography is also used for the determination of fatty acid composition of oil. Triglycerides contain various reactive positions that can act as a starting point in different reactions, that is, ester groups, C=C double bonds, the α-position of ester groups, and allylic positions. Due to this reason, it is expected that triglycerides play a significant role in synthesizing polymers from renewable sources (Table 6.4).

TABLE 6.3 Absorption Bands of Various Functional Groups in Vegetable Oil Using FTIR.

Absorption band (cm^{-1})	Functional groups
3468	-OH stretching vibration
2856–2924	C-H stretching vibration
1744	C=O stretching vibration
1655	C=C stretching vibration
1456	C-H bending vibration
1162	C-O-C stretching vibration of esters
719	Methylene rocking vibration

TABLE 6.4 ^{1}H-NMR Shift Values of Vegetable Oil.

^{1}H-NMR shift (ppm)	Proton groups
0.87–0.89	Protons of terminal methyl groups
1.60	Protons of internal –CH$_2$- groups
2.01–2.05	Allylic protons of CH$_2$
2.30–2.32	Alpha protons of ester groups
2.75–2.78	-CH$_2$ of double allylic protons
4.15–4.28	Protons of glyceride moieties
5.32–5.35	Protons of the –CH=CH-

6.3 APPLICATIONS OF VEGETABLE OIL

Vegetable oil is an essential component of various manufactured products. Vegetable oils are an admirable source of renewable materials for different applications. Due to their essential properties, triglyceride oil-based polymers have been used in various fields (Fig. 6.2)

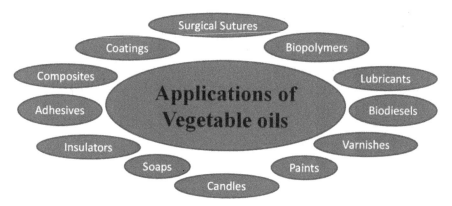

FIGURE 6.2   Applications of vegetable oils.

6.4   CONCLUSION

As the world is arriving closer to the decline of our fossil fuel resources, a great concern is increasing about continuity of our way of life. Plant seed oil and vegetable oil provide an important renewable source of such materials. Modified triacylglycerides can be used to produce polymers with significant properties. These oil-based polymers have found to be an essential source for the variety of monomers and polymers. It is an awfully challenging field of research with infinite future prospects. Researchers can focus their consideration on these renewable sources for the manufacturing of various monomers and polymers to keep the environment cleaner and greener in the coming centuries.

KEYWORDS

- **renewable sources**
- **natural oil**
- **triglycerides**
- **polymers**
- **polymerization**

REFERENCES

1. Islam, M. R.; Beg, M. D. H.; Jamari, S. S. *J. Appl. Polym. Sci.* **2014,** *40787,* 1–13.
2. Adekunle, K. F. *Open J. Polym. Chem.* **2015,** *5,* 41–46.
3. Chandra, R.; Rustgi, R. *Prog. Polym. Sci.* **1998,** *23,* 1273–1335.
4. Okada, M.; Okada, Y.; Tao, A.; Aoi, K. *J. Appl. Polym. Sci.* **1996,** *62,* 2257–2265.
5. Alam, M.; Akram, D.; Sharmin, E.; Zafar, F.; Ahmad, S. *Arab. J. Chem.* **2014,** *7,* 469–479.
6. Singh, T.; Shreaz, S.; Khan, L. A.; Hashmi, A. A.*Mater. Res. Innov.* **2014,** *16,* 204–212.
7. Singh, T.; Ambreen, S.; Shreaz, S.; Hashmi, A. A. *J. Polym. Environ.* **2013,** *21,* 81–87.
8. Bharthi, N. P.; Khan, N. U.; Alam, M.; Shreaz, S.; Hashmi, A. A. *J. Inorg. Orgnomet. Polym. Mater.* **2010,** *20,* 839–846.
9. Singh, T.; Shreaz, S.; Hashmi, A. A. *Int. J. Polym. Mater. Polym. Biomater.* **2013,** *62,* 653–662.
10. Singh, T.; Khan, N. U.; Shreaz, S.; Hashmi, A. A. *Polym. Eng. Sci.* **2013,** *53,* 2650–2658.
11. Asadaukas, S.; Erhan, S. Z. *J. Am. Oil Chem. Soc.* **2001,** *71,* 1223–1226.
12. Kumar, A; Sharma, A.; Upadhyaya, K. C., *Curr. Genom.* **2016,** *17,* 230–240.

CHAPTER 7

Renewable Resource-Based Environmental Friendly Waterborne Polymeric Anticorrosive Nanocomposite Coatings

MOHD IRFAN[1,2*], HALIMA KHATOON[2], RABIA KOUSER[2], ABU DARDA[3], SHAHIDUL ISLAM BHAT[2], and SAJID IQBAL[2]

[1]*Ajanta Caves and Conservation Research Laboratory, Aurangabad, India*

[2]*Materials Research Laboratory, Department of Chemistry, Jamia Millia Islamia, New Delhi, India*

[3]*Analytical Research Laboratory, Department of Applied Sciences and Humanities, Faculty of Engineering and Technology, Jamia Millia Islamia, New Delhi, India*

Corresponding author. E-mail: irfanc73@gmail.com

ABSTRACT

Vegetable oil (VO)-based waterborne (WB) polymeric nanocomposite coatings emerged as important alternatives to the petro-based polymers due to increasing pressure to limit detrimental health and environmental effects. In this approach, the toxic and expensive volatile organic solvents are replaced by water as an environmentally benign solvent, resulting in minimal volatile organic contents. WB polymeric (WBP) nanocomposites have attracted great attention in the field of anticorrosive coating materials due to the synergistic effect of polymeric matrix and nanofillers. This chapter deals with the classification of WBPS, various synthetic approaches, advantages, and shortcomings. In addition, the chapter also elaborated on the various VO-based WBPS in the field of anticorrosive nanocomposite coatings. The future perspectives in the development of WB polymers are also discussed.

7.1   INTRODUCTION

Nowadays the green chemistry revolution has provided an alternative route for chemists and industrialists to develop polymers through eco-friendly route, to diminish the production of hazardous volatile organic compounds (VOC), which are dangerous for health and environment.[1] Various environmental and Clean Air Act regulation has forced the scientists to synthesize WB polymers (WBPS) using renewable natural resources and green technology.[2] The term waterborne (WB) is used for those polymeric materials that contain water as the main solvent for dispersion of polymer resin or in a blend of water and green solvent. The polymers, which are synthesized in water without any organic solvents are called water-soluble polymers while those prepared in the blends of water and green solvents are called water-reducible polymers. In case of solvent-borne polymeric materials, hazardous VOC are released into the atmosphere which deplete the atmosphere and result in air pollution.[3] In comparison with solvent-borne coatings, WB coatings possess several advantages such as easy cleaning, low VOC, and are less toxic.[4] Rising alarms regarding the environment and the effect of VOC has resulted in the replacement of many solvent-borne systems with WB systems.[5] WBPS find versatile applications in coatings, adhesives, paints, fax machines, primer, defoamer, biomaterial typewriters, and wood coating.[6] WBPS solutions are applied on the surface of the substrate like metal, plastic, wood, glass, and others by spray, dip, or brush techniques to develop their coatings.[7] These coatings are either cured at ambient temperature or require elevated temperature for curing. The WB contains approximately 70–80% water as a solvent along with cosolvents such as ethanol, methanol, and glycol ether.[8] The cosolvents act as coalescing agents and as diffusion promoters in WBPS.[9,10] They primarily soften the polymer particles, reduce the viscosity, enhance the film-forming ability, prevent skinning, and help in the change of rheology to improve the quality of drying of WBPS coating materials.[11] Due to the diverse effects on human health and the environment, there is an urgent need to develop environmentally friendly coating systems.[12] There are following two possible methods for the formulation of environmentally friendly coating:

1. By the use of green solvents that are nonhazardous to health and the environment.[3]
2. By the development of WB coatings, high solid coatings, and solvent less coatings focusing on the partial or complete elimination of the solvents.[13,14]

On account of this, this chapter first classifies the WBPS then various synthetic approaches are discussed in detail. Their advantages, shortcomings, and the application of WBPS nanocomposites are covered. In addition, future perspectives in the development of WBPS are also embedded.

7.2   CLASSIFICATION OF WB COATINGS

Almost all types of polymeric resin are available in the formulation of WBPS. In general, WBP generally classified into four types as discussed below:

7.2.1   WATER-REDUCIBLE POLYMERS

Generally, copolymers are used in the formulation of water-reducible polymers because the polar groups of these polymers are easily accessible to water. The copolymers are formed by the polymerization reactions that occur in water-miscible or in green organic solvents such as alcohols (green solvents).[15] The coatings formulated by these polymers have a high gloss, clarity, good pigment-wetting, and dispersion properties. Water-reducible polymers include polyesteramide, alkyds, polyurethane, polyepoxies, and others.

7.2.2   WATER-SOLUBLE POLYMERS

Water-soluble polymers are dispersed or dissolved in water and can be made WB by the introduction of an anionic and cationic group.[16] The anionic and cationic group can be introduced by the presence of polyether or hydroxyl groups on the backbone of polymers.

7.2.3   WATER-DISPERSIBLE POLYMERS

Water-dispersible polymers are dispersed in water along with a small amount of green organic solvents, which acts as a coalescing agent. These polymers are called as colloidal-dispersion polymers and they swell in water.[17] The common examples of such polymers include styrene-butadiene copolymers, acrylate-methacrylate copolymers, vinyl acetate copolymers, and vinyl propionate copolymers.

7.2.4 WATER-EMULSION POLYMERS

In this case, polymeric resin is dispersed in water and it can be stabilized by an emulsifier. The high molecular weight emulsion polymers comprise of discrete spherical shape particles that are present in dispersed form and get swelled in water. This results in the formation of their colloidal solution which increases the molecular weight of such type of polymers.[18,19] The coatings formulated by water emulsion enhances the properties like good toughness, chemical, and water resistance. The resins commonly employed in such cases include polyvinyl acetate, copolymers, styrene-butadiene, acrylics, polystyrene, and alkyds.

7.3 VARIOUS SYNTHETIC TECHNIQUES USED FOR WBP COATING

WBPS are synthesized by the incorporation of various moieties into the backbone of the polymer matrix. These moieties can be made WB by means of appropriate modifications as discussed below:

The polymers can be made WB either by formation of salt or by the introduction of amino or carboxylic group in the backbone of the polymeric chain. These moieties result in the conversion of cations or anions by acid–base reactions (Fig. 7.1). The polymers which were synthesized by using carboxylic group have acid value in between 40 and 160.[20] These polymers are prepared through a single step polycondensation reaction.

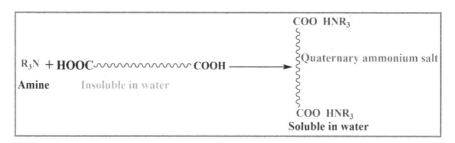

FIGURE 7.1   Formation of WBPS by the salt formation.

Figure 7.2 shows the formation of WBPS by the introduction of polyols or polyethers (nonionic groups) in the backbone of polymers. The polymers synthesized by this approach have low molecular weight and can remain water-sensitive.

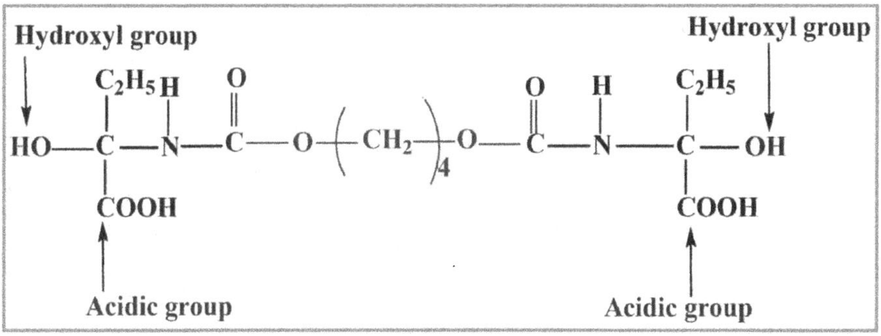

FIGURE 7.2 Formation of WBPS by the introduction of polar group.

Polymers can also be made WB by the way of intermediate zwitterions.[10] Zwitterion sometimes referred to as inner salts, which have two or more functional groups. Zwitterion is a neutral molecule with a negative and a positive electrical charge (Fig. 7.3). The common examples of such polymers include polyesteramide, alkyd, and polyepoxies.

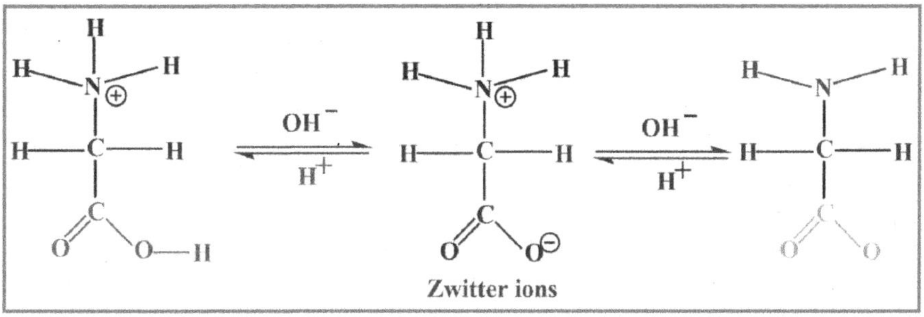

FIGURE 7.3 Formation of WBPS by the introduction of zwitterions intermediate.

7.4 MECHANISM FOR THE FORMULATION OF WBPS COATINGS

In the formulation of WBPS, different components are used for the formulation of WB polymer like binder and solvents (water or green organic solvents) and various other additives such as surfactant, stabilizers, corrosion inhibitors, and coalescing agent are used for the improvement of corrosion, rheology, wetting, and UV resistance properties.[21]

Cosolvents are added into the WBPS for the improvement in coalescence, leveling, wetting, and to the expedite release of water. They also improve drying time of coatings.[6,10]

Surfactants are mainly used for the decrease in the surface tension of the anticorrosive coating, which induce wetting of the substrate, latex stabilization, and to sustain the dispersion of pigments.[22]

Rheological modifiers are used to build up the viscosity especially in latex paints. Most commonly used rheology modifiers include a cellulosic polymer such as methylcellulose, carboxymethyl cellulose, hydroxyethyl cellulose, hydroxypropyl cellulose, polyacrylates, etc.[11,23]

7.5 ADVANTAGE OF WBPS COATING

WB coatings are considered as a feasible alternative to the solvent-borne coatings. They meet the requirement of various acts and legislations related to environments. WBPS possess low or zero VOCs (considered as environment friendly),[24] exhibit good physico-mechanical and corrosion resistance performance.[25] WBPS have attracted much attention in the industry due to several advantages such as low viscosity at high molecular weight, easy formulation, low cost, eco-friendly, and nontoxic nature.[4] Other advantages are their low flammability and no hazardous emissions. WBPS generally do not require additives, thinners, and hardeners as compared to their solvent-based counterparts.

7.6 SHORTCOMING OF WBPS COATING

Coatings formulated from WBPS usually take a longer time for dry-to-touch and dry-to-hard. They do not dry at ambient temperature rather they require high temperatures for curing.[5] Another disadvantage of these coatings is the appearance of foam in the coating material, which reduces gloss and the performance properties of these coatings.[10] One of the major disadvantages of WB coating is that the relative humidity should be less than 80%. The above-mentioned shortcomings were improved by the dispersion of various nanofillers into the vegetable oil (VO)-based polymer matrix.

7.7 VEGETABLE OILS

VOs are triglycerides of glycerol and fatty acids.[26] Most VOs exist in liquid form at room temperatures. Triglycerides have three fatty acids linked with the center known as glycerol center.[27] VOs are insoluble in water but are soluble in organic solvents like ethanol, acetone, and hexane. Fatty acids may be saturated or unsaturated. Unsaturated fatty acids are those which contain double bond while saturated fatty acids have no double bonds (lack of unsaturation).[28] A general structure of triglycerides is shown in Figure 7.4.

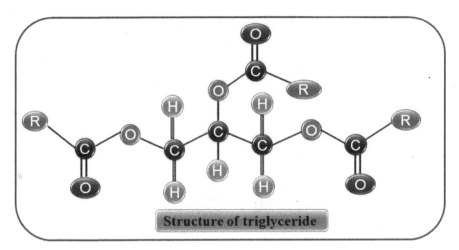

Structure of triglyceride

FIGURE 7.4 Structure of triglyceride.

Polymers obtained from living resources like plant, algae, and trees are known as bio-based polymers. WBPS were synthesized from various renewable resources as shown in Figure 7.5. Among these, VO is considered as valuable bio-renewable resources for the preparation of polymeric resin due to their easy availability, biodegradability, and versatile applications.[29] Presently, VO are considered to be extensive feedstock's used worldwide in paint and coating technology because of the following reasons[1]:

- cost effectiveness,
- ease of availability,
- functional attributes,
- inherent fluidity, and
- biodegradability

Fatty acids present in VOs comprises of long straight-chain compounds with an even number of carbons and zero to three double bonds per fatty acid, and double bond in these fatty acids adopt cis-configuration while some of them are in trans form. Ricinoleic and vernolic acids present in some fatty acid chains possess functional groups such as hydroxyl and epoxy groups (Fig. 7.6).[28] These functional groups provide an opportunity to carry out numerous modifications such as the introduction of new functionalities and crosslinking sites for curing agents. Different types of VOs are discussed in the proceeding section.

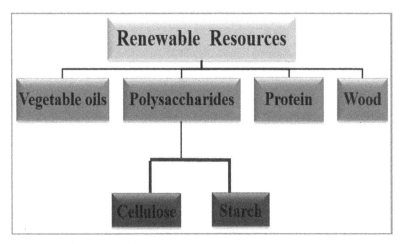

FIGURE 7.5   Various renewable resources.

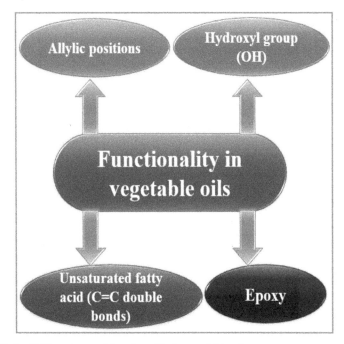

FIGURE 7.6   Different type of functionality in vegetable oils.

7.7.1 WATERBORNE POLYURETHANES (WPU)

Polyurethane (PU) was first developed by Otto Bayer in 1937. PU is synthesized by the polyaddition reaction of polyol (alcohol-containing several hydroxyl groups) and an isocyanate (R–N=C=O).[30] PU is one of the most important polymeric resins owing to their extensive applications, properties, and versatility.

$$ROH + R'NCO \longrightarrow R'NH\text{-}CO\text{-}OR$$

Due to environmental protection acts and regulations, solvent-borne PU was replaced by WB, which is synthesized by means of anionic and cationic dispersions due to their environmentally friendly nature and have zero volatile organic contents and hazardous air pollutants.[31] This WPU exhibited potential application in the field of coatings, inks, adhesives, foams, sealants, and elastomers.[10] WPU possesses low viscosity and high molecular weight.

However, WPU is associated with several drawbacks such as relatively low heat-resistance, low adhesion in the moist environment and mechanical properties, which displays the broad application of WPU. To overcome these drawbacks these polymers are modified by the incorporation of various metals, metal oxide-based nanofillers.

7.7.2 WB EPOXY

Epoxy resins are also referred to as polyepoxies and have more than one oxirane ring per molecule.[14] Epoxy resins are three-membered ring in which oxygen atom is bonded with a carbon atom.[14] It is widely used as a protective coating against corrosive ions and is considered as a potential candidate in other applications like laminates, paints adhesives, and composites.[32]

Recently, WB epoxy has gained considerable importance because of consumer awareness and legislations related to clean atmosphere.[9] Epoxy resins can be used for the preparation of WB coatings simply by the modification of epoxy resins through the introduction of polar groups in their backbone.

7.7.3 WB ALKYDS

WB alkyds (Fig. 7.7) have received a lot of importance in the past few years due to their synthesis via eco-friendly routes. Alkyd resins are defined as polyesters of moderate molecular weight, which are obtained by polymerization of (1) polyalcohols, (2) anhydrides or dicarboxylic acids, and (3) unsaturated long fatty acids chain of triglycerides.[33] In general, monoglycerides of fatty acids serve as the precursor molecules for the synthesis of various alkyd resins. In the previous case, alcoholysis of VO with glycerol resulted in the formation of monoglyceride. Then, the hydroxyls of the monoglyceride are transesterified by a polyacid. Alkyd resins obtained by this way have high viscosity, good drying, and hardening properties.

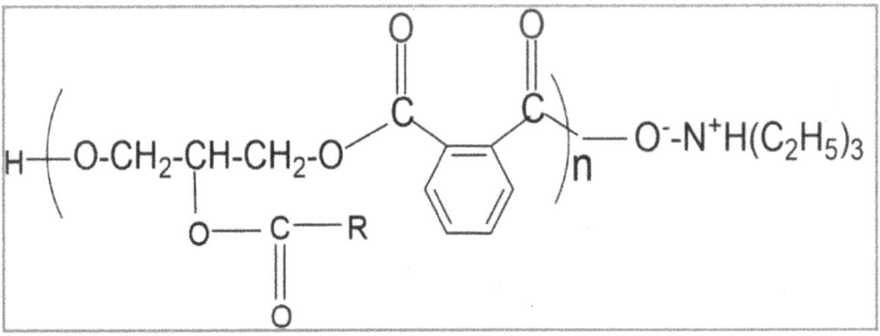

FIGURE 7.7   WB alkyd.

7.7.4   WB POLYESTERAMIDE

VO-based WB polyesteramide (Fig. 7.8) are amide modified alkyds, which is synthesized by the esterification reaction between VO, amide diol, and an anhydride or acid that enhanced the properties over normal alkyd. Polyesteramide comprised of repeating ester and amide linkage on to their backbone.[34] The preparation of polyesteramide occurs in two steps:

1.   Preparation of N,N-bis(2-hydroxyethyl) oil fatty amide from different VO with diethanolamine in the presence of sodium methoxide as a catalyst to generate amide diols.[35]
2.   By polycondensation reaction between hydroxyl groups with a dibasic acid or anhydride which results in the formation of polyesteramide.[36]
3.   Then water as a green solvent is added into it.[34]

7.7.5   WB ACRYLATE

The term "acrylic" is derived from those products that contain acrylic acid or methacrylic acid. Acrylic resins are generally polymer or copolymers of vinyl monomer such as acrylic acid, styrene, methacrylic acid, acrylonitrile, hydroxyl ethylmethacrylate,[21,37] etc. WB acrylic polymers resins are used in the preparation of WB acrylic coatings.[38] WB acrylic coatings are also known as acrylic latex coatings. Acrylic resins are used in the application of coatings and adhesives and show excellent durability and weather resistance property.[39]

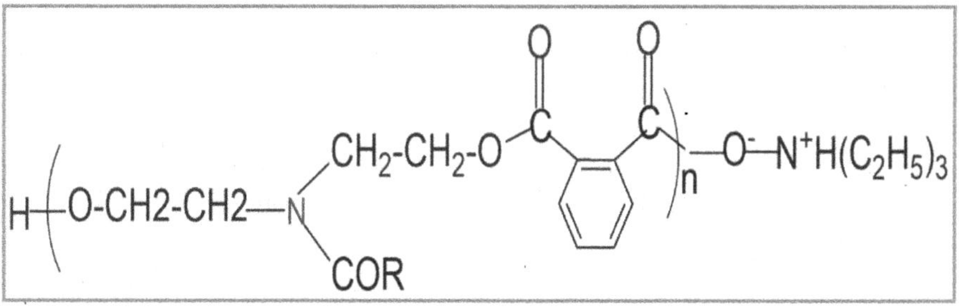

FIGURE 7.8   WB polyesteramide.

7.8   WB COATINGS FOR CORROSION PROTECTION

During the last few years, there has been a significant advancement in the development of WBPS for their anticorrosive coating applications.[32] Coating may be defined as the thin or thick layer of polymeric materials covering the surface of a substrate, which strongly adheres to the metal substrate. Generally, coatings are applied on the metal substrate in order to enhance the properties like wettability, adhesion, corrosion resistance, wear resistance, barrier properties, and dielectric properties.[40] Corrosion is the undesirable deterioration or degradation of a metal surface due to aggressive species in the environment.[41] A general approach for corrosion protection is the use of organic, inorganic, and metallic coatings to protect the underlying metal substrate. For corrosion resistance, coatings are applied to the steel substrate by using different techniques such as dip coating, brush technique, air spraying, electrostatic spraying, and airless spraying with improved thermal stability, physico-mechanical, and abrasion resistance properties[42,43] Outstanding corrosion resistance is one of the most attractive properties of high-performance WB anticorrosive coatings.[44] WBP coatings have potential as compared to solvent-borne coatings because of low or zero volatile organic content. WBPS find versatile applications in the coating industry such as powder coatings, high solid coatings, UV curable coating, and wood coatings.[31]

From the past few years, WBPS such as alkyd, epoxy, and polyurethane have gained considerable importance in high-performance anticorrosive nanocomposite coatings with excellent thermal stability, mechanical, and barrier properties.[43,45] For the formulation of nanocomposite coatings, various nanoparticles like ceramic nanoparticles, carbon nanotube, graphene oxide, reduced graphene oxide, and organic–inorganic nanoparticle were dispersed in various WBPS matrices, which bring drastic changes in their properties.[46] Different components used for the formulation of WBPS nanocomposites are shown in Figure 7.9. Irfan and coworker reported the physico-mechanical and electrochemical corrosion resistance performance of reduced graphene oxide dispersed WB polyesteramide.[34] The results revealed that there is enhanced improvement in thermal stability, bending ability, impact resistance, scratch hardness, and high corrosion protection performance of the coating material. Xiao et al.[47] synthesized WB epoxy acrylate silica hybrid materials and studied their UV curing behavior. Their results showed that upon addition of nano-silica sol within epoxy acrylate matrix improved the thermal and fire-retardant properties.

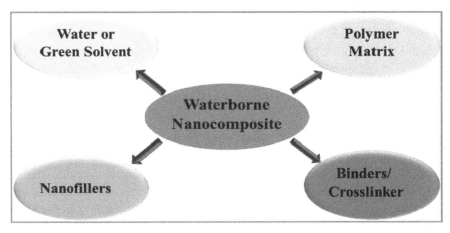

FIGURE 7.9   Different components involve in the formulations of coating.

7.9   FUTURE PERSPECTIVES

WBPS are gaining significant attention in industry and academia because of their outstanding characteristics. WB coatings formulated from VO are generally prepared in water along with green organic solvents. In this chapter,

VOs-based WBPS are considered as potential candidates for environment-friendly materials and have less deleterious and hazardous impact on the environment. The VO-based resin is modified with various nanoparticle to overcome drawbacks associated with WBPS and directly helps to renew and diversify the composite industry worldwide in the future. WBPS find potential applications in different fields such as paints, coatings, adhesives, sensors, antibacterial, and biomedical. They have a bright future in different areas like antifouling, coatings for glasses, catalysis, sensors, biomedical applications, and light-responsive materials.

7.10 CONCLUSION

This chapter illustrates WBPS and their role in the restrictions on the use of hazards volatile organic solvents. The utilization of organic solvents in the formulation coating has been strictly limited due to both human health and environmental concerns. In light of the shortcoming, WBPS adore a good position in the present era. But the real challenge for researchers and industrialists is to design sustainable resources based WB polymeric products. This chapter describes the various synthetic techniques used for the preparation of WBPS, their classification, and mechanism of formulation of WBPS. This chapter also described the development of the different types of VO-based WBPS and their anticorrosive coatings. VOs are the main precursor for the preparation of different types of polymers like polyurethane, polyesteramide, alkyd, epoxy, and polyacrylate. This is an alternate route that has the potential to replace petroleum-based polymers. The WBPS show a promising approach and it is considered as a potential alternative for the replacement of solvent-borne polymers.

KEYWORDS

- **waterborne polymers**
- **vegetable oil**
- **green chemistry**
- **anticorrosive coatings**
- **nanocomposite**

REFERENCES

1. Pathan, S.; Ahmad, S. s-Triazine Ring-Modified Waterborne Alkyd: Synthesis, Characterization, Antibacterial, and Electrochemical Corrosion Studies. *ACS Sustain. Chem. Eng.* **2013,** *1,* 1246–1257. DOI: 10.1021/sc4001077.

2. Anastas, P.; Eghbali, N. Green Chemistry: Principles and Practice. *Chem. Soc. Rev.* **2010,** *39,* 301–312. DOI: 10.1039/B918763B.

3. Sharmin, E.; Zafar, F.; Akram, D.; Alam, M.; Ahmad, S. Recent Advances in Vegetable Oils Based Environment Friendly Coatings: A Review. *Ind. Crops Prod.* **2015,** 215–229. DOI: 10.1016/j.indcrop.2015.06.022.

4. Pathan, S.; Ahmad, S. Green and Sustainable Anticorrosive Coating Derived from Waterborne Linseed Alkyd Using Organic-inorganic Hybrid Cross Linker. *Prog. Org. Coat.* **2018,** *122,* 189–198. DOI: 10.1016/j.porgcoat.2018.05.026.

5. Singh, A.P.; Suryanarayana, C.; Baloji, R.; Naik, G. Gunasekaran, Development of Hyperbranched Polyester Polyol-based Waterborne Anticorrosive Coating. *J. Coat. Technol. Res.* **2016,** *13,* 41–51. DOI: 10.1007/s11998-015-9720-1.

6. Verma, G.; Dhoke, S. K.; Khanna, A. S. Polyester Based-siloxane Modified Waterborne Anticorrosive Hydrophobic Coating on Copper. *Surf. Coat. Technol.* **2012,** *212,* 101–108. DOI: 10.1016/j.surfcoat.2012.09.028.

7. Nine, M. J.; Cole, M. A.; Johnson, L.; Tran, D. N. H.; Losic, D. Robust Superhydrophobic Graphene-based Composite Coatings with Self-cleaning and Corrosion Barrier Properties. *ACS Appl. Mater. Interf.* **2015,** *7,* 28482–28493. DOI: 10.1021/acsami.5b09611.

8. Honarkar, H.; Barmar, M.; Barikani, M. Synthesis, Characterization and Properties of Waterborne Polyurethanes Based on Two Different Ionic Centers. *Fiber Polym.* **2015,** *16,* 718–725. DOI: 10.1007/s12221-015-0718-1.

9. Yao, M.; Tang, E.; Guo, C.; Liu, S.; Tian, H.; Gao, H. Synthesis of Waterborne Epoxy/Polyacrylate Composites via Miniemulsion Polymerization and Corrosion Resistance of Coatings. *Prog. Org. Coat.* **2017,** *113,* 143–150. DOI: 10.1016/j.porgcoat.2017.09.008.

10. Panda, S. S.; Panda, B. P.; Nayak, S. K.; Mohanty, S. A Review on Waterborne Thermosetting Polyurethane Coatings Based on Castor Oil: Synthesis, Characterization, and Application. *Polym.—Plast. Technol. Eng.* **2018,** *57,* 500–522. DOI: 10.1080/03602559.2016.1275681.

11. Madbouly, S. A.; Xia, Y.; Kessler, M. R.; Rheological Behavior of Environmentally Friendly Castor Oil-based Waterborne Polyurethane Dispersions. *Macromolecules* **2013,** *46,* 4606–4616. DOI: 10.1021/ma400200y.

12. Sharmin, E.; Zafar, F.; Akram, D.; Alam, M.; Ahmad, S. Recent Advances in Vegetable Oils Based Environment Friendly Coatings: A Review. Ind. Crop Prod. **2015.** DOI: 10.1016/j.indcrop.2015.06.022.

13. Naik, R. B.; Ratna, D.; Singh, S. K. Synthesis and Characterization of Novel Hyperbranched Alkyd and Isocyanate Trimer Based High Solid Polyurethane Coatings. *Prog. Org. Coat.* **2014.** DOI: 10.1016/j.porgcoat.2013.10.012.

14. Shah, M. Y.; Ahmad, S. Waterborne Vegetable Oil Epoxy Coatings: Preparation and Characterization. *Prog. Org. Coat.* **2012,** *75,* 248–252. DOI: 10.1016/j.porgcoat.2012.05.001.

15. Saravari, O.; Phapant, P.; Pimpan, V. Synthesis of Water-reducible Acrylic-alkyd Resins Based on Modified Palm Oil. *J. App. Polym. Sci.* **2005,** *96,* 1170–1175. DOI: 10.1002/app.21009.

16. Ma, S.; Qian, J.; Zhuang, Q.; Li, X.; Kou, W.; Peng, S. Synthesis and Application of Water-soluble Hyperbranched Polyester Modified by Trimellitic Anhydride. *J. Macromol. Sci., Part A.* **2018,** *55,* 414–421. DOI: 10.1080/10601325.2018.1453261.

17. Roy, T. K.; Raval, D. A.; Mannari, V. M. Water-dispersible Polyesteramide Resins for Surface Coating Applications. *Int. J. Polym. Mater. Polym. Biomater.* **1998,** *42,* 39–52.

18. Lu, Y.; Larock, R. C. New Hybrid Latexes from a Soybean Oil-based Waterborne Polyurethane and Acrylics via Emulsion Polymerization. *Biomacromolecules* **2007,** *8,* 3108–3114. DOI: 10.1021/bm700522z.

19. Bunker, S.; Staller, C.; Willenbacher, N.; Wool, R. Miniemulsion Polymerization of Acrylated Methyl Oleate for Pressure Sensitive Adhesives. *Int. J. Adhesion Adhesives* **2003,** *23,* 29–38. DOI: 10.1016/S0143-7496(02)00079-9.

20. Pathan, S.; Ahmad, S. Synthesis, Characterization and the Effect of the s-triazine Ring on Physico-mechanical and Electrochemical Corrosion Resistance Performance of Waterborne Castor Oil Alkyd. *J. Mater. Chem. A* **2013,** *1,* 14227–14238. DOI: 10.1039/c3ta13126b.

21. Zhu, K.; li, X.; Wang, H.; Li, J.; Fei, G. Electrochemical and Anti-corrosion Behaviors of Water Dispersible Graphene/Acrylic Modified Alkyd Resin Latex Composites Coated Carbon Steel. *J. Appl. Polym. Sci.* **2017,** *134,* 1–12. DOI: 10.1002/app.44445.

22. Naik, R. B.; Jagtap, S. B.; Ratna, D. Effect of Carbon Nanofillers on Anticorrosive and Physico-mechanical Properties of Hyperbranched Urethane Alkyd Coatings. *Prog. Org. Coat.* **2015.** DOI: 10.1016/j.porgcoat.2015.05.001.

23. Asif, A.; Shi, W.; Shen, X.; Nie, K. Physical and Thermal Properties of UV Curable Waterborne Polyurethane Dispersions Incorporating Hyperbranched Aliphatic Polyester of Varying Generation Number. *Polymer* **2005,** *46,* 11066–11078. DOI: 10.1016/j.polymer.2005.09.046.

24. Liu, B.; Wang, Y. A Novel Design for Water-based Modified Epoxy Coating with Anti-corrosive Application Properties. *Prog. Org. Coat.* **2014,** *77,* 219–224. DOI: 10.1016/j.porgcoat.2013.09.007.

25. Ding, J.; ur Rahman, O.; Peng, W.; Dou, H.; Yu, H. A Novel Hydroxyl Epoxy Phosphate Monomer Enhancing the Anticorrosive Performance of Waterborne Graphene/Epoxy Coatings. *Appl. Surf. Sci.* **2018,** *427,* 981–991. DOI: 10.1016/j.apsusc.2017.08.224.

26. Lligadas, G.; Ronda, J. C.; Galià, M.; Cádiz, V. Renewable Polymeric Materials from Vegetable Oils: A Perspective. *Mater. Today* **2013,** 337–343. DOI: 10.1016/j.mattod.2013.08.016.

27. Xia, Y.; Larock, R. C. Vegetable Oil-based Polymeric Materials: Synthesis, Properties, and Applications. *Green Chem.* **2010,** *12,* 1893–1909. DOI: 10.1039/c0gc00264j.

28. Sharmin, E.; Zafar, F.; Akram, D.; Alam, M.; Ahmad, S. Recent Advances in Vegetable Oils Based Environment Friendly Coatings: A Review. *Ind. Crop Prod.* **2015,** *76,* 215–229. DOI: 10.1016/j.indcrop.2015.06.022.

29. Alam, M.; Akram, D.; Sharmin, E.; Zafar, F.; Ahmad, S. Vegetable Oil Based Eco-friendly Coating Materials: A Review Article. *Arab. J. Chem.* **2014,** *7,* 469–479. DOI: 10.1016/j.arabjc.2013.12.023.

30. Wu, Y.; Du, Z.; Wang, H.; Cheng, X. Preparation of Waterborne Polyurethane Nanocomposite Reinforced with Halloysite Nanotubes for Coating Applications. *J. Appl. Polym. Sci.* **2016,** *133,* 43949. DOI: 10.1002/app.43949.

31. Li, Y.; Yang, Z.; Qiu, H.; Dai, Y.; Zheng, Q.; Li, J.; Yang, J. Self-aligned Graphene as Anticorrosive Barrier in Waterborne Polyurethane Composite Coatings. *J. Mater. Chem. A* **2014,** *2,* 14139–14145. DOI: 10.1039/c4ta02262a.

32. Rahman, O. U.; Kashif, M.; Ahmad, S. Nanoferrite Dispersed Waterborne Epoxy-acrylate: Anticorrosive Nanocomposite Coatings. *Prog. Org. Coat.* **2015,** *80,* 77–86.

33. Irfan, M.; Bhat, S. I.; Ahmad, S. Reduced Graphene Oxide Reinforced Waterborne Soy Alkyd Nanocomposites: Formulation, Characterization, and Corrosion Inhibition Analysis. *ACS Sustain. Chem. Eng.* **2018,** *6,* 14820–14830. DOI: 10.1021/acssuschemeng.8b03349.

34. Irfan, M.; BHAT, S. I.; Ahmad, S. Waterborne Reduced Graphene Oxide Dispersed Bio-polyesteramide Nanocomposites: An Approach Towards Eco-friendly Anticorrosive Coatings. *New J. Chem.* **2019,** *43,* 4706–4720. DOI: 10.1039/c8nj03383h.

35. Zafar, F.; Sharmin, E.; Ashraf, S. M.; Ahmad, S. Ambient-cured Polyesteramide-based Anticorrosive Coatings from Linseed Oil—A Sustainable Resource. *J. Appl. Polym. Sci.* **2005,** *97,* 1818–1824. DOI: 10.1002/app.21953.

36. Zafar, F.; Ashraf, S. M.; Ahmad, S. Cd and Zn-incorporated Polyesteramide Coating Materials from Seed Oil—A Renewable Resource. *Prog. Org. Coat.* **2007,** *59,* 68–75. DOI: 10.1016/J.PORGCOAT.2007.01.009.

37. Lv, X. L.; Caihong, L. H.; Yang, Y.; Li, H.; Huang, C. Waterborne UV-curable Polyurethane Acrylate/Silica Nanocomposites for Thermochromic Coatings. *RSC Adv.* **2015,** *5,* 25730–25737.

38. Wang, S.; Li, W.; Han, D.; Liu, H.; Zhu, L. Preparation and Application of a Waterborne Acrylic Copolymer-siloxane Composite: Improvement on the Corrosion Resistance of Zinc-coated NdFeB Magnets. *RSC Adv.* **2015,** *5,* 81759–81767. DOI: 10.1039/c5ra10851a.

39. Mishra, V.; Mohanty, I.; Patel, M.R.; Patel, K. I. Development of Green Waterborne UV-Curable Castor Oil-Based Urethane Acrylate Coatings: Preparation and Property Analysis. *Int. J. Polym. Analy. Charact.* **2015,** *20,* 504–513. DOI: 10.1080/1023666X.2015.1050852.

40. Yuan, R.; Wu, S.; Yu, P.; Wang, B.; Mu, L.; Zhang, X.; Zhu, Y.; Wang, B.; Wang, H.; Zhu, J. Superamphiphobic and Electroactive Nanocomposite Toward Self-Cleaning, Antiwear, and Anticorrosion Coatings. *ACS Appl. Mater. Interf.* **2016,** *8,* 12481–12493. DOI: 10.1021/acsami.6b03961.

41. Kordzangeneh, S.; Naghibi, S.; Esmaeili, H. Coating of Steel by Alkyd Resin Reinforced with Al2O3Nanoparticles to Improve Corrosion Resistance. *J. Mater. Eng. Perform.* **2018,** *27,* 219–227. DOI: 10.1007/s11665-017-3080-1.

42. Dhoke, S. K.; Sinha, T. J. M.; Dutta, P.; Khanna, A. S. Formulation and Performance Study of Low Molecular Weight, Alkyd-based Waterborne Anticorrosive Coating on Mild Steel. *Prog. Org. Coat.* **2008,** *62,* 183–192. DOI: 10.1016/j.porgcoat.2007.10.008.

43. Selim, M. S.; Shenashen, M. A.; Elmarakbi, A.; EL-Saeed, A. M; Selim, M. M.; El-Safty, S. A. Sunflower Oil-based Hyperbranched Alkyd/Spherical ZnO Nanocomposite Modeling for Mechanical and Anticorrosive Applications. *RSC Adv.* **2017,** *7,* 21796–21808. DOI: 10.1039/C7RA01343D.

44. Njoku, D. I.; Cui, M.; Xiao, H.; Shang, B.; Li, Y. Understanding the Anticorrosive Protective Mechanisms of Modified Epoxy Coatings with Improved Barrier, Active and Self-healing Functionalities: EIS and Spectroscopic Techniques. *Sci. Rep.* **2017,** *1,* 1–15. DOI: 10.1038/s41598-017-15845-0.

45. Harb, S. V.; Pulcinelli, S. H.; Santilli, C. V.; Knowles, K. M.; Hammer, P. A Comparative Study on Graphene Oxide and Carbon Nanotube Reinforcement of PMMA-Siloxane-Silica Anticorrosive Coatings. *ACS Appl. Mater. Interf.* **2016,** *8,* 16339–16350. DOI: 10.1021/acsami.6b04780.

46. Luo, X.; Yuan, S.; Pan, X.; Zhang, C.; Du, S.; Liu, Y. Synthesis and Enhanced Corrosion Protection Performance of Reduced Graphene Oxide Nanosheet/ZnAl Layered Double Hydroxide Composite Films by Hydrothermal Continuous Flow Method. *ACS Appl. Mater. Interf.* **2017**, *9*, 18263–18275. DOI: 10.1021/acsami.7b02580.

47. Xiao, X.; Hao, C. Preparation of Waterborne Epoxy Acrylate/Silica Sol Hybrid Materials and Study of Their UV Curing Behavior. *Colloid Surf. A* **2010**, *359*, 82–87. DOI: 10.1016/j.colsurfa.2010.01.067.

CHAPTER 8

An Approach for Development of Materials for Green Chemical Catalytic Processes: Green Catalysis

RIMZHIM GUPTA[1*], AKANKSHA ADAVAL[2], and SUSHANT KUMAR[1]

[1]*Department of Chemical Engineering, Indian Institute of Science, Bangalore, India*

[2]*Department of Metallurgical Engineering and Materials Science, Indian Institute of Technology, Mumbai, India*

ABSTRACT

Green catalysis is an important branch of green engineering as it is inherently derived from the green chemistry and possesses the capability to bring a substantial change in this field by moving a step forward toward foreseeable sustainability. Demand of energy inputs and high operating conditions for fuel generation by direct or indirect usage of fossil fuels in any chemical process can be suppressed by reducing the energy barrier to facilitate the reaction. Catalysts are known to have the ability to suppress the activation energy of the chemical processes, accelerate the reaction rate, increase the yield and selectivity, and limit the excessive usage of stoichiometric reagents by providing surface sites without being involved in the reaction. Energy inputs are as much responsible as the direct usage of chemicals in adding toxins in the environment posing serious health problems. Therefore, nonhazardous ways to deliver energy inputs and reduction of waste from the overall process is the first and foremost requirement of moving toward the green approach. In this regard, solar irradiation, potential difference across electrodes as the energy source/driving force for the reactions operating at room temperature is a kick start to demonstrate the preventive as well as the after treatment for pollution prevention measures. Photons absorbed via solar energy promote the generation of charge carriers and are widely used for fuel generation by replacing the thermocatalytic processes using high temperature and high pressure as the reaction parameters. Production of electricity using solar energy, i.e., solar cells and photovoltaics, solar driven air and water purification are few of the examples of moving forward toward greener approach. Green engineering and green chemistry provide an enormous scope to direct the research toward designing the highly efficient materials, catalysts, and processes with high environmental benefits.

8.1 ORIGIN OF GREEN CATALYSIS

Observing the continuous increase in toxicity and harmful human health effects due to the energy-intensive nature of chemical processes and their harmful byproducts, the concept of green synthesis and green chemistry was adopted by research and scientific community in no time. Since then, continuous efforts have been made to turn the processes, synthesis, and chemistry toward the greener approach to make the world a better place. Green approach promotes the pollutant-free environment along with cleaner and greener ways to produce energy. The word green catalysis originates from words "green" and "catalysis" that can be associated with environmentally friendly ways that assist a chemical process via catalysis.[1] Catalysis itself is an energy-intensive route to perform chemical reactions. Although, catalysts majorly nanoparticles or nanostructured materials do not participate in the reaction, yet it may impart its vital role in the generation of harmful byproducts and release its toxicity in the final product in case of liquid-phase reactions. Therefore, the design of catalysts is an integral part of the catalysis.

Green catalysis concept was first introduced by Prof. Robert Crabtree who gave his contribution in the field of organometallic homogeneous catalysis and used iridium-based complex for homogeneous hydrogenation.[2] Figure 8.1 shows the structure of the iridium complex as a homogeneous catalyst, this is also known as "Crabtree catalyst."

US Environmental Protection Agency defines the green chemistry and its processes in the following ways:

- Use of low energy-intensive processes (processes that do not use fossil fuels as energy inputs, rather solar energy, wind energy, and potential gradient is used as the driving force of the chemical reaction and biomass as the feedstock for deriving the fuels).
- Use of alternative reaction systems and conditions (elimination of harmful or toxic reagents/solvents and promoting the usage of natural

reducing agents, biological, and enzymatic nonharmful compounds), or processes generating low emission and chemicals improving the selectivity for less harmful products.
- Development of eco-compatible chemicals (replacing the more harmful chemicals by inherently less harmful or do not cause accidents).

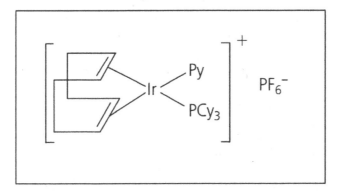

FIGURE 8.1 Iridium complex, (1,5-cyclooctadiene) pyridine (tricyclohexylphosphine) iridium (I) hexafluorophosphate.

Design of catalysts includes minimal or elimination of the solvents/ reagents while synthesis, eliminating the use of heavy metals or carcinogenic compounds with the aim of complete recovery, and reusability of the catalyst material. Design of the catalyst should be such that it produces a minimal concentration of harmful byproducts. Its elemental, surface, and physico-chemical nature decides the fate of the reactant conversion into the product. Therefore, it is of utmost importance for researchers to find ways to improve and understand how the design of the catalysts can change the selectivity of the chemical process. The greenness of the process can be estimated by "atom economy."[3] It is a widely used concept in green chemistry; however, consumption of low energy and low waste generation is also a benign part of green catalysis. Catalysis is one of the most important pillars of green chemistry. There can be a significant difference in the energy consumption between the chemical processes with and without catalysts due to extreme operating conditions in the prior case.

Two measures defining the greenness of the process are E factor and atom economy. E factor represents the amount of waste generated during the process and can be calculated by the mass ratio of the waste divided by the mass ratio of the desired product. While atom economy is calculated by the molecular weight of the desired product divided by the total sum of the molecular weights of all the products generated during the reaction in the stoichiometric equation.[4] Water is not usually included in the calculation of the E factor; however, the organic and inorganic content present in the aqueous phase or in water are included.

Principles for green chemistry were given by Anastas and Warner, that includes minimum generation of waste, maximum conversion (consumption of reactants), minimum generation of hazardous products, design of less toxic chemicals, low energy requirements (operating conditions), use of catalysts, and minimizing the accidental hazards during the operations.[5]

In the following sections, you will see how the catalysis is appropriately following the guidelines of green chemistry and the advancements to make the process and catalysts greener.

8.2   TYPES AND ROLES OF CATALYSTS

8.2.1   METAL CATALYSTS

Transition metals and noble metal catalysts, metal hybrids (metal dispersed with metal oxides supports silica, alumina, etc.) are widely explored for various reactions such as hydrogenation reaction,[6] Haber–Bosch process,[7] methanol decomposition,[8] partial oxidation,[9] steam reforming,[10] production of nitric acid, and epoxyethane.[11] The reactions and the reaction details are tabulated in Table 8.1.

However, considering the cost of noble metal catalysts, operating conditions, and turning them into environmentally benign, continuous research is being performed in the field of catalysis to explore bimetallic,[8] multimetallic,[12] and alloy-based catalysts for such applications.[13]

Metal nanoparticles have their wide applications apart from catalysis such as sensors, optical devices, electronics, etc. depending upon their particle sizes, shape, and morphology.[14] The chemical bottom-up synthesis of these metal nanoparticles majorly includes excessive usage of reducing agent and stabilizing agent mostly a polymer and metal salt. This is considered to be a cost-effective method. However, the reducing agents used in such applications are highly toxic and can cause hazards and environmental harm. Reducing agents reduce the metal ions into metal atoms and further aggregation atoms result into metal nanoparticles of different sizes attributing various interesting properties. Reducing agents mostly used are hydrazine hydrate, $NaBH_4$ for transition metal nanoparticles, H_2O_2, alcohol, and citrate reduction, usually for noble metal (Au, Ag, Pt) nanoparticles.[15] Variation in the reducing agents may result in particle sizes of nanoparticles due to metal–ligand interaction and double layer formation. Apart from the reducing agent, the residual of metal salts can incorporate excessive toxicity in the system. Chloride salts after dissociating from the metal ions form chloride ions which impart significant toxicity. Therefore, replacing reducing agents and using nitrates as metal salts may improve the parameters required for the green synthesis.

Green synthesis offers various interesting properties of metal nanoparticles along with environmental compatibility. Various reducing agents obtained from the plant extract can act as stabilizing agents as well as the reducing agents. These organic molecules containing various polyol, polyphenol, alkaloids, proteins, sugar groups, and other functional groups are capable of reducing the metals and simultaneously stabilizing them. The green synthesis via plant extract involves three steps: activation/nucleation, growth, and termination. First phase involves the dissociation of metal ions in the liquid medium and reduction of those metal ions into zerovalent atoms giving rise to the formation of nucleus. Second phase includes the association of metal atoms due to the high surface energy of atoms. During this process, the thermodynamic stability is achieved and various morphologies can be achieved as the result of metal–ligand interaction. Termination occurs when the steady-state of the synthesis is achieved.

Bacterial, fungal, algae, and other microbial species have also been used to reduce the metal salt into metal nanoparticles. Various strains of different bacteria result in different sizes, shape, and morphology of the nanoparticles. Monodispersed particles with varying morphologies are also synthesized by using fungal species as they carry a variety of intracellular enzymes. The yield of nanoparticles obtained from fungal strains is higher than that of bacterial strains.[16]

These particles in different sizes show interesting properties in terms of fluorescence and emit different wavelength of light upon excitation due to their quantized energy states owing to excellent optical properties. These particles also have their applications in drug delivery etc.

Different stabilizing and reducing agents impart different surface properties to the nanoparticles. Surface properties have a vital role in catalytic applications such as electrocatalysis and photocatalysis. Nanoparticles such as platinum, palladium, ruthenium, and various alloy nanoparticles are widely being used for electrocatalytic applications and new understanding is coming into the picture in terms of effect of shape, sizes, and morphologies for electrochemical reactions such as water splitting,[17] CO_2 reduction,[18] N_2 fixation,[19] oxygen evolution, oxygen reduction reactions, etc.[20] Metal electrocatalysts reduce the electrochemical overpotential to initiate the reaction. High surface-to-volume ratio, unsaturated surface atoms, and good electron conductivity of these materials makes them an appropriate candidate for such applications. However, there are other parameters in these nanoparticles which can result in varying catalytic properties for different shaped particles. Surrounding medium/coordination of surface atoms, geometry, nature of exposed facets, etc. play a vital role in the catalysis. Recently, alloy nanoparticles have gained enormous attention as they impart interesting properties and prevent the deactivation of the catalysts. Apart from that, charge dynamics at the metal heterojunction/interface, the lattice compression due to the growth of different metal may influence the catalytic activity. Therefore, control over such parameters is rather important.[21]

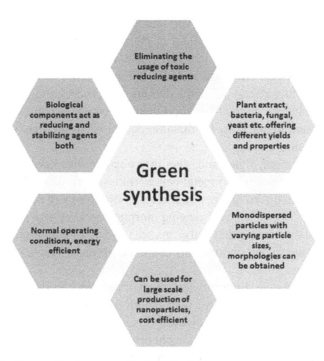

FIGURE 8.2   Key merits of green synthesis methods.

There are various reports where alloy metal nanoparticles are synthesized by green synthesis using plant extract for electrocatalytic applications.[22,23]

In Table 8.1, there are few examples of the reactions where metal nanoparticles are commercially used, or metal or alloy nanoparticles have an edge over other nanomaterials.

TABLE 8.1 Reactions Including Metal Nanoparticle as Commercial Catalysts.

Process	Reaction	Catalysts
Methanol decomposition	$CH_3OH \rightarrow CO + 2H_2$	Cu, Ni, Pd & Pt (comm.)
Steam reforming	$CH_4 + H_2O \rightarrow CO + 3H_2$	Ni (comm.)
Haber–Bosch process	$2NH_3 \rightarrow N_2 + 3H_2$	Promoted Fe (comm.)
Nitric acid production	$4NH_3 + 5O_2 \rightarrow 4NO + 6H_2O$	Pt and Ru (comm.)
Epoxyethane production	$C_6H_{14} \rightarrow$ +H_2	Silver on alumina (comm.)
Reforming of naphtha	$C_2H4 + 1/2O_2 \rightarrow$	Pt and Ru on alumina (comm.)
Photocatalysis	Antimicrobial, CO_2 reduction, organic decontamination,	Metals, metal/semiconductor hybrids
Electrocatalysis	N_2 fixation, CO_2 reduction	Metal/alloy-based catalysts

8.2.2 METAL OXIDE/CHALCOGENIDE CATALYSTS

Nanotechnology is all about controlling the fascinating properties of nanoparticles and nanohybrids by tailoring size, morphology, and composition of particles/structure which is not only restricted to metals. Metal oxides/chalcogenides nanostructures/nanocomposites with diverse morphology also exhibit properties like magnetic,[24] photoluminescence,[25] semiconducting properties,[26] etc. which makes them viable for application in the field of biomedical, electronics, optical sensing, memory storage, and microelectronics.[27] The properties of nanohybrids are completely dependent on the architecture of nanocomposites. Noble metals can be coupled to metal oxides and metal chalcogenides to yield diverse nanostructures including noble metal decorated-metal oxide nanoparticles, nanoarrays, noble metal/metal oxide core/shell, noble metal/metal oxide yolk/shell, and Janus noble metal–metal oxide nanostructures used in catalysis for various application as shown in Figure 8.3.

Metal chalcogenides have also gained interest because of their superior electronic properties in comparison to graphene, as graphene is zero-gap half metal[29] which makes them suitable for optoelectronic applications. Theoretical and experimental investigations have been done on metal chalcogenides like GaN,[30,31] GaSe,[32] InSe,[33] GaS,[34,35] and BN to find out their properties and suitable applications. Chalcogenides can also be applied as catalysts for Hydrogen generation and dye removal,[38] reduction of 4-nitrophenol,[39] photodegradation of organic dyes,[40] and photocatalysis.[41] Figure 8.4 shows the photolysis mechanism using ZnO decorated with MoS_2.[92]

Several methods have been reported for the synthesis of metal oxides and metal chalcogenides like thermal method (hydrothermal and solvothermal),[43–47] chemical method,[48,49] and microwave-assisted synthesis.[50–52] These methods sometimes involve the use of toxic chemicals which can prove to be hazardous for the environment and the person handling it.

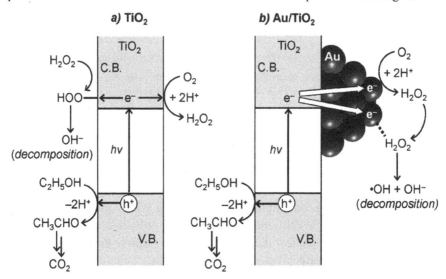

FIGURE 8.3   Scheme of synthesis and decomposition of H_2O_2 on (a) TiO_2 and (b) Au/TiO_2 photocatalysts.

Source: Reprinted with permission from Ref. [91]. © 2012 American Chemical Society.

Due to the large rate of toxic chemicals and extreme environment employed in the physical and chemical production of these nanoparticles NPs, green methods employing the use of plants, fungus, bacteria, and algae have been adopted.[53]

8.2.3   METAL ORGANIC FRAMEWORKS (MOFs)

MOFs are the inorganic–organic hybrids and a novel class of materials with various interesting properties leading to their use in various applications such as catalysis, sensing, adsorption, capture, drug delivery, magnetism, etc. They are highly porous materials with high specific surface area, tunable functionality, conductivity, and pore size and shape size selectivity.[54] Porosity of these materials can be controlled based on the organic linker, framework

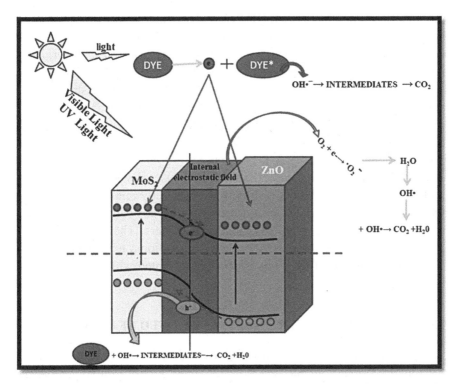

FIGURE 8.4 Schematic representation of metal oxide-metal chalcogenide nanocomposite ZnO/MoS2 for solar energy-driven photocatalytic degradation of organic dyes.

Source: Reprinted with permission from Ref. [92]. © 2019 Elsevier.

connectivity between inorganic unit and ligand. These materials can bridge the gap between the zeolites which have fixed structures and pore sizes and silicate materials. However, MOFs are reported to have lower thermal stability than the oxides.

There are various methods of synthesis for MOFs such as solvothermal method, microwave synthesis, mechanochemical method, and electrochemical and sonochemical method. However, most of these methods use excessive solvents that are toxic or harmful for the environment. In this regard, mechanochemical method uses a nontoxic approach without using organic solvents that are carcinogenic or toxic for the environment. Besides, solvent-based approaches have limitations of remaining solvents molecules in the framework that may lead to the collapse of the frameworks. It uses metals or metal oxides as the starting material that promotes solid–solid reaction. Emmerling and coworkers reported the mechanochemical synthesis of traditionally used MOF HKUST-1 for catalysis and adsorption application due to the presence of unsaturated Cu metal center. Conventional ball milling was used to synthesize the material in the presence of liquid ethanolic grinding of fine precursor materials.[55] Crystalline nature of MOFs allows us to explore various interesting properties for catalysis. Identification and distribution of active sites can be significantly improved in such materials.[56] Synthesis of MOF was reported using platinum nitrate and di and tri pyridine ligands in the presence of ethylenediamine as the precursors to develop metal-organic square and metal-organic bowl, respectively.[57]

Stability of MOFs is another concern because of which MOFs usage is limited. Stability of MOFs is directly or indirectly related to the metal–ligand strength. Chemical stability in these materials can be incorporated by using high valance metal ions as metal centers. Thermal stability of these materials can be increased by using oxy anionic terminated linkers. Mechanical strength of MOFs is inversely correlated with the porosity of the materials.[54]

8.3 GREEN ORGANIC SYNTHESIS IN THE FIELD OF CATALYSIS

Catalytic reagents have been reported to be more effective than stoichiometric reagents for green organic reactions as they improve atom efficiency and energy efficiency. For example, while converting $PhCH(OH)CH_3$ to $PhCOCH_3$ when using Jones Reagent ($CrO_3 + H_2SO_4$) the atom efficiency is only 42% but increases two-folds to about 87% while using a catalyst.[58] Over the period of time, researchers have utilized catalysts for various environmental applications like removal of NOx to improve the quality of air, minimizing the usage of volatile organic compounds, replacing the use of toxic chlorine-based processes, and reducing waste production. Researchers are focusing on establishing an understanding of the new process technologies and the existing commercialized processes. Moustafa and group have developed many green synthesis techniques in the last decade for the production of polyfunctionalized heteroaromatic substances. They have focused on implementing the green methodologies with shorter reaction times and enhanced yields with improved efficiency.[59]

8.3.1 IMMOBILIZED CATALYSTS ON SOLID SUPPORTS

As experimentalists are working on new techniques for organic synthesis, they are utilizing the new solutions and alternatives for solution-phase based synthesis. A major disadvantage of using a catalyst is its complete nonremoval from the product once the reaction is completed and thereby rendering it difficult to use it in industrial applications. The solution to that is immobilizing the metal catalyst on solid supports which makes it easier to remove the catalyst from the reaction via filtration and also allows its usage in successive runs. Palladium being the most widely used transition metal catalyst is used in many homogeneous cross-coupling reactions. Some of the polymeric and inorganic supports used are cross-linked polystyrene, mesoporous silica, alumina, aluminosilicates, zirconia, graphene, carbon nanotubes, and modified graphene. Suzuki-Miyaura cross-coupling reactions use palladium polymer on adamantane, adamantane resin, bipyridyl ligand modified porous organic polymer,[60,61] and bimetallic copolymers-based frameworks.[62,63] Several new developments in Sonogashira reactions,[63–65] Mizoroki-Heck reactions,[65,66] and Tsuji-Trost reactions[65–67] have also successfully been utilized in broad applications of synthesis of organic molecules by using polymer-bound catalysts. Apart from this, a novel class of phosphine-based organic materials is developed using sterically hindered phosphine on phospha-adamantane framework, and used it for Suzuki coupling reaction (Fig. 8.5).[93]

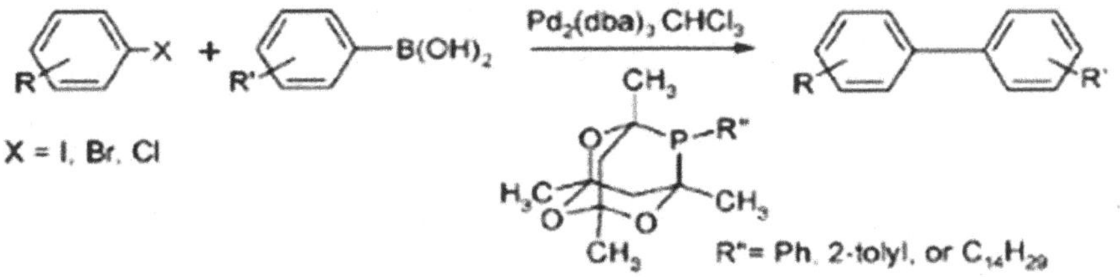

FIGURE 8.5 Use of a catalytic system with Pd2(dba)3 and 1,3,5,7-tetramethyl-2,4,8-trioxa-6-phenyl-6-phospha-adamantane to promote the Suzuki cross-coupling reaction.

Source: Reprinted with permission from Ref. [93]. © 2018 American Chemical Society.

These catalysts have shown to have high yields with easy removal through filtration and can also be reused without any loss in their catalytic activity. Rhodium and its complexes have been commonly used in several organic reactions like C-H bond activation, hydrogenation, hydroformylation, and carbonylation reactions. To follow in the steps of green chemistry, many solid supported catalysts have been developed. Petrukhina et al. have covalently anchored rhodium complex on functionalized resin for use in cyclopropanation reaction.[69] Similarly, several immobilized catalysts on solid supports have been unearthed through subsequent and relentless research for rhodium-catalyzed reactions. A recent study reported the use of rhodium and palladium complex immobilized on functionalized graphene oxide being used in hydrogenation reactions.[70,71] Ruthenium-catalyzed reactions are commonly being used for the synthesis of many organic molecules. As mentioned earlier, solid-supported ruthenium catalysts have found a breakthrough as they offer all the advantages of green chemistry, like ease of separation, reusability, and effortless handling. Grubs and Grubbs-Hoveyda ruthenium catalysts have been prepared on SBA silica and mesoporous silica supports for ring-closing metathesis reactions.[72] Magnetic beads, organic polymers, inorganic oxides, and carbonaceous materials have been used as supports for hydrogenous catalytic reactions. A report published by Ge et al. reported the use of Grubbs–Hoveyda ruthenium complex for the one-pot reaction of olefin metathesis and functionalized 2H-atrazines were synthesized by ring contraction of isoxazols using the G–H II catalyst, as shown in Figure 8.6.[94]

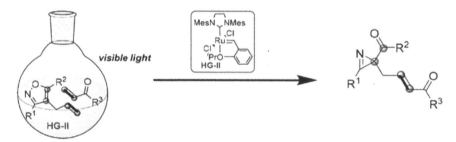

FIGURE 8.6   Usage of G–H II catalyst for ring contraction and olefin metathesis reaction.
Source: Reprinted with permission from Ref. [94]. © 2003 American Chemical Society.

Similarly, several other transition metals like cobalt,[74] iridium,[75] and copper[76] have also been immobilized on solid supports for their applications in cross-coupling reactions. The ease of separation of the solid-anchored catalyst from the reaction mixture renders it useful to be used in certain organic reactions under aqueous conditions.

8.3.2   SOLVENTS

Green solvents refer to a condition of an ideal solvent that facilitates mass transfer but does not dissolve. It has advantages of being natural, cheap, low reactivity, nontoxic, and easily available. It also follows the rule of green chemistry as it aids the reaction and provides the benefit of easy removal from the reaction and easy recyclability. Research community through its persistent efforts has established many alternatives to organic solvents which are discussed below in brief:

8.3.2.1   WATER

Water being cheap, inflammable, synthetically efficient, ease of operation, recyclable, and environment friendly is used for several organic reactions. Wittig olefination and Diels–Alder reactions have been investigated for use in aqueous mediums.[77]

8.3.2.2   IONIC LIQUIDS

Ionic liquids have gained considerable attention because of advantageous properties like low vapor pressure, excellent chemical and thermal stability, inflammable, high ionic conductivity, and their usage as a catalyst.[78] Friedel–Crafts acylation reaction actively use acidic chloroaluminate ionic liquids to generate acylium ions.

8.3.2.3   POLYETHYLENE GLYCOL (PEG)

PEG is a linear polymer which is obtained by polymerizing ethylene oxide. It has been used for phase-transfer catalysis reactions as it has the benefits of being cheap, nontoxic, biocompatible, low vapor pressure, inflammable, thermally stable, and an appropriate green alternative.

8.3.2.4   PERFLOURINATED SOLVENTS

Perfluorous liquids have properties like high density, chemical stability, low solubility in other solvents, low dielectric constant, thermal stability, nontoxicity, and environment friendly which renders them useful as catalysts in many medical applications. Perfluoroalkanes, perfluoroalkyl ethers, and perfluoroalkylamines have been intensively used with many organic solvents for various coupling reactions.

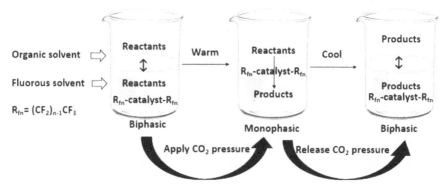

FIGURE 8.7   Organic synthesis of fluorous-based solvents.

8.3.2.5   *SUPERCRITICAL LIQUIDS*

Any substance above its critical temperature and pressure is defined as a supercritical liquid for example carbon dioxide (CO_2). CO_2 is renewable, inflammable, readily evaporating with a better ability to dissolve organic compounds, and better flowability (low viscosity), and thus has been extensively used for homogenous free radical and cationic polymerizations.

8.3.3   *NANOREACTORS*

Nanoreactors like micelles and vesicles refer to the macromolecular assemblies of noncovalently attached building units which physically confine the catalysts and render them beneficial in catalysis. Micelles have been a topic of extensive research because of their utility as a nanoreactor in green organic reactions. Depending upon the packing parameter, micelles of various morphologies like spherical, cylindrical, or worm-like can be obtained. Micelles have shown to be an effective platform in catalytic systems for various reactions like C-N coupling reactions (palladium loaded dl-α-tocopherolmethoxypolyethylene glycol succinate [TPGS-750-M]), animation of allylic ethers (polyoxyethanyl α-tocopherylsebacate [PTS]-based micelles), cross-coupling reactions (tetramethylethylenediamine [TMEDA] stabilized Pd-catalytic micelles), aldol reactions, etc.[79]

Polymersomes, also known as polymer vesicles, are self-assembly of amphiphilic block copolymers and are hollow in nature. Being synthetic in nature, their properties such as size, permeability, and stability are tunable.

Since they have multicompartments, it is easier for encapsulating or tethering a catalyst and thus in turn improving the overall catalytic activity from a green catalysis point of view.[80] Polymersomes have found their applications widely in biocatalysis in pickering emulsions, enzymatic and nonenzymatic catalytic reactions, coupling reactions, and ring-opening polymerization reactions.

8.3.4 BIOCATALYSTS

Over the years, taking inspiration from nature, enzymes, whole cells, and catalytic antibodies are being utilized for organic synthesis. Enzymes being biocompatible are the vital endogens that act as a catalyst for all the necessary in vivo metabolic reactions in the body. Enzymes such as oxidoreductases, transferases, hydrolases, lyases, isomerases, and ligases have been utilized in a multitude of applications in food, textile, and biomedical industries. Itoh and Hanefeld[81] have recently published an article emphasizing the advantages of enzyme catalysis for a sustainable and green future.

8.3.5 NANOCATALYSTS

Nanocatalysts, also termed as semiheterogenous catalysts, have high aspect ratio and surface area which results in enhanced interactions between the catalyst surface and reactants. Still, the catalyst contamination poses a problem and magnetic nanocatalysts have been found to an easy solution to it as they can be easily extracted from the reaction using an external magnetic field. This makes them a promising candidate for application in chemical industries. Various magnetic nanocatalysts have been synthesized and used in synthetic reactions like Suzuki, Sonogashira, and Heck. Silicon-zcoated nickel oxide, as a nanocatalyst, has been subsequently used to obtain tricyclic benzodiazepine with high yield.[82] Similarly, a magnetically active nanocatalyst was reported as a green alternative for hydrogenation of π bonds and the nitro groups, whereafter the reaction gets completed the catalyst can be easily removed magnetically.

8.4 DEVELOPMENT OF GREEN CATALYSIS

The escalation in the global energy demand with such high rates is responsible for the world's energy crisis. Dependence on fossil-based fuels from several decades is the reason for depletion of natural resources, excessive CO_2 emission, and increase in the global temperature.[83] Therefore, a switch toward renewable energy sources as the motive force for the generation of energy with the greener approach is a beneficial effort to address the problem of energy crisis and issues associated with it. Therefore, considering the fact that thermocatalytic pathways for CO_2 conversion and utilization, NOx conversion to ammonia or nitrogen, and N_2 to NH_3 (Haber) conversion consumes enormous energy and emits excessive CO_2 into the environment. Therefore, researchers are continuously trying to find the ways to reduce the CO_2 emission from the flue gas, consumption of CO_2 into value-added products, reduction of CO_2 from direct air, and simultaneously eliminating the usage of fuels obtained from nonrenewable sources by renewable energy. In this regard, photocatalysis and electrocatalysis can be considered as a greener approach to obtain value-added products. These processes use solar energy and potential difference as the driving force to perform chemical reactions. However, limited yield of products due to charge carrier recombination, and complexity to achieve high faradaic efficiency and low overpotential, limit their commercial viability. Therefore, entire research community stand together to overcome the issues related to these processes and trying to design more efficient catalysts appropriate for such applications.

8.4.1 STEPS TOWARD GREEN CATALYTIC PROCESSES

8.4.1.1 PHOTOCATALYSIS

Photocatalysis drives the chemical reaction by activation of catalysts via absorption of photons.[84] Photocatalysts are the substances that facilitate the active surface sites for pollutants or reactant molecules to adsorb/adhere at its surface. The photocatalysts are the photoactive materials/semiconductors that absorb in UV/visible or infrared wavelengths. Absorption of photons of energy equal to its bandgap facilitates the excitation of electrons from valence band to the empty conduction band, leaving behind a vacancy of opposite charge, that is, holes. These bands are the filled or unfilled orbitals of the elements of the respective semiconductor. Based on the location of maxima of valence and minima of the conduction band, band gaps are classified into two categories, that is, direct and indirect band gaps. The species adsorbed at the surface of catalysts are hydroxyl ion and dissolved oxygen.[85] The following schematic shows the photon absorption, recombination, and surface reactions in a semiconductor (Fig. 8.8).

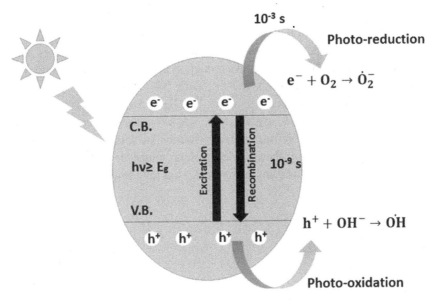

FIGURE 8.8 Schematic of radical generations in a semiconductor.

If the band edge potentials of the semiconductor cross the potentials of generation of hydroxyl and superoxide radicals, in this case, the reduction of surface adsorbed dissolved oxygen and oxidation of adsorbed hydroxyl ions generates superoxide and hydroxyl radicals, respectively. Band edge potentials of some well-known semiconductors and reduction/oxidation potential of superoxide and hydroxyl radicals are shown in Figure 8.9. Holes also possess oxidative capabilities and are responsible for direct oxidation of the contaminant. However, higher adsorption ability of the contaminant onto the catalyst encourages the direct hole attack mechanism. Low surface coverage due to similar surface charges of catalyst and the pollutant reduces the adsorption and results in a homogeneous radical reaction in the bulk solution.[86] Moreover, the role of holes, adsorbed hydroxyl radicals (bound), and free hydroxyl radicals and its pathways are debatable. This can be well understood by having information about highest occupied molecular orbital and lowest unoccupied molecular orbital of the contaminants.[87] Generally direct hole attack is also promoted by shallow surface traps or defects.[87]

Transition metal oxides are widely studied as photocatalysts due to various features such as bandgap, stability, high surface area, suitable morphology, etc. In transition metal oxides, the valance band and conduction band consist of O-2p orbitals and 3d orbitals, respectively. The ionic concentration at the surface leads to the generation of internal defects or vacancies in the system. Defect formation plays a major role in suppressing the recombination or increasing the charge separation; however, these sites sometimes act as recombination centers and result in a reduction in photocatalytic activity. Wide bandgap semiconductors such as TiO_2, ZnO, Nb_2O_5, WO_3, etc. demonstrate excellent photocatalytic activity under UV light. However, solar spectrum contains only 3–4% UV wavelength, 45–47% visible wavelength, and remaining infrared. Therefore, a larger fraction of solar spectrum can only be utilized by modifying these wide bandgap materials into lower bandgap materials, by band engineering or by designing new lower bandgap material of mixed oxides, etc.

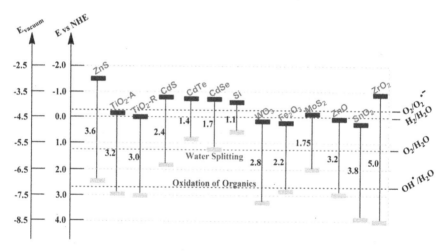

FIGURE 8.9   Band edge potentials of some well-known semiconductors and radical generation (Normal Hydrogen Electrode (NHE) and Absolute Vacuum Scale (AVS)).

Photocatalysis is a versatile technique and is used for reactions such as water splitting, CO_2 reduction, and N_2 fixation. This is also termed as artificial photosynthesis. This technique has been proven to be utilizing the most abundant sources such as water and solar light and resulting in the gen fuel. The thermocatalytic processes utilize an enormous amount of energy to produce hydrogen and emit an excessive amount of CO_2, CO, and other harmful gases. Therefore, switching from steam reforming methods to photocatalysis and electrocatalysis has been widely used. Thus, it can be considered that the research community as well as the industrial approach is heading toward obtaining the green fuel via green approach of catalysis.

8.4.1.2 ELECTROCATALYSIS

Electrocatalysis is a potential method that can be used to produce fuels for heavy transports, air vehicles, etc. This method has the potential to replace the energy and cost involved in steam reforming by natural gas; however, insufficient infrastructure, limited yields, and lower conversion and selectivity restrict its viable usage at large scale. Therefore, continuous efforts have been made in terms of designing better and stable catalysts for this application.

Along with the catalyst improvement, there are several challenges that have to be addressed for the occurrence of the reaction. Proton needed for the CO_2 reduction reaction is reduced at similar reduction potentials, limiting the kinetic feasibility of CO_2 reduction reaction. Also, the stability of the CO_2 molecule requires excess energy to facilitate the reduction reaction. Therefore, application of potential becomes rather important for driving this transformation. Here, the major challenge for the catalyst is to operate at low overpotentials and elimination of inherent kinetic limitation of the reaction. Augmentation in faradaic efficiency of the catalyst is another challenge to overcome. Suppression of unwanted competitive reactions such as HER, catalyst degradation, and solvent decomposition to reduce the energy dissipation can solve both the major challenges. Therefore, development of the photoelectrode with high partial current density is required to make use of absorption of photons to minimize the overpotential as well as accelerate the product formation.[88] In literature, electrolyte resistance has been mentioned as the responsible factor for lower efficiencies toward carbon products.[89] Therefore, efforts should be made to overcome the interfacial resistance at the interface. Electrocatalytic reactions take place at the electrode–electrolyte interface; therefore, efficiency of such systems is predicted by the structure and composition of the surface of electrode and the nature of electrolyte. Development and optimization of an efficient catalyst is a difficult task.

A major limitation of thermocatalytic methods apart from energy usage is the quality of products. The H_2 generated by the steam reforming method contains various impurities such as constituents of N, S, and C oxides. These impurities can deactivate the catalysts rapidly by poisoning the active sites and by weakening the strong metal-support interaction. Photoelectrocatalysis in this regard is the greener approach to generate H_2 because it uses direct protons to form H_2 molecule. However, the yield of H_2 obtained from this method is significantly low. Hydrolysis of reactive metal, metal hydrides, alkali, and alkaline earth metal borohydrides is also considered to be a rapid method for generation of H_2; however, the sources for the H_2 generation such as metal hydrides and borohydrides are harmful for the environment and add excessive toxicity to the environment during their synthesis procedure. Therefore, given these constraints, these methods are not recommended for the production of H_2.

Water electrolysis can be an attractive choice for H_2 generation, although the yield obtained by this method is limited yet it does not cause harmful effects to the environment. Continuous research is directed toward improving the cell performance by introducing novel approaches to explore the electrode materials. The reaction given below shows the approach for the generation of hydrogen and its energetics.

$$H_2O(l) \rightarrow H_2(g) + 1/2O_2(l) \ E_0 = 1.23 \ V \ vs. \ SHE$$

In order to minimize the energy losses during the process, this method should be taken place at highly acidic or basic conditions for the availability of the protons in acidic conditions. In the case of oxygen evolution reactions, at this pH, alloys of Ir or Ru are highly efficient and active at low overpotentials. Pt catalysts are highly efficient for hydrogen evolution reaction at low overpotentials. However, cost involved in these catalysts is the major limitation for commercial viability of the process. Therefore, efforts have been made to reduce the cost by combining these precious metals with inexpensive metals. Metal oxides and chalcogenides are also being explored to analyze their properties for such applications.[90]

KEYWORDS

- **green catalysis**
- **renewable sources**
- **solar energy**
- **wind energy**
- **potential difference**
- **toxicity**

REFERENCES

1. Kaneda, K.; Mizugaki, T.; Mitsudome, T. Green Catalysis. *Encyclopedia of Catalysis*, 2002.
2. Cavell, K.; Golunski, S.; Miller, D. Handbook of Green Chemistry-Green Catalysis. *Platinum Metals Rev.* **2010**, *54*, 233–238.
3. Li, C.-J.; Trost, B. M. Green Chemistry for Chemical Synthesis. *Proc. Natl. Acad. Sci.* **2008**, *105*, 13197–13202.
4. Sheldon, R. A.; Arends, I.; Hanefeld, U. *Green Chemistry and Catalysis*. John Wiley & Sons, 2007.
5. Anastas, P. T.; Kirchhoff, M. M.; Williamson, T. C. Catalysis as a Foundational Pillar of Green Chemistry. *Appl. Cataly. A: Gen.* **2001**, *221*, 3–13.
6. Zaera, F. The Surface Chemistry of Metal-Based Hydrogenation Catalysis. *ACS Cataly.* **2017**, *7*, 4947–4967.
7. Vu, M.-H.; Sakar, M.; Do, T.-O. Insights Into the Recent Progress and Advanced Materials for Photocatalytic Nitrogen Fixation for Ammonia (NH3) Production. *Catalysts* **2018**, *8*, 621.
8. Araiza, D. G.; Gómez-Cortés, A.; Díaz, G. Methanol Decomposition Over Bimetallic Cu-M Catalysts Supported on Nanoceria: Effect of the Second Metal on the Catalytic Properties. *Cataly. Today* **2019**, In press.
9. Christian Enger, B.; Lødeng, R.; Holmen, A. A Review of Catalytic Partial Oxidation of Methane to Synthesis Gas with Emphasis on Reaction Mechanisms Over Transition Metal Catalysts. *Appl. Cataly. A: Gen.* **2008**, *346*, 1–27.
10. Amjad, U.-E. S.; Vita, A.; Galletti, C.; Pino, L.; Specchia, S. Comparative Study on Steam and Oxidative Steam Reforming of Methane with Noble Metal Catalysts. *Ind. Eng. Chem. Res.* **2013**, *52*, 15428–15436.
11. van den Reijen, J. E.; Kanungo, S.; Welling, T. A. J.; Versluijs-Helder, M.; Nijhuis, T. A.; de Jong, K. P.; de Jongh, P. E. Preparation and Particle Size Effects of Ag/α-Al2O3 Catalysts for Ethylene Epoxidation. *J. Cataly.* **2017**, *356*, 65–74.
12. Patra, N.; Taviti, A. C.; Sahoo, A.; Pal, A.; Beuria, T. K.; Behera, A.; Patra, S. Green Synthesis of Multi-metallic Nanocubes. *RSC Adv.* **2017**, *7*, 35111–35118.
13. Kawawaki, T.; Negishi, Y.; Kawasaki, H. Photo/Electrocatalysis and Photosensitization Using Metal Nanoclusters for Green Energy and Medical Applications. *Nanoscale Adv.* **2019**.
14. Choi, J.-R.; Shin, D.-M.; Song, H.; Lee, D.; Kim, K. Current Achievements of Nanoparticle Applications in Developing Optical Sensing and Imaging Techniques. *Nano Converg.* **2016**, *3*, 30.
15. Yonezawa, T. Preparation of Metal Nanoparticles and Their Application for Materials. In *Nanoparticle Technology Handbook*; Elsevier, 2018, pp. 829–837.
16. Singh, J.; Dutta, T.; Kim, K.-H.; Rawat, M.; Samddar, P.; Kumar, P. "Green" Synthesis of Metals and Their Oxide Nanoparticles: Applications for Environmental Remediation. *J. Nanobiotechnol.* **2018**, *16*, 84.
17. Tran, T. D.; Nguyen, M. T.; Le, H. V.; Nguyen, D. N.; Truong, Q. D.; Tran, P. D. Gold Nanoparticles as an Outstanding Catalyst for the Hydrogen Evolution Reaction. *Chem. Commun.* **2018**, *54*, 3363–3366.
18. Zhao, G.; Huang, X.; Wang, X.; Wang, X. Progress in Catalyst Exploration for Heterogeneous CO^2 Reduction and Utilization: A Critical Review. *J. Mater. Chem. A* **2017**, *5*, 21625–21649.

19. Chen, X.; Li, N.; Kong, Z.; Ong, W.-J.; Zhao, X. Photocatalytic Fixation of Nitrogen to Ammonia: State-of-the-art Advancements and Future Prospects. *Mater. Horiz.* **2018**, *5*, 9–27.

20. Wang, J.; Gu, H. Novel Metal Nanomaterials and Their Catalytic Applications. *Molecules* **2015**, *20*, 17070–17092.

21. Wu, L.; Xi, Z.; Sun, S. Well-Defined Metal Nanoparticles for Electrocatalysis. In *Studies in Surface Science and Catalysis*; Elsevier, 2017, pp. 123–148.

22. Sun, L.; Yin, Y.; Wang, F.; Su, W.; Zhang, L. Facile One-pot Green Synthesis of Au–Ag Alloy Nanoparticles Using Sucrose and Their Composition-dependent Photocatalytic Activity for the Reduction of 4-nitrophenol. *Dalton Trans.* **2018**, *47*, 4315–4324.

23. Zhang, G.; Du, M.; Li, Q.; Li, X.; Huang, J.; Jiang, X.; Sun, D. Green Synthesis of Au–Ag Alloy Nanoparticles Using Cacumen Platycladi Extract. *RSC Adv.* **2013**, *3*, 1878–1884.

24. Teja, A. S.; Koh, P.-Y. Synthesis, Properties, and Applications of Magnetic Iron Oxide Nanoparticles. *Prog. Cryst. Grow. Charact. Mater.* **2009**, *55*, 22–45.

25. Tang, Y.; Yang, Q.; Wu, T.; Liu, L.; Ding, Y.; Yu, B. Fluorescence Enhancement of Cadmium Selenide Quantum Dots Assembled on Silver Nanoparticles and Its Application to Glucose Detection. *Langmuir* **2014**, *30*, 6324–6330.

26. Subramanian, V.; Wolf, E.; Kamat, P. V. Semiconductor−Metal Composite Nanostructures. To What Extent Do Metal Nanoparticles Improve the Photocatalytic Activity of TiO_2 Films? *J. Phys. Chem. B* **2001**, *105*, 11439–11446.

27. Guo, T.; Yao, M.-S.; Lin, Y.-H.; Nan, C.-W. A Comprehensive Review on Synthesis Methods for Transition-metal Oxide Nanostructures. *Cryst. Eng. Comm.* **2015**, *17*, 3551–3585.

28. Ray, C.; Pal, T. Recent Advances of Metal–metal Oxide Nanocomposites and Their Tailored Nanostructures in Numerous Catalytic Applications. *J. Mater. Chem. A* **2017**, *5*, 9465–9487.

29. Congxin, X.; Jingbo, L. Recent Advances in Optoelectronic Properties and Applications of Two-dimensional Metal Chalcogenides. *J. Semiconduct.* **2016**, *37*, 051001.

30. Zhao, J.-W.; Zhang, Y.-F.; Li, Y.-H.; Su, C.-h.; Song, X.-M.; Yan, H.; Wang, R.-Z. A Low Cost, Green Method to Synthesize GaN Nanowires. *Sci. Rep.* **2015**, *5*, 17692.

31. Bao, K.; Shi, L.; Liu, X.; Chen, C.; Mao, W.; Zhu, L.; Cao, J. Synthesis of GaN Nanorods by a Solid-state Reaction. *J. Nanomater.* **2010**, *2010*, 6.

32. Late, D. J.; Liu, B.; Luo, J.; Yan, A.; Matte, H. R.; Grayson, M.; Rao, C.; Dravid, V. P. GaS and GaSe Ultrathin Layer Transistors. *Adv. Mater.* **2012**, *24*, 3549–3554.

33. Park, K. H.; Jang, K.; Kim, S.; Kim, H. J.; Son, S. U. Phase-controlled One-dimensional Shape Evolution of InSe Nanocrystals. *J. Am. Chem. Soc.* **2006**, *128*, 14780–14781.

34. Shen, G.; Chen, D.; Chen, P.-C.; Zhou, C. Vapor−solid Growth of One-dimensional layer-structured Gallium Sulfide Nanostructures. *Acs Nano* **2009**, *3*, 1115–1120.

35. Hu, P.; Wang, L.; Yoon, M.; Zhang, J.; Feng, W.; Wang, X.; Wen, Z.; Idrobo, J. C.; Miyamoto, Y.; Geohegan, D. B. Highly Responsive Ultrathin GaS Nanosheet Photodetectors on Rigid and Flexible Substrates. *Nano Lett.* **2013**, *13*, 1649–1654.

36. Golberg, D.; Bando, Y.; Huang, Y.; Terao, T.; Mitome, M.; Tang, C.; Zhi, C. Boron Nitride Nanotubes and Nanosheets. *ACS Nano* **2010**, *4*, 2979–2993.

37. Zhi, C.; Bando, Y.; Terao, T.; Tang, C.; Kuwahara, H.; Golberg, D. Chemically Activated Boron Nitride Nanotubes. *Chem.−Asian J.* **2009**, *4*, 1536–1540.

38. Huerta-Flores, A. M.; Torres-Martínez, L. M.; Moctezuma, E.; Singh, A. P.; Wickman, B. Green Synthesis of Earth-abundant Metal Sulfides (FeS 2, CuS, and NiS 2) and Their

Use as Visible-light Active Photocatalysts for H 2 Generation and Dye Removal. *J. Mater. Sci.: Mater. Electron.* **2018**, *29*, 11613–11626.

39. Peng, W.-c.; Chen, Y.; Li, X.-y. MoS2/Reduced Graphene Oxide Hybrid with CdS Nanoparticles as a Visible Light-driven Photocatalyst for the Reduction of 4-nitrophenol. *J. Hazard. Mater.* **2016**, *309*, 173–179.

40. Yang, X.; Wu, R.; Liu, H.; Fan, H.; Zhang, H.; Sun, Y. Amorphous Molybdenum Selenide as Highly Efficient Photocatalyst for the Photodegradation of Organic Dyes Under Visible Light. *Appl. Surf. Sci.* **2018**, *457*, 214–220.

41. Liu, H.; Su, Y.; Chen, P.; Wang, Y. Microwave-assisted Solvothermal Synthesis of 3D Carnation-like SnS2 Nanostructures with High Visible Light Photocatalytic Activity. *J. Mol. Cataly. A: Chem.* **2013**, *378*, 285–292.

42. Liu, Y.; Xie, S.; Li, H.; Wang, X. A Highly Efficient Sunlight Driven ZnO Nanosheet Photocatalyst: Synergetic Effect of P-Doping and MoS2 Atomic Layer Loading. *ChemCatChem.* **2014**, *6*, 2522–2526.

43. Hayashi, H.; Hakuta, Y. Hydrothermal Synthesis of Metal Oxide Nanoparticles in Supercritical Water. *Materials* **2010**, *3*, 3794–3817.

44. Titirici, M-M.; Antonietti, M.; Thomas, A. A Generalized Synthesis of Metal Oxide Hollow Spheres Using a Hydrothermal Approach. *Chem. Mater.* **2006**, *18*, 3808–3812.

45. Yang, Q.; Lu, Z.; Liu, J.; Lei, X.; Chang, Z.; Luo, L.; Sun, X. Metal Oxide and Hydroxide Nanoarrays: Hydrothermal Synthesis and Applications as Supercapacitors and Nanocatalysts. *Prog. Nat. Sci.: Mater. Int.* **2013**, *23* 351–366.

46. Yiamsawas, D.; Boonpavanitchakul, K.; Kangwansupamonkon, K. Preparation of ZnO Nanostructures by Solvothermal Method. *J. Microsc. Soc. Thailand* **2009**, *23*, 75–78.

47. Liang, M.-T.; Wang, S.-H.; Chang, Y.-L.; Hsiang, H.-I.; Huang, H.-J.; Tsai, M.-H.; Juan, W.-C.; Lu, S.-F. Iron Oxide Synthesis Using a Continuous Hydrothermal and Solvothermal System. *Ceram. Int.* **2010**, *36*, 1131–1135.

48. Kumar, R. V.; Diamant, Y.; Gedanken, A. Sonochemical Synthesis and Characterization of Nanometer-size Transition Metal Oxides from Metal Acetates. *Chem. Mater.* **2000**, *12*, 2301–2305.

49. Pinna, N.; Niederberger, M. Surfactant-free Nonaqueous Synthesis of Metal Oxide Nanostructures. *Angew. Chem. Int. Ed.* **2008**, *47*, 5292–5304.

50. Faraji, S.; Ani, F. N. Microwave-assisted Synthesis of Metal Oxide/Hydroxide Composite Electrodes for High Power Supercapacitors–a Review. *J. Power Sourc.* **2014**, *263*, 338–360.

51. Lagashetty, A.; Havanoor, V.; Basavaraja, S.; Balaji, S.; Venkataraman, A. Microwave-assisted Route for Synthesis of Nanosized Metal Oxides. *Sci. Technol. Adv. Mater.* **2007**, *8*, 484.

52. Hamedani, N. F.; Mahjoub, A. R.; Khodadadi, A. A.; Mortazavi, Y. Microwave Assisted Fast Synthesis of Various ZnO Morphologies for Selective Detection of CO, CH4 and Ethanol. *Sens. Actuat. B: Chem.* **2011**, *156*, 737–742.

53. Agarwal, H.; Kumar, S. V.; Rajeshkumar, S. A Review on Green Synthesis of Zinc Oxide Nanoparticles–An Eco-friendly Approach. *Resource-Efficient Technologies* **2017**, *3*, 406–413.

54. Remya, V.; Kurian, M. Synthesis and Catalytic Applications of Metal–organic Frameworks: A Review on Recent Literature. *Int. Nano Lett.* **2019**, *9*, 17–29.

55. Klimakow, M.; Klobes, P.; Thünemann, A. F.; Rademann, K.; Emmerling, F. Mechanochemical Synthesis of Metal–organic Frameworks: A Fast and Facile Approach

Toward Quantitative Yields and High Specific Surface Areas. *Chem. Mater.* **2010**, *22*, 5216–5221.

56. Pascanu, V.; González Miera, G.; Inge, A. K.; Martín-Matute, B. n. Metal–organic Frameworks as Catalysts for Organic Synthesis: A Critical Perspective. *J. Am. Chem. Soc.* **2019**, *141*, 7223–7234.

57. Friščić, T. Metal-organic Frameworks: Mechanochemical Synthesis Strategies. *Encycl. Inorg. Bioinorg. Chem.* **2011**, 1–19.

58. Murzin, D. Y. *Chemical Reaction Technology.* Walter de Gruyter GmbH & Co KG, 2015.

59. Hafez, E. A. A.; Al-Mousawi, S. M.; Moustafa, M. S.; Sadek, K. U.; Elnagdi, M. H. Green Methodologies in Organic Synthesis: Recent Developments in Our Laboratories. *Green Chem. Lett. Rev.* **2013**, *6*, 189–210.

60. Wang, C.-A.; Han, Y.-F.; Y.-W. Li, Y.-W.; Nie, K.; Cheng, X.-L.; Zhang, J.-P. Bipyridyl Palladium Embedded Porous Organic Polymer as Highly Efficient and Reusable Heterogeneous Catalyst for Suzuki–Miyaura Coupling Reaction. *RSC Adv.* **2016**, *6*, 34866–34871.

61. Xia, W.; Huang, L.; Huang, X.; Wang, Y.; Lu, C.; Yang, G.; Chen, Z.; Nie, J. Main-chain NHC-palladium Polymers Based on Adamantane: Synthesis and Application in Suzuki–Miyaura reactions. *J. Mol. Cataly. A: Chem.* **2016**, *412*, 93–100.

62. Mennecke, K.; Kirschning, A. Polyionic Polymers–heterogeneous Media for Metal Nanoparticles as Catalyst in Suzuki–Miyaura and Heck–Mizoroki Reactions Under Flow Conditions. *Beilstein J. Org. Chem.* **2009**, *5*, 21.

63. You, L.; Zhu, W.; Wang, S.; Xiong, G.; Ding, F.; Ren, B.; Dragutan, I.; Dragutan, V.; Sun, Y. High Catalytic Activity in Aqueous Heck and Suzuki–Miyaura Reactions Catalyzed by Novel Pd/Ln Coordination Polymers Based on 2, 2'-bipyridine-4, 4'-dicarboxylic acid as a Heteroleptic Ligand. *Polyhedron* **2016**, *115*, 47–53.

64. Niknam, K.; Deris, A.; Panahi, F.; Nezhad, M. R. H. Immobilized Palladium Nanoparticles on Silica Functionalized *N*-propylpiperazine Sodium *N*-Propionate (SBPPSP): Catalytic Activity Evaluation in Copper-free Sonogashira Reaction. *J. Iran. Chem. Soc.* **2013**, *10*, 1291–1296.

65. Suzuka, T.; Kimura, K.; Nagamine, T. Reusable Polymer-Supported Terpyridine Palladium Complex for Suzuki-Miyaura, Mizoroki-Heck, Sonogashira, and Tsuji-Trost Reaction in Water. *Polymers* **2011**, *3*, 621–639.

66. Sarkar, S. M.; Rahman, L. M.; Yusoff, M. M. Pyridinyl Functionalized MCM-48 Supported Highly Active Heterogeneous Palladium Catalyst for Cross-coupling Reactions. *RSC Adv.* **2015**, *5*, 19630–19637.

67. Chanda, K.; Rej, S.; Liu, S. Y.; Huang, M. H. Facet-Dependent Catalytic Activity of Palladium Nanocrystals in Tsuji–Trost Allylic Amination Reactions with Product Selectivity. *ChemCatChem* **2015**, *7*, 1813–1817.

68. Adjabeng, G.; Brenstrum, T.; Wilson, J.; Frampton, C.; Robertson, A.; Hillhouse, J.; McNulty, J.; Capretta, A. Novel Class of Tertiary Phosphine Ligands Based on a Phospha-adamantane Framework and Use in the Suzuki Cross-Coupling Reactions of Aryl Halides under Mild Conditions. *Org. Lett.* **2003**, *5*, 953–955.

69. Kumar, D. K.; Filatov, A. S.; Napier, M.; Sun, J.; Dikarev, E. V.; Petrukhina, M. A. Dirhodium Paddlewheel with Functionalized Carboxylate Bridges: New Building Block for Self-assembly and Immobilization on Solid Support. *Inorg. Chem.* **2012**, *51*, 4855–4861.

70. Zhao, Q.; Chen, D.; Li, Y.; Zhang, G.; Zhang, F.; Fan, X. Rhodium Complex Immobilized on Graphene Oxide as an Efficient and Recyclable Catalyst for Hydrogenation of Cyclohexene. *Nanoscale* **2013**, *5*, 882–885.

71. Bai, C.; Zhao, Q.; Li, Y.; Zhang, G.; Zhang, F.; Fan, X. Palladium Complex Immobilized on Graphene Oxide as an Efficient and Recyclable Catalyst for Suzuki Coupling Reaction. *Cataly. Lett.* **2014**, *144*, 1617–1623.

72. Zhang, H.; Li, Y.; Shao, S.; Wu, H.; Wu, P. Grubbs-type Catalysts Immobilized on SBA-15: A Novel Heterogeneous Catalyst for Olefin Metathesis. *J. Mol. Cataly. A: Chem.* **2013**, *372*, 35–43.

73. Ge, Y.; Sun, W.; Pei, B.; Ding, J.; Jiang, Y.; Loh, T.-P. Hoveyda–Grubbs II Catalyst: A Useful Catalyst for One-Pot Visible-light-Promoted Ring Contraction and Olefin Metathesis Reactions. *Org. Lett.* **2018**, *20*, 2774–2777.

74. Iosub, A. V.; Stahl, S. S. Catalytic Aerobic Dehydrogenation of Nitrogen Heterocycles Using Heterogeneous Cobalt Oxide Supported on Nitrogen-doped Carbon. *Org. Lett.* **2015**, *17*, 4404–4407.

75. Rimoldi, M.; Nakamura, A.; Vermeulen, N. A.; Henkelis, J. J.; Blackburn, A. K.; Hupp, J. T.; Stoddart, J. F.; Farha, O. K. A Metal–organic Framework Immobilised Iridium Pincer Complex. *Chem. Sci.* **2016**, *7*, 4980–4984.

76. Gawande, M. B.; Goswami, A.; Felpin, F.-X.; Asefa, T.; Huang, X.; Silva, R.; Zou, X.; Zboril, R.; Varma, R. S. Cu and Cu-based Nanoparticles: Synthesis and Applications in Catalysis. *Chem. Rev.* **2016**, *116*, 3722–3811.

77. (a) Rideout, D. C.; Breslow R. Hydrophobic Acceleration of Diels-Alder Reactions. *J. Am. Chem. Soc.*, **1980**, *102.26*, 7816–7817. (b) Grieco, P. A.; Kiyoshi, Y.; Graner, P., Aqueous Intermolecular Diels-Alder Chemistry: Reactions of Diene Carboxylates with Dienophiles in Water at Ambient Temperature. *J. Org. Chem.* **1983**, *48.18*, 3137–3139.

78. Wasserscheid, P.; Thomas W., eds., *Ionic Liquids in Synthesis*, John Wiley & Sons, 2008, pp. 7–35.

79. De Martino, M. T.; Abdelmohsen, L. K. E. A.; Rutjes, F. P. J. T.; van Hest, J. C. M. Nanoreactors for Green Catalysis. *Beilstein J. Org. Chem.* **2018**, *14*, 716–733.

80. Peters, R. J.; Marguet, M.; Marais, S.; Fraaije, M. W.; Van Hest, J. C.; Lecommandoux, S. Cascade Reactions in Multicompartmentalized Polymersomes. *Angew. Chem. Int. Ed.* **2014**, *53*, 146–150.

81. Itoh, T.; Hanefeld, U. Enzyme Catalysis in Organic Synthesis. *Green Chem.* **2017**, *19*, 331–332.

82. Nasir, Z.; Ali, A.; Shakir, M.; Wahab, R. Silica-supported NiO Nanocomposites Prepared via a Sol–gel Technique and Their Excellent Catalytic Performance for One-pot Multicomponent Synthesis of Benzodiazepine Derivatives Under Microwave Irradiation. *New J. Chem.* **2017**, *41*, 5893–5903.

83. Coyle, E. D.; Simmons, R. A. *Understanding the Global Energy Crisis;* Purdue University Press, 2014.

84. Nahar, S.; Zain, M. F. M.; Kadhum, A. A. H.; Hasan, H. A.; Hasan, M. R. Advances in Photocatalytic CO(2) Reduction with Water: A Review. *Materials* **2017**, *10*, 629.

85. Ahmed, S. N.; Haider, W. Heterogeneous Photocatalysis and Its Potential Applications in Water and Wastewater Treatment: A Review. *Nanotechnology* **2018**, *29*, 342001.

86. Yang, S.-y.; Chen, Y.-y.; Zheng, J.-g.; Cui, Y.-j. Enhanced Photocatalytic Activity of TiO_2 by Surface Fluorination in Degradation of Organic Cationic Compound. *J. Environ. Sci.* **2007**, *19*, 86–89.

Green Polymer Chemistry and Composites

87. Maurino, V.; Minella, M.; Sordello, F.; Minero, C. A Proof of the Direct Hole Transfer in Photocatalysis: The Case of Melamine. *Appl. Cataly. A: Gen.* **2016**, *521*, 57–67.

88. White, J. L.; Baruch, M. F.; Pander III, J. E.; Hu, Y.; Fortmeyer, I. C.; Park, J. E.; Zhang, T.; Liao, K.; Gu, J.; Yan, Y. Light-driven Heterogeneous Reduction of Carbon Dioxide: Photocatalysts and Photoelectrodes. *Chem. Rev.* **2015**, *115*, 12888–12935.

89. Vesborg, P. C.; Seger, B. Performance Limits of Photoelectrochemical CO^2 Reduction Based on Known Electrocatalysts and the Case for Two-electron Reduction Products. *Chem. Mater.* **2016**, *28*, 8844–8850.

90. Anantharaj, S.; Ede, S.; Karthick, K.; Sankar, S. S.; Sangeetha, K.; Karthik, P.; Kundu, S. Precision and Correctness in the Evaluation of Electrocatalytic Water Splitting: Revisiting Activity Parameters with a Critical Assessment. *Energ. Environ. Sci.* **2018**, *11*, 744–771.

91. Tsukamoto, D.; Shiro, A.; Shiraishi, Y.; Sugano, Y.; Ichikawa, S.; Tanaka, S.; Hirai, T. Photocatalytic H2O2 Production from Ethanol/O2 System Using TiO2 Loaded with Au-Ag Bimetallic Alloy Nanoparticles. ACS Catal. 2012, 2, 599–603.

92. Krishnan, Unni, et al. "MoS2/ZnO nanocomposites for efficient photocatalytic degradation of industrial pollutants." Materials Research Bulletin 111 (2019): 212–221.

93. Ge, Y.; Sun, W.; Pei, B.; Ding, J.; Jiang, Y.; Loh, T.-P., Hoveyda–Grubbs II Catalyst: A Useful Catalyst for One-Pot Visible-Light-Promoted Ring Contraction and Olefin Metathesis Reactions. *Organic letters* **2018**, 20 (9), 2774–2777.

94. Adjabeng, George, et al. "Novel class of tertiary phosphine ligands based on a phospha-adamantane framework and use in the Suzuki cross-coupling reactions of aryl halides under mild conditions." Organic Letters 5.6 (2003): 953–955.

CHAPTER 9

Bionanocomposites, Biofuels, and Environmental Protection: A Far-Reaching Review and the Visionary Future

SUKANCHAN PALIT

Department of Chemical Engineering, University of Petroleum and Energy Studies, Dehradun 248007, India

43, Judges Bagan, Haridevpur, Kolkata 700082, India

*Corresponding author. E-mail: sukanchan68@gmail.com, sukanchan92@gmail.com, sukanchanp@rediffmail.com

ABSTRACT

The world of environmental protection and nanotechnology are today in the avenues of new scientific regeneration. The scientific domain and the civil society today stand shocked with the evergrowing challenges of global climate change, global warming, and frequent environmental disasters. Also, loss of ecological biodiversity and depletion of fossil fuel resources are veritably challenging the scientific firmament. In this paper, the author with insight and academic rigor delineates the recent and significant advances in the field of bionanocomposites, biofuels, and environmental protection with a clear vision toward true realization of environmental or green sustainability. The author also stresses the application of nanotechnology and composite science in environmental remediation and green sustainability. A larger investigation in the field of biofuels and application of biofuels are the other cornerstones of this paper. Mankind's immense scientific perseverance and determination and the scientific ingenuity of the field of environmental protection will surely be the torchbearers toward a new era in environmental sustainability today. The application of nanomaterials and engineered nanomaterials in energy engineering and environmental engineering stands as veritable pillars of this treatise. Material science and composite science are today in the process of re-envisioning and deep regeneration. This chapter will surely open new doors of innovation and scientific sagacity in the field of environmental engineering science. A new chapter of engineering science and technology will evolve as civilization confronts the issues of green engineering, green sustainability, and green chemistry. This chapter will validate these issues.

9.1   INTRODUCTION

Science and technology in the global scenario are today in the middle of vision, ingenuity, and scientific forbearance. Mankind and science are in the process of drastic changes. This is due to the burning issues of climate change, global warming, depletion of fossil fuel resources, and loss of ecological diversity. Nanotechnology and nanoengineering are the marvels of science today. In a similar manner, composite science, material science, and the interfaces of biological sciences are in the juxtaposition of scientific rejuvenation and vast scientific vision. The status of research pursuit in the field of biofuels and bioenergy are extremely bright and far-reaching. Environmental disasters in the global scenario are in an extremely dismal state and caution to the civil society, scientists, and engineers. Thus, the domain of science and engineering of environmental protection and environmental sustainability needs to be scientifically upright and far-sighted as civilization moves forward. In this treatise, the author elucidates on the science of biofuels, bionanocomposites, and its interface with environmental protection. The challenges, the targets, and the scientific vision of environmental remediation are deeply comprehended in this treatise. This treatise also uncovers the recent scientific advances in the field of biofuels and bionanocomposites with a clear and definite vision toward the larger emancipation of green and environmental sustainability. Petroleum engineering science and petroleum exploration are in the state of immense scientific catastrophe due to the depletion of fossil fuel resources. So it is a tremendous responsibility for engineers, scientists, and civil society. Scientific trustworthiness and scientific humanism will surely unfold the intricacies of environmental engineering particularly the areas of biofuels and bioenergy. Renewable energy and biomass energy are the needs and the marvels of science and technology in the global scenario. The author with vast academic rigor confronts the various issues of application of biomass energy, bioenergy, and biofuels. A new chapter in the field of renewable energy will surely evolve as scientists and engineers plunge into the unknown depths of environmental engineering science and biological sciences.

9.2   THE AIM AND OBJECTIVE OF THIS STUDY

The scientific progress in the field of environmental remediation and environmental engineering are immensely bleak and thought-provoking. Engineering and technology have practically no answers to the growing concerns of drinking water issues and climate change globally. The veritable aim and objective of this study is to target the murky and the unknown areas of bionanocomposites, biofuels, and environmental engineering science. Civilization and science are today in the process of new regeneration. Renewable energy is the need of human civilization today. Scientific forbearance, scientific humanism, and deep scientific and engineering vision are the needs of global research initiatives in renewable energy and environmental protection. Composite science, material science, and nanotechnology are similarly in the avenues of scientific ingenuity. The primary objective of this study is to unravel the applications of biofuels and bionanocomposites in environmental protection. Human scientific regeneration and the validation of the science and engineering of environmental protection will surely open up new doors of innovation, instinct, and vision in the field of environmental sustainability. Today technology has no answers to arsenic and heavy metal groundwater contamination. This chapter will surely open up new avenues in the field of environmental remediation, drinking water treatment, and industrial wastewater treatment. Technological and engineering validation in the field of biofuels and bionanocomposites are changing the vast scientific scenario globally. Energy, environmental, social, and economic sustainability are the veritable needs of human society today. Sustainable development and environmental protection are today interconnected with each other. These areas of scientific deliberation are dealt with scientific rigor in this treatise.[31,32]

9.3   THE NEED AND THE RATIONALE OF THIS STUDY

Science, technology, and engineering are today standing in the middle of vision, scientific grit, and strong determination. The need and the rationale of this study are immense and far-reaching today. The world of science and technology today are in the juxtaposition of vision, scientific forbearance, and deep scientific foresight. Climate change and global warming are today changing the face of human scientific progress, thus the imminent need for a vast deliberation in the field of nanotechnology, composite science, and biological sciences. Bionanocomposites and biofuels are today the veritable necessities of human civilization and scientific progress today. Nanocomposites and smart materials are the futuristic eco-materials of tomorrow. In a similar vision, renewable energy and biomass energy are in the path of a new vision and newer scientific ingenuity. Today the challenges and the vision of global science and technology initiatives are immense and far-reaching. Every branch of science and engineering are aligned with sustainability science and sustainability engineering. Nanotechnology is the need of civilization as well as sustainability engineering, thus the immediate need and rationale of this study. There is a tremendous scientific revolution in the field of material science, composite science, and nanotechnology. Biotechnology and biological sciences are also not lagging behind and are in the avenues of newer revolution. The need for green chemistry and green engineering in human civilization and human scientific progress are immense and in a similar vein groundbreaking. Today water purification science and industrial wastewater treatment stands in the middle of scientific contemplation and scientific forbearance. In a similar vein, science, technology, and engineering of environmental remediation are in a state of immense scientific distress. The need for a comprehensive treatise in environmental engineering and the field of renewable energy are immense, versatile, and far-reaching. The author deeply comprehends the immediate necessity of nanotechnology and renewable energy in the future scientific emancipation of human civilization. This chapter will truly unfold the intricacies and barriers of biofuels and bionanocomposites in the further realization of sustainable development.[31,32]

9.4   THE VAST SCIENTIFIC DOCTRINE OF ENVIRONMENTAL SUSTAINABILITY

Environmental and green sustainability forms the fundamental basis of the domains of environmental engineering and water purification science. Today water purification science and drinking water treatment are in the crucial juncture of immense vision and scientific comprehension. The vast scientific doctrine and the scientific provenance in the application of biofuels and bionanocomposites in human society need to be reorganized and reframed as civilization treads forward. Environmental engineering and environmental sustainability are two opposite sides of the visionary coin today. Research status in the field of environmental sustainability in a similar vein needs to be revamped as science and technology moves in the right direction. Research and development initiatives in the field of environmental sustainability are new yet far-reaching. Civilization's scientific stance and vast scientific profundity are at its zenith as mankind confronts energy and environmental sustainability. In this chapter, the authors deeply pronounce the linkage between environmental sustainability, environmental protection, and biological sciences. Without the definite vision of environmental sustainability and environmental engineering, mankind cannot move forward. This chapter is an immense glimpse on the scientific success, the deep scientific cognizance, and the scientific ingenuity in the application of bionanocomposites and biofuels in the advancement of science and human progress. Environmental and green sustainability are today revolutionizing the scientific landscape. In this chapter, the author repeatedly pronounces the need of renewable energy and the domain of nanotechnology in the true scientific realization of human mankind. Green sustainability and green engineering need to be aligned together for the betterment of mankind today. A newer visionary era in the field of green nanotechnology also needs to be envisioned as man and mankind moves forward. The status of global environment is highly derogative and disastrous. This chapter will surely open up newer futuristic thoughts and newer future recommendations in the field of environmental protection which goes far above any global research and development agenda. A newer dawn in the field of environmental protection will emerge as science and mankind trudge forward. A new scientific hope, grit, and determination will evolve as man and civilization ushers in a new era.

9.5 WHAT DO YOU MEAN BY BIONANOCOMPOSITES?

Nanocomposites are composites in which at least one of the phases shows dimensions in the nanometre range (1 nm = 10^{-9} m). Nanocomposite materials have successfully and veritably emerged as suitable alternatives to the limitations of microcomposites and monolithics, while posing preparation challenges related to the control of elemental composition and stoichiometry in the nanocluster phase. Natural nanocomposites can be characterized as bionanocomposites. Natural bionanocomposites combine a high resilience and tolerance toward failure, adaptation, modularity, and multifunctionality. Technological and engineering challenges and the vast scientific vision of composite science are unfolded in today's research and development initiatives in nanocomposites. They are originally designed and optimized for the needs of human life and to meet the needs of the surrounding environment in order to guarantee the survival of respective species they are integrated with. Structures and properties of biological polymers have been studied by biologists and biotechnologists mainly to understand and investigate their essential role in biological systems. However, the potential applications of biological molecules in the design and formation of bionanocomposites require considering them as synthetic building blocks that may distance itself from their natural environment and function. Natural materials are veritably dynamic in structure and composition. During the last few years, "bionanocomposite" has turned into a typical term to assign those nanocomposites including a naturally occurring biopolymer in the mix with organic moiety, and no less than one measurement in the nanoscale. Nanocomposites that contain naturally occurring polymers in the mix with an inorganic moiety speak to an absolutely different class of materials called bionanocomposites.

9.6   WHAT DO YOU MEAN BY BIOFUELS?

Biofuels are the utmost need of human civilization and human scientific progress today. Due to global climate change and global warming, biofuels and renewable energy are the need of the hour. Biofuels are renewable energy source, made from organic matter or wastes that play a pivotal role in reducing carbon dioxide emissions. Biofuels are one of the largest sources of renewable energy in use in mankind today. In the vast transport sector, they are blended with existing fuels such as gasoline and diesel. A biofuel is a fuel that is produced from contemporary processes from biomass, rather than a fuel produced by the very slow geological processes involved in the formation of fossil fuels such as oil. The ingenuity and profundity of renewable energy are today opening new doors of scientific innovation and scientific instinct in engineering science in decades to come. Human scientific vision and scientific provenance in the domain of biofuels and renewable energy are today in the process of newer regeneration. The author in this treatise deeply focuses on the scientific success and the scientific barriers in the application of biofuels in human society and human scientific progress.

9.7   THE VAST SCIENTIFIC RIGOR IN THE APPLICATION OF BIOFUELS IN ENVIRONMENTAL PROTECTION

Science and technology of environmental protection are today in the avenues of vast scientific vision and deep scientific acuity. Application of biofuels in environmental engineering is one of the marvels of science and technology today. The truth of science in environmental engineering, the vast scientific forbearance and the needs of environmental sustainability will surely open new doors of innovation and scientific instinct in the global scenario. Biofuels and biomass energy are the innovations of the field of renewable energy today. Technology and engineering of renewable energy domain are highly challenged today and are replete with vision and introspection. In this paper, the author deeply comprehends the scientific needs, the scientific targets, and the ingenuity of the applications of biofuels and bionanocomposites in the further emancipation of environmental engineering science. Today the science and engineering of environmental protection need to be re-envisioned as man and mankind treads forward. Science and technology are baffled with the ever-growing concerns for climate change and global warming, thus the need for innovations such as biofuels in environmental remediation. Biofuels are the marvels of the science of renewable energy. Technology and engineering science of biofuels applications are latent yet immature. In this chapter, the author reiterates the scientific success and the scientific ingenuity of both the applications of nanocomposites as well as biofuels. A new area of research and development initiative is the area of application of composite science and nanocomposites in arsenic groundwater remediation. Governments around the world are veritably baffled at the evergrowing environmental disaster of arsenic drinking water poisoning in South Asia, many developing and developed nations around the world. The question of successful water treatment technologies and innovations are the marvel of science and engineering today. The scientific and academic rigor needs to be revamped if environmental concerns are to be mitigated. This chapter veritably opens up newer futuristic thoughts and future recommendations in the field of applications of biofuels in various avenues and the applications of bionanocomposites in environmental protection.

9.8 RECENT SCIENTIFIC ADVANCES IN THE FIELD OF BIONANOCOMPOSITES

Bionanocomposites and bioenergy are relatively innovative branches of science and engineering today. The world of scientific challenges needs to be envisioned as civilization treads forward. Scientific advances in the field of bionanocomposites are in the process of surpassing vast and versatile scientific boundaries. This treatise explores the intricacies, the vision, and the fortitude of scientific research pursuit in the field of bionanocomposites.

Pande et al.[1] discussed with immense insight bionanocomposites in a detailed review. Technological and engineering vision are at its helm today with the application of composite science in environmental protection.[1] During the last few years, bionanocomposite has turned into a typical terminology to assign those nanocomposites including a naturally occurring polymer (biopolymer) in the mix with an inorganic moiety and appearing no less than one measurement on the nanoscale. Nanocomposites that contain normally occurring polymers (biopolymer) in the mix with an inorganic moiety are termed as a different class of materials called bionanocomposites. The authors in this chapter discussed with vision and lucid insight method of preparation of bionanocomposite, solution intercalation, in-situ intercalative polymerization, melt intercalation, components of · bionanocomposites, characterization of bionanocomposites, and the vast application domain.[1] Today bionanocomposites and its applications are in the process of new scientific rejuvenation. Bionanocomposites are prepared by various methods which are solution intercalation, in-situ intercalative polymerization, melt intercalation, and template synthesis.[1] The components of bionanocomposites include biomaterial, cellulose, chitosan, poly-lactic acid, starch, chitin, and polyhydroxyalkanoates. Nanoparticles include layered silicates, nanotubes, and spherical particles. Characterization of bionanocomposites involves particle size determination, surface morphology, and thermal study. Application domain of bionanocomposites are catalysts, gas-separation membranes, contact lenses, and bioactive implant materials.[1] The other areas of applications are fabrication of scaffolds, implants, diagnostics, and biomedical devices and drug-delivery systems. Chitosan nanocomposites have applications in tissue engineering, drug and gene delivery, wound healing, and bioimaging. Silica nanocomposites are used in bone applications.[1] Nanocomposite fibers are used in tissue regeneration by using the fibers as bioactive materials. Copper polymer nanocomposites are excellent and cost-effective biocide controlling or can inhibit the growth of microorganisms and prevent food-borne diseases. Bionanocomposites in the overall sense have applications in diagnostics, drug delivery, and tissue regeneration. Science of applications of bionanocomposites are today in the avenues and paths of newer scientific rejuvenation.[1] This chapter vastly reviews with scientific rigor the scientific intricacies and the vast scientific ingenuity in the applications of bionanocomposites in human society. A new dawn of science and engineering will evolve as composite science and nanotechnology enters into a new phase of research pursuit.[1]

Darwish et al.[2] discussed and elucidated with vast scientific far-sightedness functionalized nanomaterials for environmental techniques. Today nanotechnology is veritably enabling the development of new solutions to environmental issues due to the high surface area (surface to volume ratio) and associated high reactivity in the nanoscale.[2] Environmental protection and the applications of nanomaterials are today in the process of drastic challenges as science and engineering moves forward. The most attractive nanomaterials for environmental protection are derived from silica, noble metals, semiconductors, metal oxides, polymer, and carbonaceous materials. The areas of treatment tools include adsorption, photocatalytic degradation and disinfection, nanofiltration and sensing, and monitoring of various contaminants and pollutants in industrial wastewater.[2] Novel separation processes such as membrane science are the needs toward environmental remediation in the global scenario. In this treatise, the authors discussed with cogent insight the limitations of nanomaterial-based environmental techniques and how functionalization will mitigate these scientific issues.[2] The types of nanomaterials deeply discussed in this chapter are silica-based nanomaterials, carbonaceous nanomaterials, and metal-based nanomaterials. This treatise also elucidates the environmental remediation applications of functionalized nanostructures.[2] The authors discussed in minute details nanoadsorption, membranes and membrane processes, nanophotocatalysis, and nanosensing. Methods of nanomaterials' functionalization techniques are the other cornerstones of this chapter. Functionalization and applications of carbonaceous nanomaterials, silica-based nanomaterials, and functionalization of metal and metal compound nanomaterials are the other areas of investigation. In the midst of functionalization, nanomaterials may overcome the principal limitations that act as a barrier to various application areas.[2] Various types of functionalization methods were discussed mainly the areas of decoration of nanomaterials with a large number of functionalities.[2] This chapter is a water-shed treatise in the field of application of nanomaterials. The authors deeply traverses the scientific success and the deep scientific ingenuity in applications of functionalized nanomaterials.[2]

National Nanotechnology Initiative Workshop Report, USA[3] deeply deliberated nanotechnology and the environment. The authors of this report discussed nanotechnology applications for measurement in the environment, nanotechnology applications for sustainable materials and resources, applications for sustainable manufacturing processes, nanotechnology implications in natural and global processes, nanotechnology vision in health and environment, and the vast research and development frontiers.[3] Nanotechnology is the marvel and scientific innovation of global vision today. Nanotechnology has the immense potential to significantly affect environmental remediation and it has the potential to remove the finest contaminants from water supplies and air and continuously monitor industrial water and air pollution. A planning workshop was held on May 8–9, 2003 in Arlington, Virginia, USA.[3] This report formulated a vision statement for nanotechnology research and development in the next decade. The world of scientific challenges, the needs of environmental protection, and the larger vision of environmental sustainability will surely open up the innovations in nanotechnology and nanomaterial science. Nanomaterial science today is in the middle of deep vision and comprehension. This report vastly pronounces the needs of nanotechnology in the true advancements of science and technology.[3]

Shchipunov[4] dealt lucidly bionanocomposites and the green sustainable materials for the near future. Bionanocomposites are the novel and innovative class of nanosized materials. They are of biological origin and the particles with at least one dimension in the range of 1–100 nm.[4] This article is divided into two parts: bionanocomposite definition and classification along with nanoparticles and the methods of their preparation. The first approach concerns the preparation of bionanocomposites from chitosan and nanoparticles. The second avenue deals with the biomimicking mineralization of biopolymers by using a novel silica precursor. Plastics on the basis of synthetic polymers are widely used in the making of smart materials for daily human life.[4] The challenges of science and technology of plastics applications in human society are immense today. Biodegradable plastics are the scientific genre of present-day human civilization. The authors discussed in minute details nanofibers, nanotubes, nanorods, and other nanomaterials. The other areas of deep deliberations are biopolymers, polysaccharides, proteins, DNA, and polyhydroxyalkanoates. Deep scientific imagination and the scientific truth of bionanocomposites are the areas of immense research and development initiatives globally.[4]

The status of scientific research in developing and developed nations around the world in the field of biofuels and bionanocomposites are surpassing scientific and engineering frontiers. The scientific transcendence and the scientific provenance in the field of renewable energy are the utmost concerns of science and civilization today.

9.9 RECENT SCIENTIFIC ADVANCEMENTS IN THE FIELD OF BIOFUELS

Research pursuits in the field of biofuels are in the avenues of vision, scientific far-sightedness, and deep scientific alacrity. The world of science and engineering stands mesmerized with the progress of renewable energy. Biofuels, biomass energy, and bioenergy are truly revolutionizing the scientific firmament today. In this section, the author unabashedly and with scientific rigor elucidates on the recent advances in the field of biofuels. Biofuel technology is today in the avenues of newer rejuvenation and newer scientific ingenuity. Alternate and renewable energy technology are the coin words of mankind today. These domains are the urgent needs of human scientific progress today. Nanotechnology in a similar vein is a revolutionary domain of human scientific advancement today.

Rasool et al.[5] discussed with scientific insight bioenergy and biofuels and their production in a detailed review. Bioenergy refers to energy obtained from biomass. Biomass is any organic material where sunlight is stored. This review article deals with the veritable conversion of nonedible oils to biodiesel or by the modification of the process of transesterification as well as the conversion of sugars to bioethanol.[5] In this chapter, emphasis on the production of biodiesel and bioethanol and how to modify these methods are carried out.[5] Bioenergy and biofuels are the challenges and vision of the science of renewable energy. Biofuels are alternatives to fossil fuels which are liquid or gaseous fuels that are derived from biomasses. Biodiesel and bioethanol come under the first generation green biofuels. The first-generation biofuels are produced from starch and sugars and from seed oils. The development and production of biofuels are surpassing scientific frontiers.[5] Malaysia and Indonesia are among the largest producers of palm oil in the world and together they produce 80% of the world's palm oil. Humanity's immense scientific and knowledge prowess and the vast scientific in the field of biofuels and bioenergy are ushering in a new era in science and technology. The author validates and justifies these true scientific issues.[5] Taking modern bioenergy into account, the major products of the domain of bioenergy are biodiesel and bioethanol. These two major bioenergy products can be used as a replacement for transportation fuels.[5] Biomass refers to all biological material from living or recently living organisms. In this well-researched treatise, the authors discussed in minute details biofuel development across the globe, bioenergy products, sources of bioenergy, nonedible vegetable oils as a bioenergy source, modification of the chemical process to increase the efficiency of products, and alternative methods of biofuel production.[5] The other hallmarks of this treatise are biodiesel production using the microwave process, production of bioethanol through yeast fermentation, the role of genetic engineering in bioenergy, and advantages and disadvantages of bioethanol.[5] Today biofuels are the best alternatives to petroleum-based fuels because of their enhanced combustion profile and environmentally friendly nature. Biomass energy and bioenergy are the marvels of renewable energy science and technology today.[5] Technological and scientific profundity in the field of biofuels applications are today surpassing scientific and engineering frontiers. This treatise evolves newer scientific understanding and newer scientific ingenuity in the field of bioenergy. The world of scientific challenges and scientific intricacies in the field of renewable energy are scaling newer heights today. This chapter goes deeper into the field of bioenergy and biomass energy with a strong scientific understanding of the applications of biofuels.[5]

Godbole et al.[6] reviewed the production of biofuels. Biofuels are the most renewable and biodegradable sources of energy today. It is highly important to harvest this genuine and innovative technology due to the increasing fear of the impact of petro-fuels on the global environment. This study deeply reviews some of the methods for the production of biofuels. Some of the existing and emerging technologies are highly visionary, inspiring, and thought-provoking in the scientific sense.[6] It can be seen at the end of this research endeavor that the technologies for the production of biofuels are highly environment friendly. In the biofuels production, the choice of feedstock is extremely vital. The other areas of this research pursuit are the domain of biomass catalytic pyrolysis. This research treatise also reiterates the connection of genetic engineering with the production of biofuels. The hallmarks of this treatise are the areas of large scale production of biofuels in developing and developed nations around the world. Particularly large scale production of biofuels from microalgae has been deeply analyzed. India is an emerging developing nation and the need of biofuels and bioenergy are of immense importance.[6] This concept and these tenets of genetic engineering approach in biofuels production are highlighted in details in this chapter.[6] The authors discussed in minute details existing and emerging technologies such as (1) conventional agricultural products, (2) lignocellulosic products and residues, (3) inedible feedstock for biodiesel, (4) use of brown grease, and (5) microalgae for production of biofuels.[6] The other areas dealt lucidly in this chapter are biofuel conversion which are (1) bioethanol and (2) vegetable oil as straight vegetable oil. The genetic engineering approaches for production of biofuels using microalgae are the other cornerstones of this research pursuit. The authors concluded this chapter with the area of feasibility of large scale production of biofuels in India.[6] Technological and engineering envisioning are the needs of biofuels science and bioenergy research globally today. This chapter with vast scientific truth pronounces the need of renewable energy research globally today. Biofuels are today highly effective in transforming the global renewable energy scenario.[6] Due to the global demand for biofuels and bioenergy, biofuels plants in the engineering scenario needs to be scaled up. Today genetic engineering concepts are needed to enhance the scientific and technological vision of biofuels.[6] This research review vastly enhances the success of science and engineering of biofuels and bioenergy today. A deeper scientific discernment is necessary in the application of renewable energy globally today. These issues are dealt lucidly in this research endeavour.[6]

Gashaw et al.[7] reviewed biodiesel production as an alternative fuel. Biodiesel has today become an alternate source of conventional fuel. It is today highly regarded as a substitution fuel and there is an immense scientific understanding of this fuel as a future fuel.[7] This is an alternative fuel for diesel fuel. There are four primary ways to make biodiesel: direct use and blending, microemulsions, thermal cracking, and transesterification.[7] Biodiesel has attracted global attention in the world due to its renewability, biodegradability, nontoxicity, and vastly environmental benefits. Transesterification is a highly common method of production. These avenues of research endeavor are dealt in minute details in this chapter.[7] Renewable energy science is in the path of newer scientific rejuvenation. The authors with profound scientific insight open up and unfurl the vast scientific ingenuity of biofuels production. Biofuels production and application in the global scenario are entering a new phase in human civilization. It has now become a seminal contribution of research and development initiatives in biological sciences globally.[7] In this review, the processes of biodiesel production by the process of transesterification and factors affecting biodiesel production in the global scenario are vastly addressed. Fossil fuels are veritably nonrenewable energy resources. Although these fossil fuels are contributing largely to the global energy supply, their production and usage have raised lots of questions about environmental concerns.[7] It has been reported that 98% of carbon emissions are from fossil fuel combustions.[7] Mankind's immense knowledge prowess in the domains of chemical process engineering is today concentrated on biofuels and bioenergy. Research ingenuity in fossil fuel science is in a similar vein and is highly relevant in the global scenario today. The need for energy is increasing continuously due to the rapid increase of vehicles globally.[7] Biodiesel is defined as monoalkyl esters of long-chain fatty acids originated from natural oils and fats of plants and animals and is veritably a kind of alternative for fossil fuels. The authors discussed in minute details the production of biodiesel, direct use and blending, microemulsion process, thermal cracking (pyrolysis), and factors affecting biodiesel production.

Biodiesel is an important new alternative transportation fuel.[7] It can be produced from different feedstock containing fatty acids, such as animal fats, nonedible oils, and waste cooking oils and by the products of the refining vegetable oils and algae. Today, the science and engineering of biodiesels and biofuels are entering into a newer visionary arena. The authors in this chapter deeply discussed with insight and vision the scientific intricacies and the scientific barriers in the road to scientific wisdom in the field of biofuels and alternate fuels. A new dawn of science and engineering will emerge as renewable energy science revolutionizes the global domain of science and technology.[7]

Food and Agricultural Organization Report[8] deeply discussed with immense scientific conscience and technological profundity challenges and opportunities for developing countries in the field of algae-based biofuels. This report delineates concepts for bioenergy from algae, sustainability of energy from algae, and the wide domain of biofuels.[8] Sustainability science and diverse areas of engineering sciences are today two opposite sides of the visionary coin. Civilization's vast scientific understanding in biofuels, the scientific girth, and determination will all open new doors of innovation and intuition in the field of biological sciences and biotechnology.[8] This report vastly reiterates the multitude of algae-based biofuel pathways and are determined on a case by case basis. Today there is a variety of land-based cultivation system and thus several different algal-based biofuels and other bioenergy carriers are produced.[8] Many input sources like combustion gas, saltwater, and wastewater can be used. Climatic conditions such as annual solar irradiation and temperature strongly influence the algal-based biofuels design. In conclusion, this report delineates (1) algal-based biofuels holds vast promise for developing countries, (2)All algal-based biofuels require significant capital investment, (3) various knowledge levels are required, and (4) knowledge gaps exist for several critical issues. Biotechnology and biological sciences are today in the path of newer scientific rejuvenation. Application of biotechnology in agricultural sciences is a visionary avenue today.[8] This report elucidates the scientific issues in the fusion of application and production of biofuels and the vast world of sustainability.[8] Environmental or green sustainability are today the marvels of science and engineering. This report will go a long and effective way in uncovering the truth and ingenuity of biofuels production and applications.[8]

Wang et al.[9] deeply discussed with insight and lucidity sustainability assessment of bioenergy from a global perspective. Technological and engineering motivation and transcendence are today witnessing immense challenges.[9] Application of bioenergy in human society is a visionary avenue of science and technology. Bioenergy as a renewable energy source is veritably expected to witness significant development in the future. The future of civilization, science, and engineering today stands mesmerized.[9] Today science of sustainability is a huge colossus with a definite vision of its own. A key issue in global scientific progress is the sustainability of energy. To enhance the scope of bioenergy sustainability, this chapter vastly reviews a broad range of current research on this topic and increases the literature into a multidimensional framework covering the economic, environmental and ecological, social, and land-related aspects of bioenergy sustainability. In addition to this, this review shows that crop-based bioenergy and forest bioenergy are the main sources of bioenergy. The final step which needs to be taken is the areas of electricity generation from bioenergy.[9] Climate change has been seen as a major scientific challenge for sustainable development of human society. This chapter deeply reiterates the necessity of bioenergy in tackling these global issues.[9] To address these intricate issues, most of the countries around the world are taking measures to control these carbon emissions. The authors discussed in minute details geospatial distribution of bioenergy studies and the environmental and ecological aspects. Types of bioenergy resources and its vision are the other cornerstones of this treatise.[9]

Civilization, science, and engineering are today experiencing drastic and visionary challenges. Recent scientific advancements in biofuels are challenging the vast scientific fabric of bioenergy and renewable energy science. The author in this entire treatise poignantly depicts the scientific needs and the scientific divination in the application of renewable energy in human society. Research pursuit in bioenergy and renewable energy is highly advanced and groundbreaking. This chapter lucidly investigates the success and profundity of these global issues.

9.10 THE TECHNOLOGICAL AND ENGINEERING CHALLENGES OF ENVIRONMENTAL REMEDIATION

Scientific, technological, and engineering challenges of environmental remediation are in the path of newer regeneration today. The status of the global environment today stands in the middle of vision, introspection, and vast contemplation. Global climate change, the frequent environmental catastrophes, and depletion of fossil fuel resources are today changing the face of the scientific research endeavor. The domain of chemical process engineering and the aligned domain of sustainability engineering need to be re-envisioned and reorganized globally if the true realization of environmental remediation ushers in a newer age. Water purification science and the domain of industrial wastewater treatment stand in the crucial juncture of vision and scientific transcendence. In this chapter, the author reiterates the global stance and the global scientific divination in the field of biofuels, renewable energy, and the vast domain of nanotechnology. There are today immense concerns of application of nanomaterials and nanotechnology in environmental remediation. The challenges and scientific intricacies have practically no bounds. Yet, man and mankind should take positive steps in the proper realization of environmental remediation science. Technological challenges and the scientific ardor will thus open up a new dawn of environmental remediation in years to come if adequate and proper steps are taken toward the true emancipation of green or environmental sustainability.

9.11 APPLICATION OF NANOTECHNOLOGY AND NANOMATERIALS IN ENVIRONMENTAL PROTECTION

Human civilization's scientific stance and scientific ingenuity as regards environmental protection are in a state of immense disaster. Nanotechnology is a wonder of science today. Science and technology are huge colossus with a definite vision of its own. Today human society and human civilization are technology-driven society. Application of nanomaterials and engineered nanomaterials are the needs of human society. Environmental protection science and water purification science are in a state of immense disaster and veritably deep scientific introspection. A deep scientific need today is a detailed investigation of the application of nanotechnology in environmental remediation, water purification science, and water disinfection. The authors discussed in minute details the success of nanotechnology in the proper emancipation of environmental remediation. This treatise deals in detail the success of nanomaterials in proper water and wastewater treatment. The author with vast foresight details some of the recent review papers in the field of nanotechnology and environmental protection. A new dawn in human civilization will emerge if nanotechnology reaches every nook and corner of human society.

Singh et al.[12] discussed and deliberated with scientific and engineering insight wonders of nanotechnology for remediation of polluted aquatic environments. Today science, technology, and human civilization are moving rapidly. On earth, all forms of life wholly or solely depend on the clean water for definite survival. The freshwater ecosystems are the home of a large number of organisms from microscopic and macroscopic species.[12] Thus challenges and the vision of science and engineering have become unmeasured. The problem of water pollution is worsening day by day which ultimately affects the limited freshwater resources. The scientific vision and the engineering acuity needs to be envisioned as mankind treads forward.[12] The literature cited in this chapter immensely suggests that nanotechnology could be a viable and visionary alternative. It is not selective to cleanup only organic-based pollutants but efficient to remediate heavy metals and pesticides in water and industrial wastewater.[12] Water is one of the precious gifts of human civilization today. The majority of the water (97%) on the surface of the ground is saltwater while only 3% is available as freshwater. Two-third of the 3% is in frozen form and only 1% is in the assessable form. Thus, this is an enigmatic quandary to human civilization and human scientific progress. The uncontrolled rise in population is adding more problems and more issues on the water bodies as per one visionary estimate that the population of the world will reach up to 9 billion by 2050 and this rising population will increase the water and wastewater pollution by increasing the amount of waste accumulation in water bodies.[12] The authors discussed with scientific acuity and vast contemplation adsorption, nanoadsorbents based on carbon substances, heavy metal removal, nanoadsorbents based on metals, potential application in water treatment, the vast world of nanofiltration, nanofiber membranes, nanocomposite membranes, thin-film nanocomposite membranes, photocatalysis, trace contaminant detection, and nanomaterials for adsorption of pollutants.[12] Technological and scientific validation in the field of nanotechnology are poised for newer challenges and this treatise deeply elucidates the immense success of nanotechnology applications in environmental remediation. A new era in the field of environmental engineering will emerge if there is strong emancipation in environmental protection.[12]

Gopakumar et al.[13] discussed and described with vision and scientific determination nanomaterials, the state of the art, new challenges, and opportunities. Nanotechnology is today an interdisciplinary branch of science and engineering.[13] It integrates the engineering aspects of biology, physics, and chemistry.[13] Nanotechnology today has lots of drawbacks and needs to be envisioned as science and civilization treads forward. There is a strong need for developing new approaches in manufacturing which needs to be developing to reach visionary milestones. Nanomaterials applications in water purification science are wonders of science and technology today. Similarly, nanotechnology applications in diverse areas of environmental protection are overcoming immense scientific hurdles and crossing vast scientific frontiers. Today the human civilization is a technology-driven society. The authors reiterate the immense need for nanotechnology in environmental protection, water, and wastewater treatment. The area of nanoscience is of immense debate due to the health effects of the application of nanomaterials.[13] The toxic and health hazards of nanoparticles and other nanoobjects are an evergrowing scientific concern today, thus the need for deep scientific ingenuity and deep scientific and engineering acuity. The authors in this well-researched treatise discussed nanomaterials for water filtration, the application of fullerenes, carbon nanotubes, graphene, nanocellulose, nanochitin, noble metal nanoparticles, metal-based nanoadsorbents, nanodendrimers, and nanostructured membranes. Nanofiber membranes and the challenges and limitations of nanotechnology in water treatment are the other cornerstones of this chapter.[13] Nanotechnology today is envisioned deeply to enhance the performance of existing technologies and veritably develop newer technologies. The vast challenges in the field of nanotechnology, its vision, and scientific profundity are detailed in this chapter.[13] Nanomaterial properties desirable for water and wastewater applications include a high surface area for adsorption, high activity for photocatalysis, antimicrobial properties, superparamagnetism for particle separation, and other optical and electronic properties that find applications in novel separation processes and the vast array of sensors for water quality monitoring. The needs of science and engineering, the ingenuity of chemical process engineering, and environmental engineering science will today open up new doors of innovation and scientific instinct in years to come.[13]

Nasreen et al.[14] deeply discussed with lucid insight nanomaterials as solutions to water concomitant challenges. Plenty of freshwater resources are still inaccessible to human society. Rapid industrialization and the march of human civilization are veritably threatening the environment and the ecological biodiversity, thus the need for a thorough and sound understanding of environmental protection science. Calamities such as pollution, climate change, and global warming pose immense threats to the freshwater system and public health engineering.[14] Industrialization and urbanization are totally destroying the vast acumen of science and technology. The authors discussed in minute details the role of nanomaterials for effective pollutant removal of heavy metals, dyes, and antibacterial activity. Heavy metal removal from drinking water and wastewater is an enigmatic scientific issue. The field of nanotechnology is a marvel of science and engineering. The vision, the challenges, and the ardor of science and technology in the global scenario are today vast, varied, and ground-breaking. The authors with vision and insight target mainly heavy metal and dye removal from wastewater. This is an absolutely visionary effort of global science and engineering and will surely open vast doors of innovation and scientific intuition in the field of environmental engineering science.

Application of nanomaterials and engineered nanomaterials in environmental remediation is a veritable success of science and technology in the global scientific scenario. The author in this chapter reiterates and focuses on the deep issues of environmental and health impacts of nanomaterials applications. The vast environmental engineering ardor, the profundity of science and the needs of environmental protection will surely one day open newer visions and newer ingenuity in science and technology globally.

9.12 HEAVY METAL AND ARSENIC GROUNDWATER REMEDIATION AND APPLICATION OF COMPOSITE SCIENCE AND NANOTECHNOLOGY

Arsenic and heavy metal groundwater and drinking water contamination are challenging the entire scientific and engineering domain today. Human research pursuit in the field of water purification is aligned with the vast domain of composite science, bionanocomposites, and biological sciences. Science and engineering today have few answers to the world of challenges in the field of arsenic and heavy metal groundwater remediation. In Bangladesh, India, and many developed and developing nations around the world, scientists and engineers are puzzled due to the enormity of this arsenic drinking water contamination crisis. The scientific enigma is beyond any scientist's and engineer's comprehension. Application of bionanocomposites in environmental protection is not a new field but needs to be re-envisioned with the progress of science and technology. In a similar vein, the application of composite science and nanotechnology in environmental protection needs to revamp with research progress globally. The situation of global environment is absolutely grave. Man and mankind today have no answers to the growing concerns for arsenic drinking water contamination in many developing and developed nations around the world. Application of composites and bionanocomposites in water and wastewater treatment is a novel idea and its scientific vision and vast scientific ingenuity needs to be unfolded with scientific progress and civilization's advancement. The vision and success of science and engineering are limited as regards arsenic and heavy metal groundwater contamination. Composite science, polymer science, and material science are the needs of environmental remediation in today's scientific world.

Hashim et al.[10] discussed and deliberated with lucid insight and scientific far-sightedness remediation technologies for heavy metal contaminated groundwater. The contamination of groundwater by heavy metal is a matter of immense concern for humanity and public health engineering.[10] Civil engineering, chemical process engineering, environmental engineering, and geological sciences are the areas of research pursuit for the mitigation of global water issues and for groundwater heavy metal and arsenic contamination. Today environmental engineering is a vast colossus with a definite vision of its own. In this chapter, the authors elucidate in minute details 35 approaches for groundwater treatment have been reviewed and classified under three categories which are: (1)chemical, (2) biochemical, and (3) physico-chemical treatment techniques. A new scientific and engineering order in the field of environmental protection is emerging today. In this chapter, comparison tables have been provided at the end of each process for a much better scientific understanding of each category.[10] Selection of a suitable technology for groundwater remediation is a challenging task due to extremely complex soil chemistry and aquifer characteristics. Groundwater heavy metal contamination is a burning issue in the global scenario. Keeping the sustainability issues in mind, the technologies encompassing natural chemistry, bioremediation, and biosorption are highly recommended to be adopted in appropriate cases.[10] Human civilization's immense scientific prowess and stance and the world of scientific challenges will all be the torchbearers toward a newer visionary era in environmental remediation and environmental engineering. These scientific and engineering stances in the field of environmental protection are deeply elucidated in this paper. Heavy metal is a general term, which applies to the group of metals and metalloids with an atomic density greater than 4000 kg m^{-3} or five times more than water. Although some of them are nutrients for living organisms, at higher concentrations they are highly poisonous. The authors of this chapter discussed deeply and with lucid insight heavy metals in groundwater and their sources, chemical property and speciation, technologies for the treatment of heavy metal contaminated groundwater, biological treatments, and physic-chemical treatments of wastewater.[10]

Hassan[11] discussed with vision and insight arsenic in groundwater and the poisoning and risk assessment. The status of global research and development initiatives in arsenic groundwater remediation is highly grave and immediate action needs to be taken. The author in this chapter discussed and deliberated in details arsenic poisoning through ages, the global scenario of groundwater arsenic catastrophe, groundwater arsenic discontinuity, spatial mapping, spatial planning and public participation, arsenic-induced health and social hazard, and the vast domain of arsenic poisoning in Bangladesh and the legal issues of responsibility.[11] Bangladesh today is in a state of immense disaster due to the arsenic drinking water contamination. The author in this book deeply elucidates this utmost concern.

Bangladesh and the state of West Bengal, India are today immensely crisis-prone. Science and engineering have few answers to arsenic and heavy metal drinking water and groundwater contamination. This is the main vision of this chapter along with a deep investigation on biofuels and bionanocomposites.

9.13 APPLICATION OF BIOFUELS AND BIONANOCOMPOSITES IN ENVIRONMENTAL PROTECTION

Application of biofuels and bionanocomposites in environmental protection are far-reaching and needs to be reframed as mankind treads forward. The success of scientific imagination and sagacity in the field of environmental protection also needs to be re-envisioned as technology and engineering move forward surpassing one visionary frontier over another. Today biological sciences and the domain of biotechnology are in the path of new scientific rejuvenation. Biofuels applications and the domain of renewable energy are connected to each other. Bionanocomposites are the next generation smart materials. Adsorption of recalcitrant pollutants in wastewater on bionanocomposites is an immensely innovative research pursuit. Here the application of material science and composite science comes into play. Today, the present-day human civilization is a technology-driven society. Renewable energy science is today ushering in a new era in the field of biofuels and biomass energy. In a similar vein, bionanocomposites are the hallmarks toward a newer scientific perspective globally today.

9.14 NANOMATERIALS IN ENVIRONMENTAL REMEDIATION

Nanomaterials and engineered nanomaterials are the next-generation wonders of science and engineering today. Nanotechnology and nanoengineering are in a similar vein in the avenues of scientific profundity and vast scientific ingenuity. Thus the applications of nanomaterials in environmental protection are the new scientific order globally. Drinking water and industrial wastewater treatments are the immediate needs of civilization, science, and technology today. Today is the world of technological challenges in the field of nano-materials and engineered nanomaterials. Graphene and fullerenes are revolutionizing the world of science and engineering. The status of environmental remediation globally is absolutely grave and needs immediate attention. Here comes the importance of biological sciences, nanocomposites, and nanotechnology. In this chapter, the author deeply stresses the scientific success, the deep scientific progeny, and the world of scientific ingenuity in the field of bionanocomposites applications in environmental protection. Drinking water issues are today in the state of immense catastrophe. In developing and developed nations around the world, arsenic and heavy metal groundwater contamination are challenging the entire scientific firmament.[15-20]

9.15 THE STATUS OF RESEARCH IN NANOTECHNOLOGY AND BIOFUELS TODAY

The status of research in nanotechnology and biofuels are drastically changing the scientific fabric globally today. Biofuels and bioenergy are the utmost needs of the hour as renewable energy scenario assumes immense importance. Today nanotechnology is integrated with diverse areas of science and engineering which includes environmental protection and environmental sustainability. The science of environmental and energy sustainability needs to be envisioned with scientific progress.[21-24] There should be farsightedness of science and engineering in the domains of environmental protection and nanotechnology. Research necessities are the mother of invention in present-day human civilization. The status of research in biofuels and alternate energy resources are today targeted toward the feasibilities of the renewable energy tools which include bioenergy and biomass energy. Integrated water resource management, wastewater treatment management, and water purification science need to be thoroughly revamped as the world stands in the midst of unending environmental catastrophes. Arsenic and heavy metal groundwater contamination in many developed and developing nations around the globe are of immense concern to the scientists and engineers. This treatise with deep scientific truth unfolds the research needs, the evergrowing concerns, and the global solutions to the water treatment and industrial wastewater treatment scenario. Civilization's scientific girth and perseverance will surely unfold as regards environmental protection in the distant future.[25-30]

9.16   FUTURE RESEARCH NEEDS AND THE FUTURISTIC FLOW OF THOUGHTS

Future research needs of science and engineering in present-day human civilization are the areas of environmental engineering, nanotechnology, water science and technology, and biological sciences. Biological sciences and the vast area of biotechnology are in the process of a new beginning, a new vision, and a new scientific divination. The futuristic flow of scientific thoughts needs to be readdressed and re-envisioned as science, civilization, and mankind moves forward at a rapid pace. The status of research and development initiatives in environmental engineering, chemical process engineering, nanotechnology, and biological sciences are in the avenues of ingenuity and profundity. The futuristic flow of scientific thoughts should target the scientific intricacies and the scientific hurdles of the laboratory to shop-floor applications. The world of environmental remediation and biofuels are highly challenged. Similarly, the domain of green or environmental sustainability. This chapter widens the scientific understanding and the areas of scientific discernment in the domains of biofuels and bioenergy. The futuristic vision of biofuels and bioenergy should definitely target the applications domain. The future research and initiatives need to be revamped and realigned as growing concerns for climate change and loss of ecological biodiversity mounts in a vicious manner. The future also belongs to material science and composite science. This treatise will surely reach the environmental engineers and nanotechnologists and open new doors of scientific innovation, ardor, and instinct in years to come.

9.17   FUTURE RECOMMENDATIONS OF THIS STUDY

Future recommendations of this detailed investigation need to be re-envisioned and restructured as environmental protection science, water purification science, and industrial wastewater treatment ushers in a new era. Biofuels and renewable energy scenario globally are immensely bright and inspiring. Biomass energy is in the path of new scientific divination and vast and varied scientific ingenuity. Future of the field of renewable energy is slowly surpassing visionary scientific frontiers. The areas of research endeavor in the field of nanocomposites should be targeted toward application areas in environmental engineering and chemical process engineering. Chemical process engineering, environmental engineering, and nanotechnology are today connected with each other. The author with immense academic and scientific rigor elucidates on the scientific success, the provenance of environmental protection, and the vast scientific doctrine and ingenuity in the application areas of green sustainability. The future of science and engineering of biofuels and bionanocomposites and its applications are immensely bright and far-reaching. This treatise targets the scientific needs of humanity such as environmental protection and biological sciences. Without vision and without ingenuity, science and technology cannot move forward. This chapter widens these definite visions of environmental remediation and water purification science with the interfaces and application areas of biofuels and bionanocomposites. It is absolutely sure that the world of challenges in environmental remediation will witness a new dawn and the scientific issues of climate change, global warming, and depletion of fossil fuel resources will be thoroughly mitigated and resolved. A new beginning, a newer scientific doctrine, and sagacity will emerge in the scientific horizon as civilization moves forward.

9.18 CONCLUSION AND FUTURE SCIENTIFIC PERSPECTIVES

The world of biological sciences, biofuels, and environmental protection are in the middle of vision, scientific ingenuity, and immense scientific farsightedness. Today the domain of nanotechnology, material science, and composites are gearing forward toward new challenges. The domain of biofuels and bionanocomposites are in the process of new scientific rejuvenation. The future environmental and scientific perspectives need to be redrawn as civil society and the scientific domain moves forward. Technological and engineering challenges are immense, vital, and visionary as environmental remediation science treads forward. Green and environmental sustainability in a similar vein are surpassing scientific frontiers. The future scientific perspectives in the field of environmental engineering and green sustainability need to be re-envisioned if the world of science and engineering are to confront global climate change and loss of ecological biodiversity. Industrial systems engineering and technology management needs to be linked with integrated water resource management and wastewater management with the progress of science, mankind, and scientific forbearance. In this chapter, the author veritably pronounces the scientific intricacies, the scientific trustworthiness, and the vast world of scientific ingenuity in the field of bionanocomposites and biofuels. Renewable energy engineering and technology are the coin words of scientific research pursuit in modern civilization. The issues of application of biofuels and bionanocomposites in present-day human civilization are the cornerstones of this research endeavor. The future recommendations and the futuristic vision of environmental protection and environmental engineering need to be connected to the vast and versatile areas of nanoscience and nanotechnology. Thus new dawn of science and engineering will evolve and mankind's immense scientific prowess will be realized with the progress of environmental engineering science and green sustainability. Technological advancements and the world of scientific validation in environmental engineering and chemical process engineering thus will witness a new beginning as civilization moves forward.

KEYWORDS

- water
- nanotechnology
- nanomaterials
- vision
- arsenic
- heavy metal

REFERENCES

1. Pande, V. V.; Sanklecha, V. M. Bionanocomposite: A Review. *Austin J. Nanomed. Nanotechnol.* **2017,** *5* (1), 1–3.
2. Darwish, M.; Mohammadi, A. Functionalized Nanomaterial for Environmental Techniques, Chapter-10, *Nanotechnology in Environmental Science*; Hussain, C. M., Mishra, A. K., Eds.; Wiley-VCH Verlag GmbH &Co.: Weinham, Germany, 2018; pp 315–349.
3. *National Nanotechnology Initiative Workshop Report, USA*, Nanotechnology and the Environment, 2003, Arlington, Virginia, USA, May 8-9 2003.
4. Shchipunov, Y. Bionanocomposites: Green Sustainable Materials for the Future. *Pure Appl. Chem.* **2012,** *84* (12), 2579–2607.
5. Rasool, U.; Hemalatha, S. A Review on Bioenergy and Biofuels: Sources and Their Production. *Braz. J. Biol. Sci.* **2016,** *3* (5), 3–21.
6. Godbole, E. P.; Dabhadkar, K. C. Review on Production of Biofuels. *IOSR J. Biotechnol. Biochem.* **2016,** *2* (6), 62–69.
7. Gashaw, A.; Getachew, T.; Teshita, A. A Review on Biodiesel Production as an Alternative Fuel. *J. For. Prod. Ind.* **2015,** *4* (2), 80–85.
8. *Food and Agricultural Organization Report, Rome, Italy*, Algae-Based Biofuels: A Review of Challenges and Opportunities for Developing Countries, May 2009.

9. Wang, J.; Yang, Y.; Bentley, Y.; Geng, X.; Liu, X. Sustainability Assessment of Bioenergy from a Global Perspective: A Review, *Sustainability* **2018**, *10* (2739), 1–19.

10. Hashim, M. A.; Mukhopdhyay, S.; Sahu, J. N.; Sengupta, B. Remediation Technologies for Heavy Metal Contaminated Groundwater. *J. Environ. Manage.* **2011**, *92*, 2355–2388.

11. Hassan, M. M. *Arsenic in Groundwater: Poisoning and Risk Assessment*; CRC Press, Taylor and Francis Group: Boca Raton, Florida, USA, 2018.

12. Singh, V. D.; Bhat, R. A.; Dervash, M. A.; Qadri, H.; Mehmood, M. A.; Dar, G. H.; Hameed, M.; Rashid, R. Wonders of Nanotechnology for Remediation of Polluted Aquatic Environs. *Fresh Water Pollution Dynamics and Remediation*; Qadri, H. et al., Eds; Springer Nature Singapore Pte Ltd.: Singapore, 2020.

13. Gopakumar, D. A.; Pai, A. R.; Pasquini, D.; Leu, S.-Y.; Abdul Khalil, H. P. S.; Thomas, S. Nanomaterials- State of Art, New Challenges and Opportunities, Chapter-1. *Nanoscale Materials in Water Purification*, 2019, https://doi.org/10.1016/B978-0-12-813926-4.00001-X.

14. Nasreen, S. A. A. N.; Sundarrajan, S.; Nizar, S. A. S.; Ramakrishna, S. Nanomaterials: Solutions to Water-Concomitant Challenges. *Membranes* **2019**, *9* (40), 1–21.

15. Palit, S.; Hussain, C. M. Nanocomposites in Packaging: A Groundbreaking Review and A Vision for the Future. *Bio-Based Materials for Food Packaging*; Ahmed, S, Ed.; Springer Nature Singapore. Pte Ltd., 2018; pp 287–303.

16. Palit, S. Advanced Environmental Engineering Separation Processes, Environmental Analysis and Application of Nanotechnology—A Far-Reaching Review, Chapter-14. *Advanced Environmental Analysis—Application of Nanomaterials, Volume-1*; Hussain, C. M., Kharisov, B., Eds.; The Royal Society of Chemistry: Cambridge, United Kingdom, 2017; pp 377–416.

17. Hussain, C. M.; Kharisov, B. *Advanced Environmental Analysis—Application of Nanomaterials, Volume-1*; The Royal Society of Chemistry: Cambridge, United Kingdom, 2017.

18. Hussain, C. M. Magnetic Nanomaterials for Environmental Analysis, Chapter-19. *Advanced Environmental Analysis—Application of Nanomaterials, Volume-1*; Hussain, C. M., Kharisov, B. The Royal Society of Chemistry: Cambridge, United Kingdom, 2017, pp 3–13.

19. Hussain, C. M. *Handbook of Nanomaterials for Industrial Applications*; Elsevier: Amsterdam, Netherlands, 2018.

20. Palit, S.; Hussain, C. M. Environmental Management and Sustainable Development: A Vision for the Future. *Handbook of Environmental Materials Management*; Hussain, C. M, Ed.; Springer Nature Switzerland A.G., 2018; pp 1–17.

21. Palit, S.; Hussain, C. M. Nanomembranes for Environment. *Handbook of Environmental Materials Management*; Hussain, C. M., Ed.; Springer Nature Switzerland A.G., 2018; pp 1–24.

22. Palit, S.; Hussain, C. M. Remediation of Industrial and Automobile Exhausts for Environmental Management. *Handbook of Environmental Materials Management*; Hussain, C. M., Ed.; Springer Nature Switzerland A.G., 2018; pp 1–17.

23. Palit, S.; Hussain, C. M. Sustainable Biomedical Waste Management. *Handbook of Environmental Materials Management*; Hussain, C. M., Ed.; Springer Nature Switzerland A.G., 2018; pp 1–23.

24. Palit, S. Industrial vs Food Enzymes: Application and Future Prospects. *Enzymes in Food Technology: Improvements and Innovations*; Kuddus, M., Ed.; Springer Nature Singapore Pte. Ltd.: Singapore, 2018; pp-319–345

25. Palit, S.; Hussain, C. M. Green Sustainability, Nanotechnology and Advanced Materials—A Critical Overview and A Vision for the Future, Chapter-1. *Green and Sustainable Advanced Materials, Volume-2, Applications*; Ahmed, S., Hussain, C. M., Eds.; Wiley Scrivener Publishing: Beverly, Massachusetts,USA, 2018; pp 1–18.

26. Palit, S. Recent Advances in Corrosion Science: A Critical Overview and A Deep Comprehension. *Direct Synthesis of Metal Complexes*; Kharisov, B.I., Ed.; Elsevier: Amsterdam, Netherlands, 2018; pp 379–410.

27. Palit, S. Nanomaterials for Industrial Wastewater Treatment and Water Purification. *Handbook of Ecomaterials*; Springer International Publishing, AG: Switzerland, 2017; pp 1–41.

28. Hussain, C. M.; Mishra, A. K. *New Polymer Nanocomposites for Environmental Remediation*; Elsevier: Amsterdam, Netherlands, 2018.

29. Hussain, C. M.; Mishra, A. K. *Nanotechnology in Environmental Science*; John Wiley and Sons: New York, USA, 2018.

30. Hussain, C. M.; Mishra, A. K. *Nanocomposites for Pollution Control*; CRC Press: Boca Raton, Florida, USA, 2018.

31. www.wikipedia.com (accessed Nov 25, 2019).

32. www.google.com (accessed Nov 25, 2019).

CHAPTER 10

Perspectives for Electronic Nose Technology in Green Analytical Chemistry

T. SONAMANI SINGH, PRIYANKA SINGH, and R. D. S. YADAVA*

Department of Physics, Sensors & Signal Processing Laboratory, Institute of Science, Banaras Hindu University, Varanasi 221005, Uttar Pradesh, India

Corresponding author. E-mail: ardius@bhu.ac.in; ardius@gmail.com

ABSTRACT

Electronic nose (eNose) is an analytical methodology for detection and quantification of chemical analytes in vapor phase analogous to mammalian nose. The analogues of olfactory receptor neurons in nasal epithelium, neural network in olfactory bulb and olfactory cortex in mammalian smell sensing are replaced respectively by an array of chemical vapor sensors, signal processing unit and pattern recognition system in an eNose. The eNose technology is extensively developed for detection and identification of a large variety of volatile organic compounds (VOCs) in many application domains such as detection of explosives and narcotics for security, monitoring of hazardous industrial emissions, detection of environmental pollutants, monitoring of food and beverage quality, detection of food freshness/spoilage and health monitoring and disease diagnostics by analysis of body odor. The principles of green chemistry dictate achieving sustainable development through chemistry related activities by minimizing harmful influences on human health and environment. This can be achieved by selecting less hazardous chemicals for input ingredients, optimization of process technology for maximum yield with minimum hazard, minimization of waste generation, automated monitoring and process control. The traditional analytical methods for chemical analysis are largely based on chromatography, mass spectrometry and infrared spectroscopy principles. These methods invariably depend on sample collection, chemical extraction and enrichment methods.

In addition, these methods are not real time, hence are not appropriate for in-line process monitoring and automated process control applications. The eNoses based on chemical sensors have proven capabilities for real-time detection of volatile organics with detection limits ppb to ppt range. In this chapter we shall briefly describe basics of an eNose system, review status of eNose technology, and present perspectives for their applications in green chemistry activities.

10.1   INTRODUCTION

The principles of green chemistry provide philosophical framework for chemistry-related activities for achieving sustainable development.[1] The design, development, and production of chemical products invariably involve generation of harmful chemical reaction products and wastes that get released in environment either directly or through various disposal pathways. The safety of human health in direct contact and of all life forms, and environment in the long run is the primary concern that these principles of green chemistry address. The efforts in green chemistry and green chemical engineering are to optimize all the steps involved in the manufacturing of chemical products starting from the selection of raw materials, reaction pathways, solvents, and reagents to the consumption, recycling, and disposal in such a way that their harmful impacts are eliminated or minimized. The assessment and monitoring of various chemical activities for their negative implications need fast and accurate analytical methods. Most of the analytical procedures, however, involve sample collection and treatment processes like the extraction from target matrix and the separation and/or enrichment of analytes, which necessitate the use of different harmful chemical solvents and reagents that finally enter into environment and human life via various disposal, transportation, and recycling routes. This has led to significant growth of greening analytical chemistry in recent years.[2,3] The objectives of green analytical chemistry focus on the greenness of analytical methods also in addition to meeting the other technical requirements like the goals of analysis, types of analytes and host matrix, real-time capability, cost-effectiveness, availability, and so on. The greenness refers to the quantified evaluation of harmful impact on environment.[3,4] An appraisal of ongoing efforts and concerns for developing appropriate metrics for quantifying greenness can be found in Ref. [5].

A recent summary of various green analytical methods and their development status is presented in Ref. [3, Table 15.2]. It is noteworthy that most of the presently developed methods utilize gas chromatography (GC) or liquid chromatography principles in some form for the analytes separation, and spectrometry (mass or ion mobility) and spectroscopy (absorption or emission involving photons or electrons) principles for their detection. These methods invariably involve sample pretreatment and analyte extraction or enrichment procedures prior to the analysis; hence, their greenness depends largely on the greenness and efficiency of sample pretreatment methods.[6–8] Besides, most of these methods are offline type requiring significant time for sample collection and analysis. There are two other important green analytical approaches that use principles of electrochemistry[9–13] and immunochemistry.[14–17] Among various analytical methods, these are perhaps the most green for the reasons that these are miniature sensors involving relatively simple sample pretreatment (hence small reagents volume), electrical or optical readout (hence efficient information collection), and small analysis time (hence large throughput). These methods are particularly suited for detection of chemical and biological analytes in liquid media. With advances in modern micro and nanotechnology, the analytical instrumentation based on these methods are being developed in microsystem and lab-on-a-chip (LOC) formats which are direct sample collection and detection systems integrated with electronic/optical readout and data analysis interface.[18–21] The miniaturization and integration technologies impart fast response and portability to these analytical systems. Besides, these are mass producible technologies based on standard microfabrication and lithography techniques, hence have the potential for being available in large numbers at low cost. The miniaturization of traditional chromatography, spectrometry, and spectroscopy-based methods also has been under focus for long, and successful demonstrations have been made.[22,23] However, current advances in micro and nanomaterials, microfluidics, and microelectromechanical system (MEMS) and nanoelectromechanical system (NEMS) sensor technologies could be seen as enablers for future generation intelligent analytical tools having features, such as real-time application, field deployable, and low-cost operation, hence overall helpful to the greening efforts for analytical methods.

In this chapter, we focus on another green analytical methodology that makes use of chemical-sensor-based electronic nose (eNose) technology. The eNose technology is fairly advanced for detection of organic vapor analytes in air for a variety of application fields like explosive detection, food-quality monitoring, industrial emission monitoring, and air pollution monitoring; it can be seen through several recent reviews.[24–31] However, its potentiality as green analytical methodology has not received much attention in green chemistry literature. There are several reasons for that. The most important reason is that the chemical sensors are not highly selective for molecular level recognition of analytes. This limitation is inherent to the chemical interaction between analytes and sensing material (usually a polymer). The fundamental mechanisms by which chemical molecules interact with the sensors are hydrogen bonding and polar and nonpolar dispersive (Van der Waals and London) types simultaneously present in various combinations of their strengths. The stronger interaction mechanisms like covalent and ionic types are not exploited for eNose chemical sensors design because then the sensors become slow responding and nonreversible.[24] The lack of high-molecular specificity is not desirable for applications where the monitoring and control of chemical reactions are involved. The analyte discrimination and identification in an eNose system are achieved by employing an array of sensors where the individual sensors are broadly selective (preferential affinities) toward different chemical analytes in an application scenario and then interpreting the sensor array response by pattern recognition methods. The quantification and calibration of analyte concentration are the other points for worry. The analyte separation in eNoses is achieved in virtual space (e.g., different analytes at separate output nodes in an artificial neural network) and the calibration is done under controlled laboratory conditions by employing computational intelligence techniques (see Refs. [24 (chapter 6), 32]). These are susceptible to variations in background chemicals in real practical conditions. Another drawback of eNose technology is that for each application, a separate set of sensor array and data processing algorithm must be optimized. These apparent limitations of eNose technology however are largely overcome by improving sensing materials design, sensor selection and optimization strategies, and intelligent chemometrics. The eNose chemical sensors with ppm–ppb level detection capabilities for volatile organic compounds (VOCs) are already reported, and some of them available as commercial instruments.[33–38] The recent advances and trends in development of nanomaterials (multifunctional and quantum dots) and nanosensors (carbon nanotubes and nanocantilevers) technologies strongly suggest that these boundaries will be pushed further, and ppb–ppt level detection capabilities may become common in coming years.[39–44]

Further, the eNose technology provides several unique features that are not available with traditional analytical methods. The eNose technology is a fully noninvasive and direct analysis methodology that requires almost no sample pretreatment (a strong plus for greenness). The eNose sensors are fast responding (second to minutes time scale) miniature (micro to nanometer dimension) electronic devices compatible with integrated circuit and microfluidic technologies.[18,19,45-50] These features make the eNose sensors mass producible and smart, and most appropriate for real-time online process monitoring and automated control applications. The trends in miniaturization and artificial intelligence applications provide impetus for development of smart eNoses in microsystem[39,40,51] or in chip[18,19,45,46] formats. These aspects also make eNose sensors suitable for wireless connectivity, hence for emerging remote sensing,[52] robotics,[53,54] and internet-of-things technologies.[55] In short, by managing cross-selectivity, integration with standard microelectronics, microfluidics, and communication technologies, eNoses may prove enablers for future generation green chemistry analytics. In this chapter, we shall first briefly outline basics of eNose systems and associated sensor technologies and then focus on their real-time chemometric aspects in an attempt to present perspectives for applications in green chemistry activities.

10.2 ELECTRONIC NOSE

eNose is a chemical sensing methodology inspired by odor discrimination ability of mammalian nose. The first eNose model was proposed and experimentally demonstrated in 1982 by Persaud and Dodd.[56] Since then, it has emerged into a vast research and development activity with many commercial instruments rolled out and still struggling to meet standards of spectrometry and spectroscopy methods in analytical accuracy,[51,57-59] and to impact commercialization of modern LOC or micro total analytical system technologies for low-cost, real-time, field-deployable, online monitoring applications.[18,19,45-50,60-63] The volume of literature available in the references cited above, and many more, gives an idea of how this field has expanded by interweaving different traditional disciplines (semiconductor, piezoelectric, electrochemistry, optics, solid mechanics, fluid mechanics, electronics, microfabrication, chemistry, and artificial intelligence) to provide alternate and efficient analytical chemistry tools for wide range of application domains (strategic safety and security, environment, industrial hazard and effluents, food quality, disease diagnostics, health, indoor air, water, fragrance, and cosmetics). In this section, we present a brief overview of the existing eNose technology as backdrop to identify and focus on those aspects that need further development for making eNoses appropriate for monitoring and control of harmful reaction byproducts in chemical activities.

10.2.1 OPERATIONAL PARADIGM

Initially, the eNose architecture was visualized to be fully analogous to the biological nose as shown schematically in Figure 10.1. It consists of four basic subparts: air sampler, chemical sensors array, electrical signal conditioning and data acquisition unit, and data processing or pattern recognition unit. The air sampler is analogous to inhalation that brings odorant molecules in contact with the nasal epithelium. The epithelium contains several types of odorant receptors located on olfactory receptor neurons; each receptor type can sense a limited number of odorants. When activated by odorant molecules, the receptor neurons send signals directly to the olfactory bulb where the signals received from each type of receptor neurons converge to separate nodes (called glomeruli). The cumulative responses at nodes fire mitral cells, which transmit signals as response patterns to cortex part in brain for odor recognition.[64] In an eNose system, the olfactory receptor neurons are replaced by selective chemical sensors array where individual sensors need not be highly specific rather respond to limited number of chemical analytes. We can call it chemical selectivity spectrum. The selectivity spectra of the sensors in array may overlap but must be centered on different chemicals in the target sample. The processing in olfactory bulb involves some filtering and reinforcement of received signals. This is done by signal conditioning part in eNose. The convergence of signals to different mitral cells and firing is the data preprocessing, feature extraction, and classification parts in the pattern recognition system. The set of outputs from mitral cells or classifier is uniquely encoded odor identity. The identification or decision activity in the brain is based on the prior experience. The brain responds to stimuli from the olfactory bulb and takes a decision about the identity and strength of the inhaled odor. This is accomplished by training the eNose in laboratory by exposing to various samples of known identity and concentration.[65]

The selection of sensor array and efficiency of data processing is at the heart of eNose operation. The overlap between chemical selectivity spectra of sensor units in the array must be small and their centers well separated. This helps in reducing the complexity and enhancing the accuracy of the pattern recognition system. The functioning of all subsystems in the eNose is implemented in a programmed manner for sniffing and purging the sensors array chamber and activating data acquisition and pattern analysis systems. The time taken from sniffing to purging after the decision out completes one cycle of analysis. The smallest level of analyte concentration that can be detected depends on the sensitivity of sensors and the associated noise. Therefore, low noise and high sensitivity sensing platforms are crucial.

(a) Human Nose

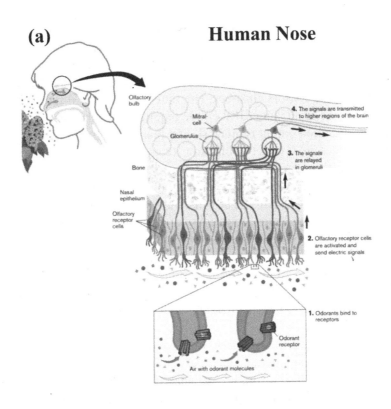

(b) Electronic Nose

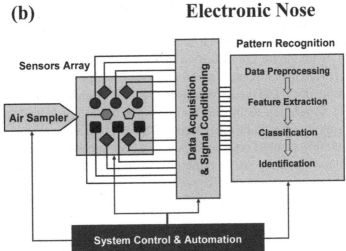

FIGURE 10.1 Schematic representation of human nose and sensors array-based eNose. The source for biological nose picture is a press release. NobelPrize.org. Nobel Media AB 2019. Nov Sat. 23, 2019. https://www.nobelprize.org/prizes/medicine/2004/press-release/.

The cleaning of sensors outputs by filtering or denoising for parasitics and interferences is done partly at the signal conditioning stage and partly by the preprocessor at pattern recognition stage. To improve limit of detection, a chemical filtering and/or preconcentrator unit can be fitted between the sampler and the sensors chamber. The time scale for "real-time" operation depends on the application. If the time required from the air sniffing to the pattern recognition output is much shorter than the characteristic time scales of the changes in a target application (e.g., the reaction times in typical chemical synthesis procedures or the time scale over which environmental changes occur) the eNose can be considered real time. The response times of modern micro/nano-electronic sensors are on milliseconds to seconds scale. The data processing times are usually much shorter. The response times of eNoses are therefore limited, mainly by the air sampling and purging cycle time. These are typically manageable on seconds scale with modern electrical, chemical, electromechanical, and mechanical parts (such as solenoid valves, chemical scrubbers, mixers, flow meters, switches, etc.). Therefore, eNose systems for most applications can be considered to be real-time analytical systems.

In view of some difficulties that come across while optimizing selectivities of eNose sensors, an alternate approach has also researched and developed. In this, certain features of traditional analytical methods are combined with fast and high sensitivity response of sensors. As an example, Electronic Sensor Technology (USA) has developed fast (GC) by combining advantage of chemicals, differentiation feature of chromatography with fast and high sensitivity response of a surface acoustic wave (SAW) sensor.[66] The SAW sensor is temperature controlled to condense VOCs eluted from the GC column. Different compounds are condensed sequentially in accordance with their elution times, and the SAW sensor responds by changes in its resonance frequency in relation to incremental changes in the condensed mass. The sensor array is replaced by GC column. The parts per billion level detection limits with few seconds response time have been reported for several VOCs of environmental concern. In principle, this approach can be adapted with other GC/sensor combinations as well.

10.2.2   SENSORS AND SENSORS ARRAY

There are many excellent books and reviews published on chemical vapor sensors and their applications in eNose designs as cited in the preceding, for example Refs. [19,24,28,35,42,50,57,61,67–70,71,72]. An account of basic principles of their operation can be seen in Refs. [24,61], for earlier chemiresistive, electrochemical, and gravimetric sensors and in Refs. [19,36–38,42,71,72] for later entries based on micro and nanotechnologies. The status on commercial eNoses can be found in many of these reviews, particularly see Refs. [25,31,73].

The sensor response generation in all types of sensors is a two-step process: chemical interaction and signal transduction. The categorization according to chemical interaction can be done in two ways: electrochemical type and chemical affinity interaction type. In the electrochemical category, the chemical analytes interact with the sensing device via some redox process and brings out change in the electrical property (resistance, voltage, or current) for measurement. In the affinity interaction category, the interaction between chemical analytes and the sensing device occurs via some weak chemical forces that produce adsorption or chemisorption effects and bring out changes in the physical properties such as resistance, volume, mass, and/or stress. Usually, this category of sensors involves a second-stage transduction to convert the changes in physical properties into electrical signals (voltage or frequency). Alternately, these sensors are also categorized according to the type of sensing platform (or device). This is the most commonly used categorization in the chemical sensor literature, and is as follows:

- Electrochemical (EC) sensors (redox-based potentiometric and amperometric)
- Metal–oxide semiconductor (MOS) sensors (redox-based chemiresistive)
- Intrinsic conducting polymer (ICP) sensors (chemisorption-based chemiresistive)
- Composite conducting polymer (CCP) sensors (chemisorption-based chemiresistive)
- Metal-oxide-semiconductor field-effect transistor (MOSFET) sensors (adsorption at gate metal produced channel conductance, also referred to as ChemFET)
- Quartz crystal microbalance (QCM) sensors (mass loading produced frequency change)
- SAW sensors (mass loading produced frequency change)
- MEMS/NEMS sensors (surface stress, stiffness, and/or mass loading produced displacement or frequency change)
- Carbon nanotube (CNT) sensors (adsorption produced charge transfer and conductance change)

The details of these sensors are repeatedly described in literature and can be found in abovementioned references. In many works, different features of these sensors have been combined to create new hybrid sensors with objectives for improving sensors' performances. For example, electrostatically driven MEMS cantilever device has been integrated as overhanging gate electrode in MOSFET structure to control surface potential and output current. When sensitized for analyte capture on cantilever surface, this creates new category of ChemFET sensors;[47,50,74,75] CNTs dispersed in polymer matrix have been reported to enhance characteristics of CCP sensors in Ref. [42]; CNTs dispersed on acoustic wave propagation path in SAW device have been shown to enhance VOC sensitivity of SAW sensors by several orders of magnitude[76]; working electrodes in electrochemical sensors have been modified by nanoparticles of metal oxides or CNTs for enhancing their catalytic activity for gas sensing.[77]

The traditional GC is the most commonly used analytical method for chemical analysis in vapor phase. It measures some attributes of the vapor molecules directly—after spatial separation in a column the vapor species are identified and quantified by mass spectrometry, flame photometry, photoionization, or electron capture detectors. In comparison, the chemical sensors measure changes in some physical property of active materials they are constituted of, and upon exposure to vapor samples, they indicate the presence of chemicals indirectly. As examples, a SAW, QCM, or MEMS resonator responds by change in frequency. A vapor sample in practical condition invariably contains several types of molecules, some specific to analysis target, and many more from the surrounding environment called interferents. The nonspecific nature of chemical sensors does not allow unique correlations to be defined between sensors output and vapor identities. This limitation of individual sensors is overcome by employing an array of sensors for generating multiple responses and applying multivariate statistical methods for extraction of unique vapor identities as a set of mathematical descriptors (or vapor signature).[32] The set can be thought of being the responses from a virtual sensor array with minimal overlap of information (or maximally orthogonal).

The set of measured responses of the individual sensors in an array can be thought of as a vector in multidimensional data space where the dimensions are represented by different sensors in the array. The responses of individual sensors are cumulative effect of all subresponses produced by different chemical species in the vapor sample. If the sensor responses are linear, the cumulative effect is simply the sum of all molecular level subresponses. Let there be n-elements sensor array generating response data for m vapor samples, each having an arbitrary number of chemical species. If we label the vapor samples by index k, the sensors by index j and the molecular species in a vapor sample by index i then the response of k-th sensor can be expressed as a vector.

$$\mathbf{R}_k = \sum_j b_{kj}\hat{s}_j = \sum_j \left(\sum_i c_{ik}a_{ij} \right)\hat{s}_j \qquad (10.1)$$

where b_{kj} denotes the magnitude of jth sensor response represented by the unit vector $\hat{s}_j$. The bracketed factor in the second equality is the expression for b_{kj} in terms of molecular level subresponses. Here, c_{ik} denotes the fractional concentration of ith chemical species in kth sample and a_{ij} denotes the sensitivity of jth sensor for ith chemical species. The measurements for m samples can be written in matrix form:

$$\begin{pmatrix} R_1 \\ R_2 \\ ... \\ R_m \end{pmatrix} = \begin{pmatrix} b_{11} & b_{12} & ... & b_{1n} \\ b_{21} & b_{22} & ... & b_{2n} \\ ... & & & \\ b_{m1} & b_{m2} & ... & b_{mn} \end{pmatrix} \begin{pmatrix} s_1 \\ s_2 \\ ... \\ s_n \end{pmatrix} \qquad (10.2)$$

10.2.3 SENSORS ARRAY CHEMOMETRICS

An eNose is a supervised learning machine. It needs to be trained *a priori* before being applied for extraction of chemical information about unknown samples. The eNose training means optimizing all data processing procedures and algorithms by adjusting all parameters needed in calculations for efficient cognitive and quantitative prediction ability. To do this, a large data set is generated in the format of Equation 10.2 by using samples of known identity and concentration. In generating training dataset, it is important to visualize actual scenes of application with various possibilities for target chemicals and interferents. The calibrated training samples must be used with diverse compositional makeup to match anticipated analytes and surrounding. The training procedures involve basically three types of algorithms. At the first stage called "preprocessing" the raw data are cleaned for noise and outliers, and rescaled and normalized.[24,32,78-81] The main idea behind these steps is to prepare data for the multivariate analysis at the second stage such that analyte identifying attributes could be maximally discriminated. At the second stage, the preprocessed data are transformed to the virtual sensor space by applying algorithms like principal component analysis (PCA), independent components analysis (ICA), linear discriminant analysis (LDA), and so on.[82-88] The algorithms at this stage extract or build unique vapor descriptors (called "feature extraction") and present that to the following third stage for classification (called "recognition"). In many applications where the interest lies only in identifying the presence of an object, the eNose instrument is complete up to this stage. However, it is important to establish how small the amount of chemicals can be detected. This is determined by the noise level in the system. Usually, the minimum detectable amount of analyte is defined to be that which produces signal at least three times the noise. This is known as "limit of detection."

Initially, the development of eNose systems was driven by this kind of need in many application domains such as explosive detection, food-quality monitoring, environmental pollution monitoring, and so on. However, to fulfill needs of many applications where quantitative determination of the concentrations of specific chemical compounds is required and also to compete with the accuracy of other established analytical methods, the eNose outputs must be adequately calibrated. That is, quantitative relations between sensors array features and input chemicals must be established. Substantial work has been done this aspect. These are all software activities commonly termed as "intelligent chemometrics and pattern recognition." An overview of various chemometric tools, their potentialities, limitations, and applications can be found in some recent reviews.[87–93]

There are basically three main categories of mathematical paradigms employed for quantitative chemometrics. These are artificial neural network (ANN),[94–98] fuzzy inference system (FIS),[99,100] and partial least squared regression (PLSR).[90,101–103] Also, these base methods are combined among themselves with others to build hybrid systems for enhanced accuracy and reliability.[104–108] An ANN is a multilayer neuronal architecture. The individual layers are constituted of a set of neurons (called nodes). The neurons in different layers are interconnected and each interconnection is assigned some weight. The network stores and processes information analogous to human brain. The network intelligence lays in weights, biases, and activation functions assigned to these interconnections and nodes. The system is made to learn these by training with known data. The training process involves determining the number of layers, the number of neurons in each layer, and adjusting the weights and biases to yield most accurate input-to-output relationship in training data. Once a vapor sample is identified, its concentration is determined by a separate network. There are as many second stage networks as the number of target vapors, each calibrated for a separate vapor. The quantification network qualifies the previously identified vapor into a number concentration levels and treats them as classification. An FIS models the input-to-output relationship by a set of fuzzy if-then rules. These rules are constructed on the basis of empirically observed instances of system output for known inputs. The sensor array responses define the input space, and the associated vapor concentrations define the output space. The optimization of inference rule (coefficients of polynomial) is done by using a training data and applying an algorithm for minimization of error between FIS output and target result.

The regression analysis is to model mathematical relationship between sensor array responses as independent and vapor concentrations as dependent variables in such a form that by making observations of the sensor array output one could predict vapor concentrations. The regression analysis is to achieve best fit to the model as per some criteria for the "goodness of fit." There are several methods of multivariate regression analysis: classical-least-square regression, inverse-least-square regression, multiple linear regression, principle components regression, and PLSR.[101] All these are linear regression models. The PLSR is the most successful in diverse conditions of analysis. All these linear regression methods have been used in several studies on sensor array-based quantitative vapor recognition.[101,109–112] In cases of sensor responses being nonlinear, these linear regression methods incur error. Several nonlinear regression models have also been developed.[113–116] The regression methods have also been combined with other nonlinear methods like ANN, FIS, discriminant analysis, and so on for utilizing the best features of regression and classification for efficient quantitative recognition.[117–123] In Ref. [124] simultaneous determination of vapor identity and concentration problem is modeled as multi-input/multi-output (MIMO) function approximation problem, and is solved by decomposing MIMO approximation task into multiple many-to-one tasks where single approximation tasks could be a multivariate logarithmic and/or quadratic regression, a multilayer perceptron or a support vector machine.

10.2.4 SENSORS SELECTION

The quality of feature extraction and accuracy of chemometric calibration depends on the set of sensors used in the array for data generation—more specificity and smaller overlap will make a better set. The number of sensors also needs to be optimized because too many of them generate degenerate or redundant information and make the feature extraction and classification tasks difficult, and too less may miss important information leading to false classification. A lot of work has been done on this aspect.[92,127,134] A usual procedure is to consider a large number of commercially available sensors, use them in combinations of smaller sets to define provisional eNoses, carry out a target analysis task by each and select those sensors from each combination that contribute most in the class discrimination task. The selected subset can be used to make new combinations, and the process can be repeated iteratively until an optimal set is obtained. Several algorithms have been developed for determining the significance of a sensor in a selection exercise. These algorithms are based on computational intelligence methods such as PCA, LDA, genetic algorithm (GA), ANN, support vector machine (SVM), fuzzy clustering, and so on, and their combinations like kernel PCA and ANN, neuro-fuzzy, and so on for data transformation, and then applying some criterion for selection of sensors like highest loads on principal components, separation distance in clustering (hierarchical or fuzzy), fitness in genetic evolution of GA, separation distance from mean in LDA, and so on.[87,124–129,130–135] Several examples of practical implementations in contexts of VOCs discrimination tasks are presented in these references, and extensively reviewed in Refs. [127,135]. This approach of sensor selection involves massive measurement and evaluation work. An exhaustive search for optimal set of sensors is practically unmanageable if the number of prospective sensors is large. As an example, a search for 5 sensors from a list of 25 prospective sensors would require measurement and analysis of 53,130 sensors arrays. The whole task becomes highly time-consuming and expensive, not friendly for developing low cost eNoses.

For making sensor selection process efficient and cost effective, an alternate approach has been suggested recently for polymeric sensors (sensors that depend on vapor-polymer solvation interaction for selective vapor capture such as SAW, CCP, and MEMS sensors)[146] The method suggests that sensors selection can be done prior to their fabrication by short listing a small set of polymers from the large list of commercially available prospective polymers. The information for vapor discrimination is contained in solvation interactions between vapor molecules and polymer matrix. The proposed methodology is based on data mining techniques of computational intelligence applied to vapor–polymer interaction data available in literature. The equilibrium partitioning of vapor species between gaseous and polymer phase is defined by partition coefficient $K = C_p / C_V$ where C_p and C_V denote molar concentration of vapor species in polymer and vapor phases, respectively. It is described by linear solvation energy relationship (LSER)[137,138]:

$$\log_{10} K = c + rR_2 + s\pi_2^* + a\sum \alpha_2^H + b\sum \beta_2^H + l \log L^{16} \qquad (10.3)$$

where each term on the r.h.s. is contribution from one type of vapor–polymer interaction. The set of parameters (c, r, s, a, b, l) and $(R_2, \pi_2^*, \sum \alpha_2^H, \sum \beta_2^H,$

log L^{16},) are complimentary solvation parameters associated with solvent (polymer) and solute (analyte), respectively. The r and R_2 are measures of interaction via dispersion forces, s and π_2^* due to dipole interactions, a hydrogen bond basicity of polymer and $\sum \alpha_2^H$ hydrogen bond acidity of vapor, b hydrogen bond acidity of polymer and $\sum \beta_2^H$ hydrogen bond basicity of vapor, l and log L^{16} combined effects of dispersion interaction and cavity formation in a reference matrix called hexadecane, and c is a regression constant that represents residual effects not covered by the other terms. The set of solvation parameters are characteristic descriptors for the solute and the solvent identities. A particular chemical analyte with its own specific set of solvation parameters will generate different outputs from the set of sensors in array functionalized by different polymers.

The rationale of polymer selection is just reverse of the vapor recognition problem. The polymers are treated as objects for recognition by the vapor analytes. The partition coefficient data is analyzed by statistical multivariate analysis for clustering polymers and selecting one polymer from each cluster. This will dramatically reduce the prospective number of polymers. The validation and fine-tuning can be done by performing real experiments. A prior short listing based on data mining of existing data may however substantially reduce time and cost for development of target-specific eNoses. The validation of this approach has been demonstrated for a number of VOC targets through sensor modal based simulation experiments.[139–143]

10.2.5 HYBRID SENSORS ARRAY AND DATA FUSION

Commonly, eNose systems are designed by using one type of sensor array (homogeneous). The information for chemical recognition lies in the differential loading of different sensors in the array due to variations in redox potential or polymer selectivity for target vapor molecules as mentioned in Section 2.2. This is primarily driven by developed methods for feature extraction which are mostly linear (PCA, SVM, LDA, etc.). The accuracy and robustness of chemical identification can be enhanced by bringing diversity in the information generation process through heterogeneous sensors array and by applying data or information fusion techniques. The heterogeneous or hybrid multisensor platforms can be designed by drawing from the commercially available pool of MOS, CCP, and CNT chemiresistors; QCM, SAW, and MEMS resonators; and also other sensor types like electrochemical and fiber-optic sensors. The selection of hybrid-sensing platform can be expected to reduce information degeneracy because different transduction processes are maximally sensitive to different types of chemical–sensor interactions. For example, SAW sensors respond to total mass loading irrespective of sorption mechanism, whereas CCP chemiresistors respond to volume swelling of polymer matrix due to vapor solvation. An overview of this approach can be obtained from Refs. [144–148]. The data fusion involves combining data from different kinds of sensors in a proper way to achieve detection and identification of analytes with high level of certainty. The data fusion techniques were developed originally for defense applications for automated detection of military targets, early threats, and friend–foe–neutral identification by combining radar, infrared sensors, and surveillance information. However, soon it found application in other areas like machine vision, robotics, weather forecasting, remote sensing, medical imaging, terrain mapping, and so on. A brief review of these can be seen in Ref. [149], and two books[150,151] provide some elaboration. An idea about applications of data/information fusion approach be in chemical sensing domain can be obtained from Refs. [152–156].

A new information fusion approach was reported recently for enhancing recognition efficiency of homogeneous sensors array-based eNoses.[157,158] The method exploits multiplicity of feature extraction methods in combination with radial basis function neural network (RBF). Each feature extractor processes data from a different perspective of data structure, for example, PCA transforms data to seek largest variance directions as principal components, ICA seeks orthogonal directions of maximum non-Gaussianity for statistically independent components, LDA seeks maximum interclass and minimum intraclass scatters, and so on. Individual feature extractors are used with separate RBF networks to generate different sets of class likelihood. The method creates a set of virtual experts in which individual expert members are defined by a different combination of a feature extractor and an RBF neural network classifier. In this work, the outputs from five different linear feature extraction methods: PCA, ICA, singular value decomposition (SVD), LDA, and PLSR, are fed separately as inputs to five different RBF neural networks. The parameters defining each RBF network are optimized separately by training them as independent decision-makers. Since a given feature extractor processes raw data with specific perspective about the data structure, and RBF network generates a set of class likelihood values, the set of virtual experts generate alternate sets of class likelihood values. The latter are finally combined by Bayesian fusion rule[157] or Dempster–Shafer fusion rule[158] for pattern recognition.

10.3 APPLICATIONS

The eNose is generic technology. There is no one universal eNose machine that can analyze all volatile chemical compounds. However, commercially off-the-shelf availability of chemical vapor sensors varieties and data processing software can be used to develop low-cost eNoses customized for specific applications. It will be unfair to expect eNoses to be competitive analytical chemometric machine in all chemical activities of greenness concern. However, eNose methodology itself being green compared to other analytical methods, and being miniature machines with fast response and portability could be a big advantage in many situations of interest. In particular, for inline hazard monitoring at workplaces and for offline continuous monitoring of air pollution, the eNose systems could outperform other traditional methods. For inline monitoring and control of chemical reactions, however, this technology does not appear to be grown enough. In those situations perhaps electrochemical sensors based electronic tongue technology could be more effective. In the following, we take a view at some demonstrated applications in literature.

10.3.1 ENVIRONMENTAL MONITORING

The volatile chemicals in environment are in general a complex mixture of various inorganic and organic compounds. The harmful chemicals in it constitute only a part of whole and originate from varied sources like emissions from oil refineries, industrial manufacturing plants (metallurgical, food processing, pharmaceuticals, etc.), fuel combustion, biodegradation of waste, and so on. The environmental monitoring application may local (at point of sources) and global. The analytical tool employed for monitoring must cater for limit of detection, resolution, response speed, and dynamic range of target species. The detection of inorganic pollutants like nitrogen dioxide, sulfur dioxide, ozone, and carbon monoxide can be most appropriately detected by redox reaction based sensors (EC or MOS type), whereas detection of VOCs is most suited for polymeric sensors (SAW, CCP, MEMS, etc.). However, the condition of sensors operation (harsh outdoor, industrial point source, or indoor air) may preclude application of certain types of sensors despite their good performance. This is probably the reason for most eNose reports being based on MOS sensors. Among one of the earliest reported applications MOS sensor array was used for detection of building rot due to fungal infection (VOCs: 3-octanone, 3-octanol, and 1,3-octen-3-ol), wastewater at input of sewage plant (sulfurous compounds), and different types of fire.[159] For the latter conducting polymer sensors were employed. An array of six QCM sensors for monitoring of 10 hazardous compounds (ammonia, propanone, hexane, acetic acid, toluene, methanol, tetrachloromethane, chloroform, ethanol, and dichloromethane) in an industrial solvents storage room was reported in Ref. [160]. The detection of CO, NO_2, O_3, and SO_2 for monitoring outdoor air quality has been reported in Ref. [161] by using MOS and EC amperometric sensors array. In Ref. [162], soil pollution by petroleum fuel (detection of hydrocarbons) using an MOS eNose. In Refs. [163,164], MOS sensors array-based e-Nose was used for online monitoring headspace of wastewater treatment bioreactor, and successful discrimination of various stages during treatment like deepening of anaerobic condition, restoration of aerobic condition, process disruption and odor nuisance were demonstrated. In Ref. [165], polypyrrole ICP sensor-based eNose was used for odor monitoring in sewage works and odor abatement units. The efficient biodegradation of farm pesticides in soil is very important as they are highly toxic to human and livestock and their prolonged persistence in soil may pollute the food and water chain to hazardous levels. A fast and simple method for detection and monitoring of their concentration in different phases (air, soil, water, fruits, and vegetables) is very important. In Refs. [166,167], an eNose based on conducting polymer sensors array has been used to detect microbial and metal contaminations and pesticides in potable water by analyzing culture headspace air. In Ref. [168], MOS sensor-based eNose has been shown to useful for monitoring benzene pollution in city air. A panoramic view of usefulness of eNose technology in various environment monitoring applications of local and global interests can be found in Refs. [28,29,169].

10.3.2   *FOOD PROCESSING AND PRODUCTS MONITORING*

The application of properly customized eNoses for various applications in food sector such as quality control and selection of raw materials, food processing, products, packaging, storage, shelf life, and freshness or spoilage in place of traditional analytical chemistry methods could be of great help in enhancing the greenness of food industry. Traditionally, the headspace GC has been used.[170–172] The drawback of eNose approach is that for different stages of food monitoring differently optimized eNoses are required because types of characteristic volatiles vary. In the selection of raw materials, the presence of bacterial and fungal infection, pesticide residues, and quality odor indicators for flavor, freshness, and spoilage need to be monitored. The quality markers are specific to the food material; however, the spoilage markers are invariably biogenic amines (BAs). The BAs are basic organic nitrogenous compounds formed by decarboxylation (removal of carboxyl group) of amino acids or proteins in foods by microbial decarboxylase activity. Free BAs in fruits and vegetables at low concentrations determine the taste and aroma, and their presence at relatively high concentrations is associated with spoilage.[173]

BAs are generated in course of microbial, vegetable, and animal metabolisms. Different types of food and beverage contain various BAs that are formed during food processing or storage and can indicate the degree of spoilage by microbial activity.[173–175] By using traditional analytical methods of chemistry, the BAs have been detected and measured in various food products like fish,[176–178] meat,[179–181] sausages,[182] milk,[183] cheese,[184] vegetable products,[181] wine,[186] and beer.[187] BAs are sources of nitrogen and precursors for the synthesis of hormones, alkaloids, nucleic acids, and proteins. Many food-processing operations involve fermentation and release volatile amines in headspace. The measurements of these in headspace can be used to control and automate food-processing operations. Some commonly occurring BAs in food are histamine, putrescine, cadaverine, tyramine, tryptamine, β-phenyl ethylamine, spermine, and spermidine.[171] The analysis and control of BAs is important because of their toxicity, and also because of their usage as indicators of the degree of freshness or spoilage of food. Among these, trimethylamine has been commonly recognized as prominent indicator of biodegradation of food products in storage.[188–192]

The lipid oxidation is another cause of food spoilage. The lipids are commonly present in most food products such as meat, poultry, fish, dairy products, edible oils, and so on. These get oxidized during storage. The lipid oxidation generates various harmful VOCs of aldehyde and ketone families. In a typical food-processing method, some sequence of heat treatment and addition of flavor and preservatives are involved.[193] The incomplete combustion and pyrolysis of organic matter during food processes like grilling, roasting, and smoking generates large amount of harmful polycyclic aromatic hydrocarbons causing pollution in air and in food matrices.[194] The variations in storage condition occur due to refrigeration, exposure to sun, and environmental pollutants (allergens and microorganisms).[196,197] The dry and packaged food requires airtight package to maintain its quality and ensures food safety. The leaks in package or improper manufacturing increase the risk of moisture entering into food package leading to softening of food, reducing taste, and growth of harmful microorganisms. The shelf life of food products is determined by natural causes inherent to the harvested or slaughtered produce, microbial spoilage, and chemical deterioration. The food packaging materials like paper and polymer contain raw material that may release volatile compounds into food matrix and deteriorate quality.[198] The food packaging materials and processes also need to be monitored for volatile emissions.[199,200]

Several applications of eNoses in agriculture and food sector have been reported. The latest status of this field can be found in some recent review papers,[201–206] and the timeline of these developments can be traced through some earlier reviews in Refs. [207–210]. For some specific examples, one can see Refs. [141,202,208] for milk and fish,[213] for meat,[210–212] for fruits and vegetables like apple, tomatoes, pineapples, broccoli,[214–216] for beverages like beer, tea, coffee,[217] for cereals, and for flavor[206] and spoilage of other agricultural produce. The specifications for these instruments however may vary according to the purpose and point of use. For example, for monitoring of bacterial load and their proliferation in food materials during storage and transportation high sensitivity and low limit of detection are needed even if the analysis time runs into few hours. However, if a food supplier or a consumer needs to ascertain the freshness or suitability of a food product for safe ingestion the analysis time must be not longer than at the most a few minutes.

10.3.3   MEDICAL DIAGNOSTICS AND HEALTH MONITORING

The use of analytical chemistry for bioanalysis is among one of the major areas of concern for green chemistry. The analytical tools are used for all steps of pharmaceutical, clinical diagnostics, and healthcare activities. An idea about the concern for greenness in this domain can be obtained from two recent editorials of the special issues of *Bioanalysis*[218] and *J. Clin. Bioanal. Chem.*,[219] and articles therein. The chemicals used in drug development and manufacturing, disease diagnosis, physiological monitoring generate health hazard in the workplace through direct emissions in air and the environmental hazard through waste disposal. The selection of materials and process chemistry should be done with green chemistry goals. The reduced size of samples for analyte extraction and simpler sampling procedure is of great help in reducing hazardous load. The choice of analytical method that requires minimum sample preparation is another major factor that must be optimized with objective for minimizing chemical hazard. The pharmaceutical manufacturing involves variety of VOCs and flavoring additives that are emitted in environment and also appear as residues in products.[220] The major thrust in this domain has been on developing efficient sampling procedures and enhancing greenness of the traditional GC and liquid chromatography methods.[221] The variety of chemicals of both organic and inorganic types are used in clinical and research laboratories for various test protocols and functions; for example, oxidizing salts (nitrates, perchlorates, etc.), acids (nitric, perchloric, hydrogen peroxide, etc.), solvents (xylene, ethanol, toluene, methanol, benzene, dimethylformamide, acetonitrile, hexane, pyridine, etc.), cyanides (sodium cyanide, potassium cyanide, calcium cyanide), sulfides (lead sulfide, iron sulfide, etc.), disinfectants (alcohols, aldehydes, phenols, etc.). Their numbers run into hundreds and their hazardous characters include carcinogens, toxins, irritants, corrosives, sensitizers, hepatotoxins, nephrotoxins, neurotoxins, etc.).[222,223]

The application of customized eNoses in bioanalytical laboratories can be done for three types of purposes: safety at workplace, support for standard method, and alternate to established standard procedures. The task of monitoring chemical hazards emitted from bioanalytical procedures is similar to environment monitoring but it must be optimized for specific chemicals under safety scanner (e.g., toxic reagents involved in particular tests). In support, eNoses can screen samples quickly and shortlist samples under suspicion for detailed analysis by standard procedures. This might be very helpful in situations where a large number of samples are to be analyzed as in the case of outbreak of some infectious disease or for the screening of population for certain diseases in remote locations. To work successfully as an alternate method of analysis the specificity, accuracy, and threshold limits of eNoses must meet the required standard with high reliability. For example, clinical diagnosis of a disease for medical remediation must employ an eNose of specificity and limit of detection for target biomarkers (either directly or through headspace of cell culture).

There are several recent experimental reports exploring appropriateness of eNoses for detection of diseases like lung cancer,[224,225] cardiometabolic diseases,[227] renal failure,[228] gastrointestinal infection,[229] pneumonia,[226] and wound infection.[230] The VOCs for detection in these diseases are mostly compounds of alkane and alcohol families like ethane, pentane, isoprene, acetone, ethanol, acetaldehyde, and so on. These compounds are produced in the normal humans also via various metabolic processes inside body. However, in diseased conditions their concentration levels undergo dramatic increase. There are also some disease specific VOCs emissions like chlorinated hydrocarbons in lung cancer, sulfur compounds in liver disorder, nitrogenous compounds in renal disease,[231] dimethyl sulfide in cirrohsis.[232] The status of eNose applications for disease diagnosis by body odor (particularly breath) can be found in some recent reviews.[73,233,234,245]

10.4   POTENTIALITIES AND CONCLUDING REMARKS

The basic nature of eNose is to detect and identify odor. The odor needs not be a single compound. In general, it is a mixture of several compounds. The odor signature of a particular chemical process or an object targeted for detection depends on the composition of intrinsic constituent compounds as well as on the chemically active other nonintrinsic compounds in background (referred to as interferents), that is, an odor signature alters with compositional changes in interferents. The implication of this is that while optimizing an eNose for some specific application in laboratory training phase, it is important to envisage and account for composition of interferents in actual testing site. The training samples must have compositional diversity to mimic real situations. The apparent simplicity and greenness of eNose approach thus comes at the cost of some uncertainty and reliability. This does not make eNose systems totally worthless. Their unique features of high sensitivity, low limit of detection, real-time response, miniature size, and portability can provide most prudent solutions for many problems where only the target detection and identification is important and not the quantitative accuracy of chemical composition; for example, threat to life or health by leakage or excess emission of hazardous chemicals in surrounding at workplace or in an industrial manufacturing plant needs real-time warning system with capability to respond above safety threshold, monitoring of processing stage and flavor in automated food-processing units, screening of raw materials for adulteration food-processing or chemical product manufacturing, screening of food products for freshness and spoilage, and so on.

The potentialities of eNoses for quantitative detection of specific chemical compounds of environmental and pathological concerns have been demonstrated in numerous studies as mentioned in the preceding section. Most of these studies are in research laboratory conditions. The conversion of this knowledge into commercial instrumentation for actual site applications is an ongoing process. The emerging nanosensor technology and LOC capabilities integrated with advanced computational paradigms are continuously upgrading the performance status of these systems and setting unprecedented standards. Although commercialization and acceptance of eNoses for high-accuracy and reliability applications like disease diagnostics and process automation appear lacking today, the trails of over three decades of development as summarized in several recent review articles mentioned above is reassuring of its positive future.

It must be acknowledged that no one technology can fulfill all analytical needs and be adequately green. The biggest advantages of eNose technology in regard to greenness are the direct sampling or minimal presampling procedures, small volume samples, and online real-time analysis. These eliminate many intermediary processes and use of harmful chemicals. The selective application of established analytical methods like chromatography and spectroscopy after screening and shortlisting by eNoses can be of great value in achieving green chemistry objectives.

KEYWORDS

- electronic nose
- green analytical chemistry
- micro/nano cantilever sensors
- environmental monitoring
- food processing
- medical diagnostics

REFERENCES

1. Anastas, P. T.; Warner, J. C. *Green Chemistry: Theory and Practice*; Oxford University Press: Oxford, 1998.
2. Armenta, S.; Garrigues, S.; de la Guardia, M.; Turrillas, F. A. E. Green Analytical Chemistry. *Trends Anal. Chem.* **2008**, *27* (6), 497–511.
3. Wasylka, J. P.; Namieśnik, J., Eds. *Green Analytical Chemistry: Past, Present and Perspectives*; Springer Nature: Singapore, 2019, Chs. 1, 12, and 15.
4. Armenta, S.; Garrigues, S.; de la Guardia, M.; Turrillas, F. A. E. Green Analytical Chemistry. In *Encyclopedia of Analytical Science*, 3rd ed.; Worsfold, P., Poole, C., Townshend, A., Miró, M., Eds.; Elsevier: Amsterdam, Netherlands, 2019; pp 356–361.
5. Lapkin, A.; Constable, D. J. C., Eds. *Green Chemistry Metrics: Measuring and Monitoring Sustainable Processes*; Wiley: Singapore, 2009.
6. Keith, L. H.; Gron, L. U.; Young, J. L. Green Analytical Methodologies. *Chem. Rev.* **2007**, *107*, 2695–2708.
7. Tobiszewski, M.; Mechlińska, A.; Namieśnik, J. Green Analytical Chemistry—Theory and Practice. *Chem. Soc. Rev.* **2010**, *39*, 2869–2878.
8. Welch, C. J.; Wu, N.; Biba, M.; Hartman, R.; Barkovic, T.; Gong, X.; Helmy, R.; Schafer, W.; Cuff, J.; Pirzada, Z.; Zhou, L. Greening Analytical Chromatography. *TrAC Trends Anal. Chem.* **2010**, *29* (7), 667–680.
9. Brett, C. M. A. Novel Sensor Devices and Monitoring Strategies for Green and Sustainable Chemistry Processes. *Pure Appl. Chem.* **2007**, *79* (11), 1969–1980.

10. Sedeño, P. Y.; Pingarrón, J. M.; Hernández, L. Greening Electroanalytical Chemistry. In *Handbook in Electroanalytical Chemistry*; de la Guardia, M., Garrigues, S., Eds.; Wiley: New Jersey, USA, 2012; Ch. 12.

11. Zappi, D.; Caminiti, R.; Ingo, G. M.; Sadun, C.; Tortolini, C.; Antonelli, M. L. Biologically Friendly Room Temperature Ionic Liquids and Nanomaterials for the Development of Innovative Enzymatic Biosensors. *Talanta* **2017,** *175,* 566–572.

12. Vargas, G. H.; Hernández, J. E. S.; Hernandez, S. S.; Rodríguez, A. M. V.; Saldivar, R. P.; Iqbal, H. M. N. Electrochemical Biosensors: A Solution to Pollution Detection with Reference to Environmental Contaminants. *Biosensors* **2018,** *8* (2), 21 pp.

13. Sedeño, P. Y.; Campuzano, S.; Pingarrón, J. M. Electrochemical (Bio)sensors: Promising Tools for Green Analytical Chemistry. *Curr. Opin. Green Sustain. Chem.* **2019,** *19,* 1–7.

14. Marco, M.-P.; Gee, S.; Hammock, B. D. Immunochemical Techniques for Environmental Analysis. II. Antibody Production and Immunoassay Development. *Trends Anal. Chem.* **1995,** *14* (8), 415–425.

15. Franek, M.; Hruska, K. Antibody Based Methods for Environmental and Food Analysis: A Review. *Vet. Med. Czech.* **2005,** *50* (1), 1–10.

16. Koivunen, M. E.; Krogsrud, R. L. Principles of Immunochemical Techniques Used in Clinical Laboratories. *Labmedicine* **2006,** *37* (8), 490–497.

17. Li, Y.-F.; Sun, Y.-M.; Beier, R. C.; Lei, H.-T.; Gee, S.; Hammock, B. D.; Wang, H.; Wang, Z.; Sun, X.; Shen, Y.-D.; Yang, J.-Y.; Xu, Z.-L. Immunochemical Techniques for Multianalyte Analysis of Chemical Residues in Food and the Environment: A Review. *Trends Anal. Chem.* **2017,** *88,* 25–40.

18. Dittrich, P. S.; Tachikawa, K.; Manz, A. Micro Total Analysis Systems: Latest Advancements and Trends. *Anal. Chem.* **2006,** *78,* 3887–3907.

19. Waggoner, P. S.; Craighead, H. G. Micro and Nanomechanical Sensors for Environmental, Chemical, and Biological Detection. *Lab-on-a-Chip* **2007,** *7,* 1238–1255.

20. Rong, G.; Robert Corrie, S.; Clark, H. A. In Vivo Biosensing: Progress and Perspectives. *ACS Sens.* **2017,** *2* (3), 327–338.

21. Pol, R.; Céspedes, F.; Gabriel, D.; Baeza, M. Microfluidic Lab-on-a-Chip Platforms for Environmental Monitoring. *TrAC Trends Anal. Chem.* **2017,** *95,* 62–68.

22. Armenta, S.; de la Guardia, M. *Green Analytical Chemistry: Theory & Practice,* Elsevier: Amsterdam, 2011; Ch. 7.

23. Ghosh, A.; Vilorio, C. R.; Hawkins, A. R.; Lee, M. L. Microchip Gas Chromatography Columns, Interfacing and Performance. *Talanta* **2018,** *188,* 463–492.

24. Pearce, T. C.; Schiffman, S. S.; Nagle, H. T.; Gardner, J. W., Eds. *Handbook of Machine Olfaction*; Wiley-VCH: Weinheim, 2003.

25. Röck, F.; Barsan, N.; Weimar, U. Electronic Nose: Current Status and Future Trends. *Chem. Rev.* **2008,** *108,* 705–725.

26. Wilson, A. D.; Baietto, M. Applications and Advances in Electronic-Nose Technologies. *Sensors* **2009,** *9,* 5099–5148.

27. Sekhar, P. K.; Brosha, E. L.; Mukundan, R.; Garzon, F. H. Chemical Sensors for Environmental Monitoring and Homeland Security. *Electrochem. Soc. Interface* **2010,** *19* (4), 35–40.

28. Wilson, A. D. Review of Electronic-Nose Technologies and Algorithms to Detect Hazardous Chemicals in the Environment. *Procedia Technol.* **2012,** *1,* 453–463.

29. Capelli, L.; Sironi, S.; Rosso, R. D. Electronic Noses for Environmental Monitoring Applications. *Sensors* **2014,** *14,* 19979–20007.

30. Gutiérrez, J.; Horrillo, M. C. Advances in Artificial Olfaction: Sensors and Applications. *Talanta* **2014,** *124,* 95–105.

31. Eusebio, L.; Capelli, L.; Sironi, S. Electronic Nose Testing Procedure for the Definition of Minimum Performance Requirements for Environmental Odor Monitoring. *Sensors* **2016,** *16* (9), 1548 (17 pp).

32. Jurs, P. C.; Bakken, G. A.; McClelland, H. E. Computational Methods for the Analysis of Chemical Sensors Array Data from Volatile Analytes. *Chem. Rev.* **2000,** *100* (7), 2649–2678.

33. Hu, W.; Wan, L.; Jian, Y.; Ren, C.; Jin, K.; Su, X.; Bai, X.; Haick, H.; Yao, M.; Wu, W. Electronic Noses: From Advanced Materials to Sensors Aided with Data Processing. *Adv. Mater. Technol.* **2019,** *4,* 1800488 (38 pp).

34. Olguín, C.; Miró, N. L.; Pascual, L.; Breijo, E. G.; Mañez, R. M.; Soto, J. An electronic Nose for the Detection of Sarin, Soman and Tabun Mimics and Interfering Agents. *Sens. Actuators B* **2014,** *202,* 31–37.

35. Devkota, J.; Ohodnicki, R. P.; Greve, W. D. SAW Sensors for Chemical Vapors and Gases. *Sensors* **2017,** *17* (4), 801 (28 pp).

36. Eom, K.; Park, H. S.; Yoon, D. S.; Kwon, T. Nanomechanical Resonators and Their Applications in Biological/Chemical Detection: Nanomechanics Principles. *Phys. Rep.* **2011,** *503,* 115–163.

37. Zougagh, M.; Ríos, A. Micro-Electromechanical Sensors in the Analytical Field. *Analyst* **2009,** *134,* 1274–1290.

38. Goeders, K. M.; Colton, J. S.; Bottomley, L. A. Microcantilevers: Sensing Chemical Interactions via Mechanical Motion. *Chem. Rev.* **2008,** *108,* 522–542.

39. Jian, R.-S.; Huang, Y.-S.; Lai, S.-L.; Sung, L-Y.; Lu, C.-J. Compact Instrumentation of a µ-GC for Real Time Analysis of Sub-ppb VOC Mixtures. *Microchem. J.* **2013,** *108,* 161–167.

40. Martin, O.; Gouttenoire, V.; Villard, P.; Arcamone, J.; Petitjean, M.; Billiot, G.; Philippe, J.; Puget, P.; Andreucci, P.; Ricoul, F.; Dupré, C.; Duraffourg, L.; Bellemin-Comte, A.; Ollier, E.; Colinet, E.; Ernst, T. Modeling and Design of a Fully Integrated Gas Analyzer Using a µGC and NEMS Sensors. *Sens. Actuators B* **2014,** *194,* 220–228.

41. Zhu, Z. An Overview of Carbon Nanotubes and Graphene for Biosensing Applications. *Nano-Micro Lett.* **2017,** *9* (3), 25 (24 pp).

42. Chatterjee, S.; Castro, M.; Feller, J. F. An e-Nose Made of Carbon Nanotube Based Quantum Resistive Sensors for the Detection of Eighteen Polar/Nonpolar VOC Biomarkers of Lung Cancer. *J. Mater. Chem. B* **2013,** *1,* 4563–4575.

43. Johnson, B. N.; Mutharasan, R. Biosensing Using Dynamic-Mode Cantilever Sensors: A Review. *Biosens. Bioelectron.* **2012,** *32,* 1–18.

44. Chomoucka, J.; Drbohlavova, J.; Masarik, M.; Ryvolova, M.; Huska, D.; Prasek, J.; Horna, A.; Trnkova, L.; Provaznik, I.; Adam, V.; Hubalek, J.; Kizek, R. Nanotechnologies for Society. New Designs and Applications of Nanosensors and Nanobiosensors in Medicine and Environmental Analysis. *Int. J. Nanotechnol.* **2012,** *9,* 746–781.

45. Fan, Z. H. Chemical Sensors and Microfluidics. *J. Biosens. Bioelectron.* **2013,** *4* (1), e117 (2 pp).

46. Lange, D.; Hagleitner, C.; Hierlemann, A. Complementary Metal Oxide Semiconductor Cantilever Arrays on a Single Chip: Mass-Sensitive Detection of Volatile Organic Compounds. *Anal. Chem.* **2002,** *74,* 3084–3095.

47. Lopez, J. L.; Verd, J.; Teva, J.; Murillo, G.; Giner, J.; Torres, F.; Uranga, A.; Abadal, G.; Barniol, N. Integration of RF-MEMS Resonators on Submicrometric Commercial CMOS Technologies. *J. Micromech. Microeng.* **2009,** *19* (1), 13–22.

48. Li, C.-S.; Hou, L.-J.; Li, S.-S. Advanced CMOS-MEMS Resonator Platform. *IEEE Electron Device Lett.* **2012,** *33* (2), 272–274.

49. Pachkawade, V.; Li, M.-H.; Li, C.-S.; Li, S.-S. A CMOS-MEMS Resonator Integrated System for Oscillator Application. *IEEE Sens. J.* **2013,** *13* (8), 2882–2889.

50. Uranga, A.; Verd, J.; Barniol, N. CMOS–MEMS Resonators: From Devices to Applications. *Microelectron. Eng.* **2015,** *132,* 58–73.

51. Boerek, P. On 'Electronic Nose' Methodology. *Sens. Actuators B* **2014,** *204,* 2–17.

52. Chansongkram, W.; Nimsuk, N. Development of a Wireless Electronic Nose Capable of Measuring Odors in Both Open and Closed Systems. *Procedia Comput. Sci.* **2016,** *86,* 192–195.

53. Jimenez, J. G.; Monroy, J. G.; Blanco, J. L. The Multi-Chamber Electronic Nose—An Improved Olfaction Sensor for Mobile Robotics. *Sensors* **2011,** *11,* 6145–6164.

54. Gongora, A.; Monroy, J.; Jimenez, J. G. An Electronic Architecture for Multipurpose Artificial Noses. *J. Sensors* **2018,** Article ID 5427693, 9 pp.

55. Choi, J.; Chang, S. J.; Bang, J.-H.; Park, J. S.; Lee, H. R. The Miniaturized IoT Electronic Nose Device and Sensor Data Collection System for Health Screening by Volatile Organic Compounds Detection from Exhaled Breath. In *Proc. SoICT* 2018; pp 405–409.

56. Persaud, K.; Dodd, G. Analysis of Discrimination Mechanisms in the Mammalian Olfactory System Using a Model Nose. *Nature* **1982,** *299,* 352–355.

57. Albert, K. J.; Lewis, N. S.; Schauer, C. L.; Sotzing, G. A.; Stitzel, S. E.; Vaid, T. P.; Walt, D. R. Cross-Reactive Chemical Sensor Arrays. *Chem. Rev.* **2000,** *100,* 2595–2626.

58. Wilson, A. D. Applications of Electronic-Nose Technologies for Noninvasive Early Detection of Plant, Animal and Human Diseases. *Chemosensors* **2018,** *6,* 45 (36 pp).

59. Szulczyński, B.; Gębicki, J. Currently Commercially Available Chemical Sensors Employed for Detection of Volatile Organic Compounds in Outdoor and Indoor Air. *Environments* **2017,** *4,* 21 (15 pp).

60. Pinnaduwage, L. A.; Gehl, A. C.; Allman, S. L.; Johansson, A.; Boisen, A. Miniature Sensor Suitable for Electronic Nose Applications. *Rev. Sci. Instrum.* **2007,** *78,* 055101 (4 pp).

61. James, D.; Scott, S. M.; Ali, Z.; O'Hare, W. T. Chemical Sensors for Electronic Nose Systems. *Microchim. Acta* **2005,** *149,* 1–17.

62. Li, M.; Myers, E. B.; Tang, H. X.; Aldridge, S. J.; McCaig, H. C.; Whiting, J. J.; Simonson, R. J.; Lewis, N. S.; Roukes, M. L. Nanoelectromechanical Resonator Arrays for Ultrafast, Gas-Phase Chromatographic Chemical Analysis. *Nano Lett.* **2010,** *10,* 3899–3903.

63. Floresa, A. R.; McConnell, L. L.; Hapeman, C. J.; Ramirez, M.; Torrents, A. Evaluation of an Electronic Nose for Odorant and Process Monitoring of Alkaline-Stabilized Biosolids Production. *Chemosphere* **2017,** *186,* 151–159.

64. Buck, L.; Axel, R. A Novel Multigene Family May Encode Odorant Receptors: A Molecular Basis for Odor Recognition. *Cell* **1991,** *65,* 175–187.

65. Nagle, H. T.; Schiffman, S. S.; Osuna, R. G. The How and Why of Electronic Noses. *IEEE Spectrum* **1998,** *35* (9), 22–34.

66. Watson, G. W.; Staples, E. J.; Viswanathan, S. Performance Evaluation of a Surface Acoustic Wave Analyzer to Measure VOCs in Air and Water. *Environ. Prog.* **2003,** *22* (3), 215–226.

67. Janata, J. *Principles of Chemical Sensors*; Springer: New York, 2019.

68. Arshak, K.; Moore, E.; Lyons, G. M. Harris, J. Clifford, S. A Review of Gas Sensors Employed in Electronic Nose Applications. *Sensor Rev.* **2004,** *24* (2), 181–198.

69. Ho, C. K.; Robinson, A.; Miller, D. R.; Davis, M. J. Overview of Sensors and Needs for Environmental Monitoring. *Sensors* **2005,** *5*, 4–37.

70. Snopok, B. A.; Kruglenko, I. V. Multisensor Systems for Chemical Analysis: State-of-the-Art in Electronic Nose Technology and New Trends in Machine Olfaction. *Thin Solid Films* **2002,** *418*, 21–41.

71. Nazemi, H.; Joseph, A.; Park, J.; Emadi, A. Advanced Micro- and Nano-Gas Sensor Technology: A Review. *Sensors* **2019,** *19*, 1285 (23 pp).

72. Zaporotskova, I. ·V.; Boroznina, N. P.; Parkhomenko, Y. N.; Kozhitov, L. V. Carbon nanotubes: Sensor Properties. A Review. *Modern Electronic Mater.* **2016,** *2* (4), 95–105.

73. Sánchez, C.; Santos, J. P.; Lozano, J. Use of Electronic Noses for Diagnosis of Digestive and Respiratory Diseases through the Breath. *Biosensors* **2019,** *9*, 2019, 35 (20 pp.).

74. Forsen, E.; Abadal, G.; Nilsson, S. G.; Teva, J.; Verd, J.; Sandberg, R.; Svendsen, W.; Murano, F. P.; Esteve, J.; Figueras, E.; Campabadal, F.; Montelius, L.; Barniol, N.; Boisen, A. Ultrasensitive Mass Sensor Fully Integrated with Complementary Metal-Oxide-Semiconductor Circuitry. *Appl. Phys. Lett.* **2005,** *87* (4), 043507 (3 pp.).

75. Teh, W. H.; Crook, R.; Smith, C. G.; Beere, H. E.; Ritchi, D. A. Characteristics of a Micromachined Floating-Gate High-Electron-Mobility Transistor at 4.2 K. *J. Appl. Phys.* **2005,** *97* (11), 114507 (17 pp.).

76. Penza, M.; Antolini, F.; Antisari, M. V. Carbon Nanotubes-Based Surface Acoustic Waves Oscillating Sensor for Vapour Detection. *Thin Solid Films* **2005,** *472*, 246–252.

77. Plashnitsa, V. V.; Elumalai, P.; Miura, N. Zirconia-Based Electrochemical Gas Sensors Using Nano-Structured Sensing Materials Aiming at Detection of Automotive Exhausts. *Electrochim. Acta* **2009,** *54* (25), 6099–6106.

78. Osuna, R. G.; Nagle, H. T. Method for Evaluating Data Preprocessing Techniques for Odor Classification with an Array of Gas Senso. *IEEE Trans. System, Man Cybern.: B* **1999,** *29* (5), 626–632.

79. Jha, S. K.; Yadava, R. D. S. Preprocessing of SAW Sensor Array Data and Pattern Recognition. *IEEE Sensors J.* **2009,** *9* (10), 1202–1208.

80. Jha, S. K.; Yadava, R. D. S. Denoising by Singular Value Decomposition and Its Application to Electronic Nose Data Processing. *IEEE Sensors J.* **2011,** *11* (1), 35–44.

81. Jha, S. K.; Yadava, R. D. S. Power scaling of SnO_2 and Conducting Polymer Composite Sensor Array Improves Odor Classification. *J. Pattern Recogn. Res.* **2011,** *9* (1), 65–74.

82. Smith, L. I. A Tutorial on Principal Component Analysis, 2002, www.cs.otago.ac.nz.

83. Diamantaras, K. I. Neural Networks and Principal Component Analysis. In *Handbook of Neural Network Signal Processing*; Hu, Y. H., Hwang, J. N., Eds.; CRC Press: New York, 2002; pp 8.1–8.7.

84. Yadava, R.D.S.; Chaudhary, R. Solvation, Transduction and Independent Component Analysis for Pattern Recognition in SAW Electronic Nose. *Sens. Actuators B* **2006,** *113*, 1–21.

85. Hyvärinen, A.; Karhunen, J.; Oja, E. *Independent Component Analysis*; Wiley Interscience, 2001.

86. Yum, H.; Yang, J. A direct LDA Algorithm for High-Dimensional Data with Application to Face Recognition. *Pattern Recogn.* **2011,** *34*, 2067–2070.

87. Nowotny, T.; Berna, A. Z.; Binions, R.; Trowell, S. Optimal Feature Selection for Classifying a Large Set of Chemicals Using Metal Oxide Sensors. *Sens. Actuators B* **2013,** *187,* 471–480.

88. Gao, D.; Wei, C. Simultaneous Estimation of Odor Classes and Concentrations Using an Electronic Nose with Function Approximation Model Ensembles. *Sens. Actuators B* **2007,** *120,* 584–594.

89. Kumar, N.; Bansal, A.; Sarma, G. S.; Rawal, R. K. Chemometrics Tools Used in Analytical Chemistry: An Overview. *Talanta* **2014,** *123,* 186–199.

90. Wold, S.; Sjöström, M.; Eriksson, L. PLS-regression: A Basic Tool of Chemometrics. *Chemometr. Intell. Lab. Syst.* **2001,** *58,* 109–130.

91. Gromski, P. S.; Correa, E.; Vaughan, A. A.; Wedge, D. C.; Turner, M. L.; Goodacre, R. A Comparison of Different Chemometrics Approaches for the Robust Classification of Electronic Nose Data. *Anal. Bioanal. Chem.* **2014,** *406,* 7581–7590.

92. Johnson, K. J.; Pehrsson, S. L. R. Sensor Array Design for Complex Sensing Tasks. *Annu. Rev. Anal. Chem.* **2015,** *8,* 14.1–14.24.

93. Jha, S. K.; Yadava, R. D. S.; Hayashi, K.; Patel, N. Recognition and Sensing of Organic Compounds Using Analytical Methods, Chemical Sensors, and Pattern Recognition Approaches. *Chemometr. Intell. Lab. Syst.* **2019,** *185,* 18–31.

94. Riedmiller, M. Advanced Supervised Learning in Multi-Layer Perceptrons—From Backpropagation to Adaptive Learning Algorithms. *Comput. Stand. Interfaces* **1994,** *16* (3), 265–278.

95. Niebling, G.; Schlachter, A. Qualitative and Quantitative Gas Analysis with Non-Linear Interdigital Sensor Arrays and Artificial Neural Networks. *Sens. Actuators B* **1995,** *26–27,* 289–292.

96. Sutter, J. M.; Jurs, P. C. Neural Network Classification and Quantification of Organic Vapors Based on Fluorescence Data from a Fiber-Optic Sensor Array. *Anal. Chem.* **1997,** *69,* 856–862.

97. Lu, Y.; Bian, L.; Yang, P. Quantitative Artificial Neural Network for Electronic Noses. *Anal. Chim. Acta* **2000,** *417,* 101–110.

98. Gao, D.; Zeping, Y.; Jianli, S. *Modular Neural Networks for Estimating Odor Concentrations.* In IEEE, Int. Joint Conf. Neural Networks, IJCNN, 2008; pp 3941–3948.

99. Guillaume, S. Designing Fuzzy Inference Systems from Data: An Interpretability Oriented Review. *IEEE Trans. Fuzzy Syst.* **2001,** *9* (3), 426–443.

100. Takagi, T.; Sugeno, M. Fuzzy Identification of Systems and Its Applications to Modeling and Control. *IEEE Trans. Syst. Man Cybern.* **1985,** *15,* 116–132.

101. Draper, N.; Smith, H. *Applied Regression Analysis*; Wiley: New York, 1981.

102. Geladi, P.; Kowalski, R. Partial Least Squares Regression (PLS): A Tutorial. *Anal. Chim. Acta* **1986,** *185,* 1–17.

103. Carey, W. P.; Beebe, K. R.; Sanchez, E.; Geladi, P.; Kowalski, B. R. Chemometric Analysis of Multisensor Arrays. *Sens. Actuators* **1986,** *9,* 223–234.

104. Ishibuchi, H.; Nii, M. Fuzzification of Neural Networks for Classification Problems. In *Hybrid Methods in Pattern Recognition*; Bunke, H., Kandel, A., Eds.; World Scientific: Singapore, 2002.

105. Yea, B.; Osaki, T.; Sugahara, K.; Konishi, R. The Concentration Estimation of Inflammable Gases with a Semiconductor Gas Sensor Utilizing Neural Networks and Fuzzy Inference. *Sens. Actuators B* **1997,** *41,* 121–129.

106. Šundić, T.; Marco, S.; Perera, A.; Pardo, A.; Hahn, S.; Bârsan, N.; Weimar, U. Fuzzy Inference System for Sensor Array Calibration: Prediction of CO and CH_4 Levels in Variable Humidity Conditions. *Chemom. Intell. Lab. Syst.* **2003**, *64*, 103–122.

107. Jha, S. K.; Hayashi, K.; Yadava, R. D. S. Neural, Fuzzy and Neuro-Fuzzy Approaches for Concentration Estimation of Volatile Organic Compounds by Surface Acoustic Wave Sensor Array. *Measurement* **2014**, *55*, 186–195.

108. Gulbag, A.; Temurtas, F. A Study on Quantitative Classification of Binary Gas Mixture Using Neural Network and Adaptive Neuro-Fuzzy Interface System. *Sens. Actuators B* **2006**, *115*, 252–262.

109. Carey, W. P. Beebe, K. R. Kowalski, B. R. Multicomponent Analysis Using an Array of Piezoelectric Crystal Sensors. *Anal. Chem.* **1987**, *59*, 1529–1534.

110. Shurmer, H. V.; Gardner, I. W.; Chan, H. T. The Application of Discrimination Techniques in Alcohols and Tobacco Using Tin Oxide Sensors. *Sens. Actuators* **1989**, *18*, 361–371.

111. Weimar, U.; Schierbaum, K. D.; Kowalkowski, R.; Göpel, W. Pattern Recognition Methods for Gas Mixture Analysis: Application to Sensor Array Based Upon SnO_2. *Sens. Actuators B* **1990**, *1*, 93–96.

112. Gardner, J. W.; Shunner, H. V.; Tan, T. T. Application of an Electronic Nose to the Discrimination of Coffees. *Sens. Actuators B* **1992**, *6*, 71–75.

113. Hierold, C.; Mtiller, R. Quantative Analysis of Gas Mixtures with Non-Selective Gas Sensors. *Sens. Actuators* **1989**, *17*, 587–592.

114. Sundgren, H.; Lundstrom, I.; Winquist, F. Evaluation of a Multiple Gas Mixture with a Simple MOSFET Gas Sensor Array and Pattern Recognition. *Sens. Actuators B* **1990**, *2*, 115–123.

115. Homer, G.; Hierold, Chr. Gas Analysis by Partial Model Building. *Sens. Actuators B* **1990**, *2*, 173–184.

116. Carey, W. P. Yee, S. S. Calibration of Nonlinear Solid-State Sensor Arrays Using Multivariate Regression Techniques. *Sens. Actuators B* **1992**, *9*, 113–122.

117. Zellers, E. T.; Pan, T.-S.; Patrash, S. J.; Han, M.; Batterman, S. A. Extended Disjoint Principal-Components Regression Analysis of SAW Vapor Sensor-Array Responses. *Sens. Actuators B* **1993**, *12*, 123–133.

118. Yang, Y. M.; Yuan Yang, P.; Wang, X.-R. Electronic Nose Based on SAWS Array and Its Odor Identification Capability. *Sens. Actuators B* **2000**, *66*, 167–170.

119. Grate, J. W.; Patrash, S. J.; Kaganove, S. N. Least-Squares Modeling of Vapor Descriptors Using Polymer-Coated Surface Acoustic Wave Sensor Array Responses. *Anal. Chem.* **2001**, *73*, 5247–5259.

120. Fernández, M. J.; Fontecha, J. L.; Sayago, I. Aleixandre, M. Lozano, J. Guti´errez, J. Grácia, I. Cané, C. Horrillo, M. C. Discrimination of Volatile Compounds through an Electronic Nose Based on ZnO SAW Sensors. *Sens. Actuators B* **2007**, *127*, 277–283.

121. Khalaf, W.; Pace, C.; Gaudioso, M. Gas Detection via Machine Learning. *World Acad. Sci., Eng. Technol.* **2008**, *37*, 139–143.

122. Khalaf, W.; Pace, C.; Gaudioso, M. Least Square Regression Method for Estimating Gas Concentration in an Electronic Nose System. *Sensors* **2009**, *9*, 1678–1691.

123. Leis, J.; Zhao, W.; Pinnaduwage, L. A.; Gehl, A. C.; Allman, S. L.; Shepp, A.; Mahmud, K. K. Estimating Gas Concentration Using Microcantilever-Based Electronic Nose. *Digit. Signal Process* **2010**, *20*, 1229–1237.

124. Zellers, E. T.; Batterman, S. A.; Han, M.; Patrash, S. J. Optimal Coating Selection for the Analysis of Organic Vapor Mixtures with Polymer Coated Surface Acoustic Wave Sensor Arrays. *Anal. Chem.* **1995**, *67*, 1092–1106.

125. Corcoran, P.; Anglesea, J.; Elshaw, M. The Application of Genetic Algorithms to Sensor Parameter Selection for Multisensor Array Configuration. *Sens. Actuators* **1999**, *76*, 57–66.

126. Grate, J. W. Acoustic Wave Microsensor Arrays for Vapor Sensing. *Chem. Rev.* **2000**, *100* (7), 2627–2648.

127. Pearce, T. C.; Sánchez-Montañés, M. A. Chemical Sensor Array Optimization: Geometric and Information Theoretic Approaches. In *Handbook of Machine Olfaction: Electronic Nose Technology*; Pearce, T. C., Schiffman, S. S., Nagle, H. T., Gardner, J. W., Eds.; Wiley-VCH: Weinheim, Germany, Chapter 14, 2003.

128. Gardner, J. W.; Boilot, P.; Hines, E. L. Enhancing Electronic Nose Performance by Sensor Selection Using a New Integer-Based Genetic Algorithm Approach. *Sens. Actuators B* **2005**, *106*, 114–121.

129. Phaisangittisagul, E.; Nagle, H. T.; Areekul, V. Intelligent Method for Sensor Subset Selection for Machine Olfaction. *Sens. Actuators B* **2010**, *145*, 507–515.

130. Zhang, L.; Tian, F. C.; Pei, G. S. A Novel Sensor Selection Using Pattern Recognition in Electronic Noses. *Measurement* **2014**, *5*, 31–39.

131. Miao, J.; Zhang, T.; Wang, Y.; Li, G. Optimal Sensor Selection for Classifying a Set of Ginsengs Using Metal-Oxide Sensors. *Sensors* **2015**, *15*, 16027–16039.

132. Park, J.; Groves, W. A.; Zellers, E. T. Vapor Recognition with Small Arrays of Polymer-Coated Microsensors. *Anal. Chem* **1999**, *71*, 3877–3886.

133. Carey, W. P.; Beebe, K. R.; Kowalski, B. R. Selection of Adsorbates for Chemical Sensor Arrays by Pattern Recognition. *Anal. Chem.* **1986**, *58* (1), 149–153.

134. Vergara, A.; Llobet, E. Sensor Selection and Chemo-Sensory Optimization: Toward an Adaptable Chemo-Sensory System. *Front. Neuroeng.* **2012**, *4*, 2012, Article 19 (21 pp.).

135. Polikar, R.; Shinar, R.; Udpa, L.; Porter, M. D. Artificial Intelligence Methods for Selection of an Optimized Sensor Array for Identification of Volatile Organic Compound. *Sens. Actuators B* **2001**, *80* 243–254.

136. Yadava, R. D. S. Modeling, Simulation, and Information Processing for Development of a Polymeric Electronic Nose System. In *Chemical Sensors—Simulation and Modeling*; Korotcenkov, G., Ed.; Momentum Press: New York, 2012; Chapter 10, Section 6.

137. Abraham, M. H. Scales of Solute Hydrogen Bonding: Their Construction and Application to Physicochemical and Biochemical Process. *Chem. Soc. Rev.* **1993**, *22* (2), 73–83, 1993.

138. Grate, J. W.; Abraham, M. H. Solubility Interactions and the Design of Chemically Selective Sorbent Coatings for Chemical Sensors and Arrays. *Sens. Actuators B* **1991**, *3*, 85–111.

139. Jha, S. K.; Yadava, R. D. S. Designing Optimal Model SAW Sensor Array Electronic Nose for Body Odor Discrimination. *Sensor Lett.* **2011**, *9*, 1612–1622.

140. Jha, S. K.; Yadava, R. D. S. Data Mining Approach to Polymer Selection for Making SAW Sensor Array Based Electronic Nose. *Sens. Transducers J.* **2012**, *147* (12), 108–128.

141. Verma, P.; Yadava, R. D. S. Polymer Selection for SAW Sensor Array Based Electronic Noses by Fuzzy C-Means Clustering of Partition Coefficients: Model Studies on Detection of Freshness and Spoilage of Milk and Fish. *Sens. Actuators B* **2015**, *209*, 751–769.

142. Verma, P.; Yadava, R. D. S. A Data Mining Procedure for Polymer Selection for Making Surface Acoustic Wave Sensor Array. *Sensor Lett.* **2013,** *11,* 1903–1918.

143. Gupta, A.; Singh, T. S.; Yadava, R. D. S. MEMS Sensor Array-Based Electronic Nose for Breath Analysis—A Simulation Study. *J. Breath Res.* **2019,** *13,* 016003 (17 pp.).

144. Mitrovics, J.; Ulmer, H.; Noetzel, G.; Weimar, U.; Gopel, W. *Hybrid Modular Sensor Systems: A New Generation of Electronic Noses.* In Proc. IEEE Int. Symp. Industrial Electronics (ISIE '97), 1997; pp SS116–SS121.

145. Ulmer, H.; Mitrovics, J.; Weimar, U.; Gopel, W. Sensor Arrays with Only One or Several Transducer Principles? The Advantage of Hybrid Modular Systems. *Sens. Actuators B* **2000,** *65,* 79–81.

146. Wilson, D. M.; Garrod, S. D. Optimization of Gas-Sensitive Polymer Arrays Using Combinations of Heterogeneous and Homogeneous Subarrays. *IEEE Sensors J.* **2002,** *2,* 169–178.

147. Kramer, K. E.; Rose-Pehrsson, S. L.; Johnson, K. J.; Minor, C. P. Hybrid Arrays for Chemical Sensing. In *Computational Methods for Sensor Material Selection*; Ryan, M. A. et al., Eds.; Springer Science & Business Media: New York, 2009; Chapter 12, pp 265–298.

148. Skutin, E. D.; Podgorniy, S. O.; Zemtsov, A. E.; Gaberkorn, O. V. Expanding Analytical Potential of Hybrid Sensor Arrays. *Procedia Eng.* **2016,** *152,* 493–496.

149. Esteban, J.; Starr, A.; Willetts, R.; Hannah, P.; Cross, P. B. A Review of Data Fusion Models and Architectures: Towards Engineering Guidelines. *Neural Comput. Appl.* **2005,** *14* (4), 273–281.

150. Hall, D. L. *Mathematical Techniques in Multisensor Data Fusion*; Artech: New Jersey, USA, 2004.

151. Klein, L. A. *Sensor and Data Fusion: A Tool for Information Assessment and Decision Making*; SPIE: Bellingham, USA, 2012.

152. Penza, M.; Cassano, G.; Aversa, P.; Antolini, F.; Cusano, A.; Consales, M.; Giordano, M.; Nicolais, L. Carbon Nanotubes-Coated Multi-Transducing Sensors for VOCs Detection. *Sens. Actuators B* **2005,** *111/112,* 171–180.

153. Haupt, S. G.; Ha, J.; Rose, D. Chemical Agent Detection. US Patent No. 2005/0254996, Nov. 17, 2005.

154. Wide, P.; Winquist, F.; Bergsten, P.; Petriu, E. The Human-Based Multisensor Fusion Method for Artificial Nose and Tongue Sensor Data. *IEEE Trans. Instrum. Meas.* **1998,** *47,* 1072–1077.

155. Hauptmann, P. R. Selected Examples of Intelligent (micro) Sensor Systems: State-of-the-Art and Tendencies. *Meas. Sci. Technol.* **2006,** *17,* 459–466.

156. Boholt, K.; Andreasen, K.; den Berg, F.; Hansen, T. A New Method for Measuring Emission of Odor from a Rendering Plant Using the Danish Odour Sensor System (DOSS) Artificial Nose. *Sens. Actuators B* **2005,** *106,* 170–176.

157. Verma, P.; Yadava, R. D. S. Enhancing Sensor Array Intelligence by Bayesian Fusion of Information Multiplicity Generated by Multiple Processors. *Sens. Transducers J.* **2014,** *182* (11), 42–48.

158. Verma, P.; Yadava, R. D. S. *Evidence Generation for Dempster-Shafer Fusion Using Feature Extraction Multiplicity and Radial Basis Network.* In IEEE Int. Conf. Emerging Trends in Electrical and Computer Technology, ICETECT 2011; pp. 542–545.

159. Persaud, K. C.; Wareham, P.; Pisanelli, A. M.; Scorsone, E. 'Electronic Nose'—New Condition Monitoring Devices for Environmental Applications. *Chem. Senses* **2005,** *30* (1), i252–i253.

160. Fernandes, D. L. A.; Teresa, M.; Gomes, S. R. Development of an Electronic Nose.to Identify and Quantify Volatile Hazardous Compounds. *Talanta* **2008**, *77*, 77–83.
161. Jasinski, G.; Wozniak, L.; Kalinowski, P.; Jasinski, P. *Evaluation of the Electronic Nose Used for Monitoring Environmental Pollution.* In IEEE XV International Scientific Conference on Optoelectronic and Electronic Sensors (COE), 2018.
162. Bieganowski, A.; Józefaciuk, G.; Bandura, L.; Guz, L.; Łagód, G. Franus, W. Evaluation of Hydrocarbon Soil Pollution Using E-Nose. *Sensors* **2018**, *18* (8), 2463 (14 pp.).
163. Łagód, G.; Guz, L.; Sabba, F.; Sobczuk, H. Detection of Wastewater Treatment Process Disturbances in Bioreactors Using the E-Nose Technology. *Ecol. Chem. Eng. S.* **2018**, *25* (3), 405–418.
164. Łagód, G.; Duda, S. M.; Majerek, D.; Szutt, A. Dołhańczuk-Śródka, A. Application of Electronic Nose for Evaluation of Wastewater Treatment Process Effects at Full-Scale WWTP. *Processes* **2019**, *7* (5), 251 (15 pp.).
165. Stuetz, R. M.; Fenner, R. A.; Engin, G. Assessment of Odors from Sewage Treatment Works by an Electronic Nose, H_2S Analysis and Olfactometry. *Water Res.* **1999**, *33* (2), 453–461.
166. Canhoto, O.; Magan, N. Potential for Detection of Microorganisms and Heavy Metals in Potable Water Using Electronic Nose Technology. *Biosens. Bioelectron.* **2003**, *18*, 751–754.
167. Canhoto, O.; Magan, N. Electronic Nose Technology for the Detection of Microbial and Chemical Contamination of Potable Water. *Sens. Actuators B* **2005**, *106* (1), 3–6.
168. DeVito, S.; Massera, E.; Piga, M.; Martinotto, L.; DiFrancia, G. On Field Calibration of an Electronic Nose for Benzene Estimation in an Urban Pollution Monitoring Scenario. *Sens. Actuators B* **2008**, *129* (2), 750–757.
169. Cipriano, D.; Capelli, L. Evolution of Electronic Noses from Research Objects to Engineered Environmental Odor Monitoring Systems: A Review of Standardization Approaches. *Biosensors* **2019**, *9* (2), 75 (19 pp.).
170. Onal, A. A review: Current Analytical Methods for Determination of Biogenic Amines in Food. *Food Chem.* **2007**, *103*, 1475–1486.
171. Sarkadi, L. S. Amino Acids and Biogenic Amines as Food Quality Factors. *Pure Appl. Chem.* **2019**, *91* (2), 289–300.
172. Capillas, C. R.; Herrero, A. M. Impact of Biogenic Amines on Food Quality and Safety. *Foods* **2019**, *8*, 62 (16 pp.).
173. Doeun, D.; Davaatseren,M.; Chung, M.-S. Biogenic Amines in Foods. *Food. Sci. Biotechnol.* **2017**, *26* (6), 1463–1474.
174. Karovicova, J.; Kohajdova, Z. Biogenic Amines in Food. *Chem. Pap.* **2005**, *59* (1), 70–79.
175. Brink, B. T.; Damink, C.; Joosten, H. M. L. J.; Veld, J. H. J. H. Occurrence and Formation of Biologically Active Amines in Foods. *Int. J. Food Microbiol.* **1990**, *11* (11), 73–84.
176. Stute, R.; Petridis, K.; Steinhart, H.; Biernoth, G. Biogenic Amines in Fish and Soy Sauces. *Eur. Food Res. Technol.* **2002**, *215*, 101–107.
177. Capillas, C. R.; Moral, A. Correlation Between Biochemical and Sensory Quality Indices in Hake Stored in Ice. *Food Res. Int.* **2001**, *34*, 441–447.
178. Niculescu, M.; Nistor, C.; Frebort, I.; Pec, P.; Mattiasson, B.; Csoregi, E. Redox Hydrogel-Based Amperometric Bienzyme Electrodes for Fish Freshness Monitoring. *Anal. Chem.* **2000**, *72*, 1591–1597.
179. Karpas, Z.; Tilman, B.; Gdalevsky, R.; Lorber, A. Determination of Volatile Biogenic Amines in Muscle Food Products by Ion Mobility Spectrometry. *Anal. Chim. Acta* **2002**, *463*, 155–163.

180. Rodriguez, M.; Carneiro, C.; Feijó, M.; Júnior, C.; Mano, S. Bioactive Amines: Aspects of Quality and Safety in Food. *Food Nutr. Sci.* **2014,** *5* (2), 138–146.

181. Silva, C. M. G.; Gloria, M. B. A. Bioactive Amines in Chicken Breast and Thigh after Slaughter and During Storage at 4±1°C and in Chicken-Based Meat Products. *Food Chem.* **2002,** *78*, 241–248.

182. Coisson, J. D.; Cerutti, C. Travaglia, F. Arlorio, M. Production of Biogenic Amines in Salamini italianialla Cacciatora PDO. *Meat Sci.* **2004,** *67*, 343–349.

183. Ampuero, S.; Zesiger, T.; Gustafsson, V.; Lunden, A.; Bosset, J. O. Determination of Trimethylamine in Milk Using an MS Based Electronic Nose. *Eur. Food Res. Technol.* **2002,** *214*, 163–167.

184. Vale, S.; Gloria, M. B. A. Biogenic Amines in Brazilian Cheeses. *Food Chem.* **1998,** *63*, 343–348.

185. Kalac, P.; Svecova, S.; Pelikanova, T. Levels of Biogenic Amines in Typical Vegetable Products. *Food Chem.* **2002,** *77*, 349–351.

186. Funel, A. L. Biogenic Amines in Wines: Role of Lactic Acid Bacteria. *FEMS Microbiol. Lett.* **2001,** *199*, 9–13.

187. Slomkowska, A.; Ambroziak, W. Biogenic Amine Profile of the Most Popular Polish beers. *Eur. Food Res. Technol.* **2002,** *215*, 380–383.

188. Parlapani, F. F.; Mallouchos, A.; Haroutounian, S. A.; Boziaris, I. S. Microbiological Spoilage and Investigation of Volatile Profile during Storage of Sea Bream Fillets under Various Conditions. *Int. J. Food Microbiol.* **2014,** *189*, 153–163.

189. Bota, G. M.; Harrington, P. B. Direct Detection of Trimethylamine in Meat Food Products Using Ion Mobility Spectrometry. *Talanta* **2006,** *68*, 629–635 (*Sensors* **2007,** *7*, 2387).

190. Chan, S. T.; Yao, M. W. Y.; Wong, Y. C.; Wong, T.; Mok, C. S.; Sin, D. W. M. Evaluation of Chemical Indicators for Monitoring Freshness of Food and Determination of Volatile Amines in Fish by Headspace Solid-Phase Micro-extraction and Gas Chromatography–Mass Spectrometry. *Eur. Food Res. Technol.* **2006,** *224*, 67–74.

191. Periago, M. J.; Rodrigo, J.; Ros, G.; Rodriguez-Jerez, J. J.; Hernandez-Herrero, M. Monitoring Volatile and Nonvolatile Amines in Dried and Salted Roes of Tuna (*Thunnus thynnus* L.) during Manufacture and Storage. *J. Food Prot.* **2005,** *66*, 335–340.

192. Boziaris, I. S. Current Trends on the Study of Microbiological Spoilage of Fresh Fish. *Fish Aquat. J.* **2014,** *6* (1), 2014, 1000e115 (2 pp).

193. Roberts, T. A.; Cordier,J. L.; Gram, L.; Tompkin, R. B.; Pitt, J. I.; Gorris, L. G. M.; Swanson, K. M. J. *Micro-Organisms in Food 6: Microbial Ecology of Food Commodities*, 2nd ed.; Kluwer Academic/Plenum Publishers: New York, 2005.

194. European Food Safety Authority (EFSA). Polycyclic Aromatic Hydrocarbons in Food Scientific Opinion of the Panel on Contaminants in the Food Chain. *EFSA J.* **2008,** *6* (8), 724 (114 pp.).

195. Phillips, D. H. Polycyclic Aromatic Hydrocarbons in the Diet. *Mut. Res/Genet Toxicol. Environ. Mutagen* **1999,** *443* (1), 139–147.

196. Rouseff, R. L.; Cadwallader, K. R., Eds. *Headspace Analysis of Foods and Flavors: Theory and Practice*, Springer Science+Business Media: New York, 2001.

197. Dainty, R. H. Chemical-Biochemical Detection of Spoilage. *Int. J. Food Microbiol.* **1996,** *33* (1) 19–33.

198. Coles, R.; Kirwan, M. *Food and Beverage Packaging Technology*; Wiley-Blackwell: West Sussex, UK, 2011.

199. Labuza, T. P. Functional Foods and Dietary Supplements: Safety, Good Manufacturing Practice (GMPs) and Shelf Life Testing. In *Essentials of Functional Foods*; Schmidl, M. K., Labuza, T. P., Eds.; Aspen Press: London, 2000.

200. German, J. B. Food Processing and Lipid Oxidation. *Adv. Exp. Med. Biol.* **1999**, *459*, 23–50.

201. Rodríguez, S. D.; Barletta, D. A.; Wilderjans, T. F.; Bernik, D. L. Fast and Efficient Food Quality Control Using Electronic Noses: Adulteration Detection Achieved by Unfolded Cluster Analysis Coupled with Time-Window Selection. *Food Anal. Methods* **2014**, *7*, 2042–2050.

202. Loutfi, A.; Coradeschi, S.; Mani, G. K.; Shankar, P.; Rayappan, J. B. Electronic Noses for Food Quality: A Review. *J. Food Eng.* **2015**, *144*, 103–111.

203. Sanaeifar, A.; ZakiDizaji, H.; Jafari, A.; de la Guardia, M. Early Detection of Contamination and Defect in Foodstuffs by Electronic Nose: A Review. *TrAC Trends Anal. Chem.* **2017**, *97*, 257–271.

204. Świgło, A. G.; Chmielewski, J. Electronic Nose as a Tool for Monitoring the Authenticity of Food: A Review. *Food Anal. Methods* **2017**, *10*, 1800–1816.

205. Shi, H.; Zhang, M.; Adhikari, B. Advances of Electronic Nose and Its Application in Fresh Foods: A Review. *Crit. Rev. Food Sci. Nutr.* **2018**, *58* (16), 2700–2710.

206. Jia, W.; Liang, G.; Jiang, Z.; Wang, J. Advances in Electronic Nose Development for Application to Agricultural Products. *Food Anal. Methods* **2019**, *12* (10), 2226–2240.

207. Haugen, J. E. Electronic Noses in Food Analysis. In *Headspace Analysis of Foods and Flavors: Theory and Practice*; Rouseff, R. L., Cadwallader, K. R., Eds.; Springer Science+Business Media: New York, 2001; pp 43–58.

208. Ampuero, S.; Bosset, J. O. The Electronic Nose Applied to Dairy Products: A Review. *Sens. Actuators B* **2003**, *94*, 1–12.

209. Zhang, H. Using Electronic Noses to Assess Food Quality. In *Rapid and On-line Instrumentation for Food Quality Assurance*; Tothill, I. E., Eds.; Woodhead and CRC Press LLC: Boca Raton, 2003; pp 324–338.

210. Peris, M.; Gilabert, L. E. A 21st Century Technique for Food Control: Electronic Noses. *Anal. Chim. Acta* **2009**, *638* (1), 1–5.

211. Gomez, A. H.; Wang, J.; Hu, G.; Pereira, A. G. Monitoring Storage Shelf Life of Tomato Using Electronic Nose Technique. *J. Food Eng.* **2008**, *85*, 625–631.

212. Ezhilan, M.; Nesakumar, N.; Babu, K. J.; Srinandan, C. S.; Rayappan, J. B. Freshness Assessment of Broccoli Using Electronic Nose. *Measurement* **2019**, *145*, 735–743.

213. Wijaya, D. R.; Sarno, R.; Zulaika, E.; Sabila, S. I. Development of Mobile Electronic Nose for Beef Quality Monitoring. *Procedia Comput. Sci.* **2017**, 124, 728–735.

214. Nimsuk, N. Improvement of Accuracy in Beer Classification Using Transient Features for Electronic Nose Technology. *J. Food Meas. Charact.* **2019**, *13* (1), 656–662.

215. Rottiers, H.; Sosa, D. A.; Van de Vyver, L.; Hinneh, M.; Everaert, H.; De Wever, J.; Messens, K. Dewettinck, K. Discrimination of Cocoa Liquors Based on their Odor Fingerprint: A Fast GC Electronic Nose Suitability Study. *Food Anal. Methods* **2019**, *12* (2), 475–488.

216. Yuan, H.; Chen, X.; Shao, Y.; Cheng, Y.; Yang, Y.; Zhang, M.; Hua, J.; Li, J.; Deng, Y.; Wang, J.; Dong, C. Quality Evaluation of Green and Dark Tea Grade Using Electronic Nose and Multivariate Statistical Analysis. *J. Food Sci.* **2019**, *84* (12), 3411–3417.

217. Srivastava, S.; Mishra, G.; Mishra, H. N. Fuzzy Controller Based E-Nose Classification of *Sitophilus oryzae* Infestation in Stored Rice Grain. *Food Chem.* **2019**, *283*, 604–610.

218. Heudi, O. Green Bioanalytical Methods are Now a Reality. *Bioanalysis* **2012**, *4* (11), 1257 (1 pp.).

219. Shaaban, H. Green, Eco-Friendly Bio-Analytical Techniques for Pharmaceutical Analysis. *J. Clin. Bioanal. Chem.* **2017,** *1* (1), 3–4.

220. Schuberth, J. Volatile Organic Compounds Determined in Pharmaceutical Products by Full Evaporation Technique and Capillary Gas Chromatography/Ion-Trap Detection. *Anal. Chem.* **1996,** *68* (8), 1317–1320.

221. Płotka, J.; Tobiszewski, M.; Sulej, A. M.; Kupska, M.; Górecki, T.; Namiesnik, J. Green Chromatography. *J. Chromatogr. A* **2013,** *1307,* 1–20.

222. OSHA Fact Sheet, Hazardous Chemicals in Laboratories. www.osha.gov

223. Examples of Common Laboratory Chemicals and Their Hazard Class, Office of Research Facilities. https://www.orf.od.nih.gov/Pages/default.aspx

224. Gregis, G.; Sanchez, J. B.; Bezverkhyy, I.; Guy, W.; Berger, F.; Fierro, V.; Bellat, J. P.; Celzard, A. Detection and Quantification of Lung Cancer Biomarkers by a Micro-Analytical Device Using a Single Metal Oxide-Based Gas Sensor. *Sens. Actuators B* **2018,** *255,* 391–400.

225. Cai, X.; Chen, L.; Kang, T.; Tang, Y.; Lim, T.; Xu, M.; Hui, H. A Prediction Model with a Combination of Variables for Diagnosis of Lung Cancer. *Med. Sci. Monit.* **2017,** *23,* 5620–5629.

226. van Oort, P. M.; Povoa, P.; Schnabel, R.; Dark, P.; Artigas, A.; Bergmans, D. C.; Felton, T.; Coelho, L.; Schultz, M. J.; Fowler, S. J.; Bos, L. D. The Potential Role of Exhaled Breath Analysis in the Diagnostic Process of Pneumonia-A Systematic Review. *J. Breath Res.* **2018,** *12* (2), 024001 (11 pp.).

227. Germanese, D.; D'Acunto, M.; Magrini, M.; Righi, M.; Salvetti, O. Cardio-Metabolic Diseases Prevention by Self-Monitoring the Breath. *Sens. Transducers* **2017,** *215* (8), 19–26.

228. Essiet, I. Diagnosis of Kidney Failure by Analysis of the Concentration of Ammonia in Exhaled Breath. *J. Emerg. Trends Eng. Appl. Sci.* **2013,** *4* (6), 859–862.

229. Wilson, A. D. Application of Electronic-Nose Technologies and VOC Biomarkers for the Noninvasive Early Diagnosis of Gastrointestinal Diseases. *Sensors* **2018,** *18* (8), 2613 (29 pp.).

230. Ashrafi, M.; Bates, M.; Baguneid, M.; Rasgado, T. A.; Richardson, R. R.; Bayat, A. Volatile Organic Compound Detection as a Potential Means of Diagnosing Cutaneous Wound Infections. *Wound Repair Regen.* **2017,** *25* (4), 574–590.

231. Buszewski, B.; Kesy, M.; Ligor, T.; Amann, A. Human Exhaled Air Analytics: Biomarkers of Diseases. *Biomed. Chromatogr.* **2007,** *21,* 553–566.

232. Sehnert, S. S.; Jiang, L.; Burdick, J. F.; Risby, T. H. Breath Biomarkers for Detection of Human Liver Diseases: Preliminary Study. *Biomarkers* **2002,** *7* (2), 174–187.

233. Wojnowski, W.; Dymerski, T.; Gębicki, J.; Namieśnik, J. Electronic Noses in Medical Diagnostics. *Curr. Med. Chem.* **2019,** *26* (1), 197–215.

234. Lourenço, C.; Turner, C. Breath Analysis in Disease Diagnosis: Methodological Considerations and Applications. *Metabolites* **2014,** *4* (2), 465–498.

235. Wilson, A. D.; Baietto, M. Advances in Electronic-Nose Technologies Developed for Biomedical Applications. *Sensors* **2011,** *11* (1), 1105–1176.

CHAPTER 11

Multiphase Materials Based on Green Polymers for Electronics with Minimized Environmental Impact

ANDREEA IRINA BARZIC*

Laboratory of Physical Chemistry of Polymers, "Petru Poni" Institute of Macromolecular Chemistry, 41A Grigore Ghica Voda Alley, 700487 Iasi, Romania

Corresponding author. E-mail: irina_cosutchi@yahoo.com

ABSTRACT

The progressive contamination of our planet with waste from various industries is an actual problem that demands urgent solutions. For this reason, scientists and engineers must find new ways to produce materials that generate minimized impact on the environment. The chapter is focused on presentation of the state of art in the field of multiphase materials based on green polymers designed for electronic industry. The overall picture and current progress concerning the preparation routes of the green polymer composites are reviewed. The surface processing techniques of reinforced green polymer materials are also described. The latest trends regarding the application of the multiphase green polymer materials in electronics are presented. The environmental concerns are reflected in the developments in this domain, opening novel perspectives in the devices for future electronics.

11.1 INTRODUCTION

The large amount of waste produced by various industries raises serious environmental problems that need to be immediately addressed.[1-4] For instance, until 1970s electronics manufacturing was regarded as a "clean" industry.[1]

However, electronic devices contain a huge number of toxic compounds, like ferrous and nonferrous metals, ceramics, and polymers.[1,5] Their improper storage and/or recycling are/is generating undesired noxious effects on the health of all living beings.[6] The growing societal awareness of the pollution issues has determined a major review of this idea. Annual global meetings are made to explore the possibilities to preserve and protect our planet from industrial residues.[7] As the years have passed, each state developed its own legislation framework, but even so, it is known that this industry is operating away from the ideal. Therefore, significant changes must start from the stage of synthesis and design of the production materials and replace as much as possible the toxic elements from the devices. It is of paramount importance for researchers, engineers, and industrial agents to become responsible for the outcome of their work and, consequently, to cooperate to create products with reduced effects on the environment once they become technologically outdated or nonfunctional.

The electronic equipment composed of plastic components should be viewed as an environmental time bomb given the fact that they represent around 20% of the 50 million ton of waste resulted each year.[8] In this way, these residues will reach 110 million ton by 2050, if the nowadays consumption practices and existing business do not turn toward more eco-friendly solutions.[8] Among them, it is worth mentioning the incorporation in devices of recycled or renewable plastics or even better biodegradable plastics, if applicable. In this context, green polymers are gradually occupying an important place in electronic market since their diminished or lack of polluting effects.[9–11] It is clear that in some practical situations green polymers are not able to fulfill all demands as in the case of synthetic compounds produced through polluting methods. For example, cellulose derivatives have lower thermal stability than polyimides and this limits their use in high-power electronics. A good alternative is to introduce in such a polymer matrix certain types of fillers that enhance the desired property. Therefore, green polymer composites represent a less polluting path toward future "clean" electronics. Such multiphase materials display adaptable properties as a function of the reinforcement agent, its distribution, and amount inserted in the matrix. However, it is very important to maintain a balance between their performance and biodegradability.

The chapter aims to review the newest developments made on green polymer composites designed for electronic industry by emphasizing the currently used preparation procedures, bulk, and surface-processing methodologies. In addition, the application of these multiphase materials in various electronic devices is reviewed.

11.2 PREPARATION AND PROCESSING PROCEDURES OF MULTIPHASE MATERIALS BASED ON GREEN POLYMERS

Multiphase materials, like polymer composites, can be categorized as a function of the type of the constituent matrix or introduced filler. In the case of green polymer composites, it is desirable that at least the matrix is a green polymer and/or the reinforcement agent could be an eco-friendly compound. According to the type of material used in their construction, green polymer composites can be classified as[12–14]

- green polymer composites
- partly green polymer composites

To get a clearer picture, Figure 11.1 illustrates a potential classification of green polymer composites taking into account the type of materials used for their fabrication. In the case of the fully green polymer composites, it is imperative to use both green matrix and reinforcement agent. Green polymers can be indulgently considered to be natural polymers (proteins—collagen, gelatin, soya; polysaccharides—cellulose, starch, pectin, chitosan) or synthetic biodegradable polymers (polylactic acid, polycaprolactone [PCL], poly(glycolic acid)). Green polymers can be attained from biomass resources (cellulose, lipids), from microorganisms extraction (polyhydroxyalkanoates), or from biotechnologies. To obtain an entirely green composite, the green polymer must be reinforced with natural fibers or nontoxic particles of micro or nanodimensions. In some cases, when specific properties are aimed and they can be hardly accomplished with common eco-friendly polymers or fillers, one may use a partly green material. For instance, biodegradable polymer matrix can be doped with several types of fillers or the green reinforcement agent is mixed with a less toxic polymer.

The fabrication of such composites is mainly accomplished via a chemical or mechanical process by using one of the following techniques[15–17]:

- *Intercalation method*[18] is developed for nanoplatelets types of fillers which are further inserted into the polymer matrix. It is well known that introduction of such reinforcements inside a polymer material enhances certain bulk properties such as shrinkage, stiffness, and flammability. Intercalation is a top-down procedure and necessitates to modify the surface features of nanoplatelets to attain the desired homogeneous distribution of plate-like nanofillers in the selected matrix. Intercalated morphology is observed when polymer chains are diffusing into the gallery spacing of layered structure.

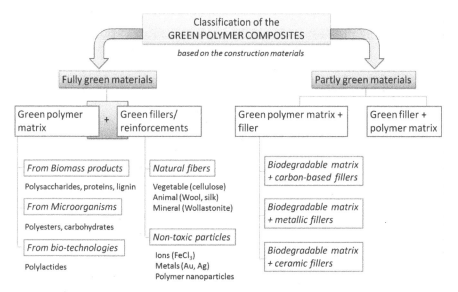

FIGURE 11.1 A schematic representation of a possible classification of green polymer composites considering the type of materials used for their fabrication.

- *In-situ polymerization*[19] requires the swelling of the reinforcement particles in the monomer solution because the solubilized low molecular weight monomer can penetrate between layers determining swelling.[20] The resulting system is then polymerized in the presence of either heat, radiation, organic initiator, or by initiator diffusion. The monomer is starting to form the polymer chain between interlayers, thereby yielding exfoliated or intercalated composites. In-situ template synthesis is an analogous procedure. Here, the clay layers are obtained in the presence of the macromolecular chains. The polymer matrix and clay reinforcement are placed in an aqueous solution leading to a gel, which is then refluxed at high temperature. The chains are blocked inside the clay layers, so in this situation, the nucleation and growth of clay layers are occurring on the polymer chains at elevated temperature. The major disadvantage of this procedure is that high temperature determines the decomposition of matrix.

- *Sol–gel method*[21] is a bottom-up procedure that relies on an opposite principle than all the previous methods. The term sol–gel is ascribed to two relations stages—sol and gel. The first one is a colloidal suspension of solid filler found in monomer solution. The second stage is the 3D interconnecting network constructed between phases.[20,21] In this approach, the tridimensional network widens throughout the liquid.[20] The polymer has the role of a nucleating agent and favors the growth of layered crystals. As they begin to grow, the macromolecular phase is seeped among layers and therefore the composite is obtained.

- *Direct mixing of polymer and nanofillers*[22] is a top-down approach of composite preparation and it consists in the breakdown of the aggregated particles during mixing process.[22] This procedure is adequate for fabricating composites with polymer matrix and it implicates two main routes of mixing the polymer and the reinforcement agent. One way is blending a polymer (without using solvents) with the fillers above the glass-transition temperature of the matrix—technique known as melt-compounding method.[20,22] The other way starts from mixing of polymer and reinforcement compound in solution in the presence of solvents—method called solvent method/solution mixing:

  - *Melt compounding method* requires the addition of the filler in the heated polymer above the vitrification temperature. In this kind of procedure, the shear stress or hydrodynamics force is generated in the polymer melt through viscous drag, and it is used to disrupt the aggregates of reinforcement material and thereby enabling a homogeneous and uniform filler distribution inside the polymer matrix.
  - *Solvent method* involves dispersing of the filler in the solvent and then the polymer is dissolved in a cosolvent. To avoid aggregates formation, the filler solution is subjected to ultrasonication.[22] The prepared composites are recovered from solution by solvent evaporation or by the solvent coagulation approach. In this technique, the shear stresses generated in the organic matrix are reduced compared to those in melt compounding.

The polymer composites achieved by one of the aforementioned methods are finally processed by common manufacturing methods such as follows:

- *Injection molding*[23] uses a molten polymer, which is subjected to high pressure and passes into a mold cavity through an opening.
- *Calendaring*[24] is generally applied to polymers in molten state to obtain rolled sheets of specific thickness and particular appearance.
- *Casting*[16] is often utilized to make flexible polymer components having the shape of a single or multilumen tube applicable in the biomedical device industry. This processing technology has the advantage that does not necessitate extrusion or injection molding, yet it readily

integrates components and characteristics traditionally produced by such processes.

- *Compression molding*[25] polymer material is squeezed into a preheated mold adopting the shape of the used mold cavity and performing curing, owing to the applied conditions of temperature and pressure.
- *Blow molding*[26,27] starts by melting down the polymer and transforming it into a parison or, in the situation of injection and injection stretch blow molding, a preform. This processing approach allows fabrication of hollow plastic parts that can be joined together. The parison is a tube-like plastic component with a hole in one end through which compressed air is able to pass. Afterward, the parison is clamped into the mold and subsequently air is blown into it. The air pressure then aids to push the plastic out (matching the mold). Once the plastic has chilled and hardened, the mold opens up and the part is removed.
- *Rotational molding*[28,29] mainly uses a heated hollow mold that is filled with a specific material. It is then slowly rotated (habitually around two orthogonal directions), making the softened polymer to spread and stick to the walls of the mold. To keep a proper uniform thickness throughout the part, the rotation of the mold is continuous during the heating stage, and to avoid deformation, rotation is sometimes kept even in the cooling phase.
- *Extrusion molding*[30] mainly relies on forcing softened polymer material to pass through a die with an opening.
- *Thermoforming*[31] is a processing approach where a polymer sheet is heated to a pliable forming temperature, modeled to a desired shape in a mold, and trimmed to attain a usable product.

Having all these aspects in view, it can be said that regardless the particularities of the aforementioned methods, the preparation of composites using green polymer matrix involves the same steps as in the case of classical polymer composites, namely:

- Selection of the matrix and the filler(s) as a function of the pursued properties that need to be improved.
- *Proper dispersion of the filler*: Uniform and homogeneous dispersion of the inserted reinforcement compounds is very important. When dealing with nanofillers, it is often noted that they tend to aggregate and constitute micron-size filler cluster that impedes the dispersion of nanoparticles in the matrix, thereby damaging the features of resulted material. Many efforts were devoted to this issue by chemical reactions, complex polymerization reactions, or surface adaptation of

filler materials. The most employed approaches for surface functionalization of filler material are[15]:

- the surface oxidation;
- the covalent coupling through oxidized materials (amidation, silylation, silanization, grafting of macromolecular compounds); and
- the noncovalent functionalization.

• Final processing stage involves drying (in case of solvents) or hardening (in the situation of molten reinforced polymers).

From the point of view of the filler distribution in the polymer, the green
composites can be divided as follows:

• *Uniformly filled polymer composites*: Generally, three ways are used
to disperse inorganic nanoparticles into polymer matrices. The first
uses the top-down methodology, which consists in the direct mixing
of the filler into a polymeric matrix in solution or in melt. The second
is based on in-situ polymerization in the presence of the nanoparticles
previously obtained or in-situ synthesis of inorganic nanoparticles in
the presence of polymer. The last one, inorganic and organic components are both formed in situ. These two last methods are defined
as bottom-up methodologies. Nevertheless, Figure 11.3 makes a
summary of the basic methods used to prepare nanocomposites.
• *Gradient polymer composites*[32]: The techniques of fabrication of
polymeric gradient composites are as follows:

- *The gravity casting* is particularly based on the action of gravity
forces without applying pressure. It is possible to obtain the
desired spatial distribution of component materials by step-by-
step deposition of two or several melts. It is essential that the first
melt is cast, and only when it begins to solidify or crosslink, then
the second melt is poured.
- *The centrifugal casting* is dependent on the nature of forming
material (paste or powder). The composite material is placed near
to the heated mold's wall. Mold with processed blend is rotated
around one or two directions, while the material is subjected
to temperatures inside the mold. The applied centrifugal force
makes one of the components to progressively sediment on
the form's walls. In the following stage, the temperature of the
form is lowered and molding is removed.[33] In previous reports
from literature, it was revealed that this technique is suitable for

polymer materials containing small particles. The gradation of physical properties is affected by the dimensions and shape of filler and by changing rate and time of centrifugation.

- *Corona discharge* is not only suitable for adapting the features of surface layer of flat products (like plates, films), but also for materials with more complex shapes (like pipes, containers). Generally, a corona-discharge apparatus contains a radiofrequency (RF) generator, high-voltage transformer, and two electrodes. One of them is ground down and wrapped with a dielectric material, and the other electrode which is made up of aluminum is bounded to the RF signal generator[34];

- *Selective laser sintering* is a process mainly applied to three-dimensional objects. Parts are obtained by fusing polymer, metal, or ceramic powder. This processing method necessitates peculiar properties of material's particles, namely the powder must be in such a form that allows to flow freely. The approach involves deposition and consolidation layer by layer, which supposes that the layer of material is selectively fused by applying laser beam. Subsequently, the next layer is deposited on the top, and the process is repeated.[35,36]

- *Pressing* can be performed uniaxially or at elevated temperatures (hot pressing, hot isostatic pressing).[37] The first processing requires the use of rigid forms that consists of matrix and punch. Pressing is done at a certain pressure and only in vertical direction enabling fabrication of density gradient composites with polymer matrix.

- *UV photopolymerization*: UV radiations are used to initiate reaction of monomers to render polymeric structures. The main advantages are the reduced temperature conditions and the fact that light can be focused onto a desired spot. The incident radiation wavelength, light intensity and characteristics of monomer, and induction, dissociation, or ionization of monomer can take place.[38] Photopolymerization is useful to attain gradient materials and their production can be imparted into two steps. First one consists in absorbance variation of UV in the medium to accomplish gradient material. The second part relies on thermal crosslinking reaction.

- *Nonuniformly filled polymer composites*: If no procedure is applied to break the filler aggregates, then the polymer will be reinforced nonuniform. In other words, simple addition of particles in the matrix without further ultrasonication or functionalization, the filler is then randomly distributed in the polymer.

Figure 11.2 presents a schematic representation of green polymer composites based on the filler type and its distribution in the matrix.

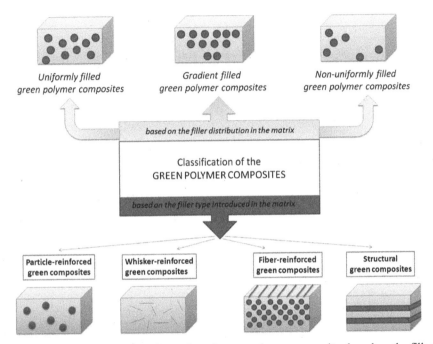

FIGURE 11.2 The schematic illustration of green polymer composites based on the filler type and its distribution in the matrix.

11.3 BULK AND SURFACE PROCESSING ROUTES FOR ALIGNMENT OF THE FILLER MATERIAL

The advantage of introduction in green polymers of fillers with high aspect ratio (like fibers or nanotubes) is that one may produce multiphase materials with directed and/or enhanced properties. There are several proposed strategies for the design and preparation of green polymer composites with complex alignment of the dispersed phase.[39-44] According to literature,[45] the alignment procedures can be divided into two large categories:

- In-situ techniques
- Ex-situ techniques or postsynthesis methods.

Figure 11.3 displays the main types of alignment approaches that can be used for both common and green polymer composites. It can be remarked that both in-situ and ex-situ procedures involve application of electrical or magnetic external fields. In addition, the alignment of the dispersed phase in the matrix can be achieved in bulk or at the surface of the reinforced material.

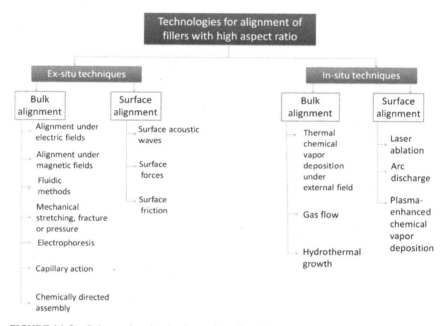

FIGURE 11.3   Scheme showing in-situ and ex-situ alignment techniques of alignment of the fillers with high aspect ratio in common or green polymer nanocomposites.

Detailed information concerning each method depicted in Figure 11.3 can be found in literature.[39–45] In-situ methods directly align anisotropic fillers during their growth process. For instance, aligned carbon nanotubes (CNT) or nanofibers (NFs) are often grown using the in-situ approaches. If a particular attention is given to alignment from thermal chemical vapor deposition (CVD) technique assisted by catalytic nanoparticles, one may distinguish the following subcategories[45]:

- thermal CVD with crowding effect
- thermal CVD growth under electric field
- thermal CVD growth under gas flow
- thermal CVD under magnetic field
- thermal CVD growth with epitaxy

Besides the already known graphoepitaxy methods, there are other in-situ alignment procedures, such as arc discharge, laser ablation, magnetic field, gas flow, hydrothermal method, electric field, and flame.[44]

In contrast, ex-situ alignment does not impose the use of a filler-fixing agent. The approaches involving electric fields often generate varied results, with a degree of filler orientation commonly less than comparable shear methods. This is mainly caused due to the use of filler liquid suspensions, where the viscous and surface tension largely influencing the degree of orientation. The linear packing density resulted from this procedure was found to be considerably dependent on the ink concentration used. Ex-situ alignment of composite samples via magnetic fields lead often relatively poor results. Usually suspended within a polymer solution or melt, the slowing of the filler paramagnetic alignment is critical, similarly to the manner in which a fixing agent makes the methods impractical for certain applications. The application of electric[46–50] and magnetic[51–54] fields has been demonstrated to be promising for fabrication of composites with oriented fillers. For instance, CNT suspensions after spin or drop casting seem to have a moderate success when using dielectrophoresis to exploit the weak filler dipole. If applying an electric field of about 10 V/μm[55–57] that was used to align solution dispersed nanotubes, a field of 1 V/μm[48,58] has been proved to orient the CNT during synthesis. Despite the advantages of using electric and magnetic fields have in filler orientation in composites with polymer matrix, the demanding electrode and catalyst-patterning steps limit scalability, both in terms of filler (CNT or NF) length and mass production. In practice, the devices must have larger scales.[42]

Among the ex-situ techniques, those that involve mechanical forces have known some interesting developments.[45] As an example, the shearing techniques have been adapted from loading at a predetermined angle, to stretching and compressing vertically aligned fillers. In this way, enhanced packing densities can be accomplished particularly at the thin film surface of the composite. Bulk assessment is somehow challenging, though alignment through such a procedure is affected by the angle of shear for the pressing technique and is very dependent on the user. In conditions of compressing from a certain and fixed axis of rotation, many degrees of misalignment can be produced in the material, but it was demonstrated that if the press is translated forward as it is rotating the orientation of the filler can be dramatically enhanced. Moreover, when using green polymer matrix that is able to develop liquid-crystalline properties under shearing, one may align fillers with high aspect ratio. For such purposes, cellulose derivatives are suitable green polymers which after removing the shearing field begin to relax, creating a banded morphology.[59-66] This presents a great interest in orientation of CNT or NF inside green polymers and not only. A typical band texture of sheared cellulosic matrix in lyotropic phase is displayed in Figure 11.4(a). In fact, literature[65] shows that sheared cellulosic solutions are forming two sets of periodic structures. The most prominent one is produced perpendicular to the shear direction, whereas the second finer set consists of bands slightly tilted from the mechanical force direction. An out-of-plane angle fluctuation (of about 9°–13°), of the sinusoidal macromolecular orientation, is also noticed.[65] Inserting fillers with high aspect ratio in such green polymer matrix was shown to be well aligned at low percentage of reinforcements,[59] as depicted in Figure 11.4(b). This could be helpful to attain green polymer composites with directed electrical or thermal conduction for electronic industry and other type of applications.

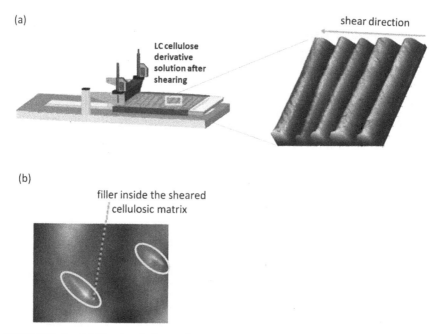

FIGURE 11.4 (a) The atomic force microscopy scans revealing the morphology of pristine cellulose derivative in lyotropic phase after shear cessation and (b) filler orientation inside the anisotropic sheared polymer matrix.

Another category of methods is focused on nanomanipulation but the disadvantage arises from the fact that the processing is serial and demands time-consuming electron microscopy approaches.[67] Relatively recent trends[45] show that a good solution is the fabrication of freestanding alternative, capable of creating significant lengths of freestanding thin composite films containing aligned nanofillers without involvement of a binder or another supporting substrate.[68] Such kind of structures are observed to be very well aligned and enable particularly enhanced linear packing densities. The practical use of reinforced green polymers in electronic industry is described in the next section of the chapter.

11.4 PRACTICAL USE OF MULTIPHASE GREEN POLYMER MATERIALS IN ELECTRONICS

Electronic products become more and more important in our lives, but there is a great importance that the technology should not leave a permanent mark on the nature even after disposal. Green polymers seem to be the key of the future in this industry. However, in the case of some specific applications, the basic requirements (flexibility, electric/thermal conductivity, gas, and water vapor barrier) cannot be fulfilled by the majority of biodegradable polymers. Therefore, the fabrication of nanocomposites by adding reinforcement compounds into biopolymers could endow them with the desired functional characteristics, while preserving the biodegradability.

In the described context, green polymers and their composites provide a valuable link between electronics, soft matter, and environment. The broad chemical design of such materials enables tunability of electronic, thermal, and mechanical characteristics. The biodegradable electronic compounds are highly desirable for current technologies and can be imparted in two categories of materials[69–72]:

- *Type I* are disintegrable materials that are characterized by partial degradation and are often used in devices made to operate for a defined time scale. These materials suffer macroscopic degradation, the molecular cleavage of the polymer chains into oligomers and monomers is facilitated by microorganisms in the environment.
- *Type II* are materials that endure complete degradation and can decompose into bioderived or natural building blocks. Such products are highly important owing to their intrinsic biodegradation and recyclability. These materials can disintegrate into known biocompatible small molecules, which have lower potential to elicit negative long-term responses.

To understand in a clearer manner the features of each type of biodegradable polymer composites for electronics, Figure 11.5 presents a schematic representation of the earlier mentioned categories. Each of them plays an important role in applications, where the lifetime of the product is essential. Type I are disintegrable from the point of view of matrix properties leading upon degradation to small molecules of the building units. On the other hand, the second type refers to composites where both matrix and filler are biodegradable and thus they have minimized impact on the environment after disposal of the product. Such devices are more easy to recycle and should be more often found on market these days to keep a clean planet and maintain the health of living beings.

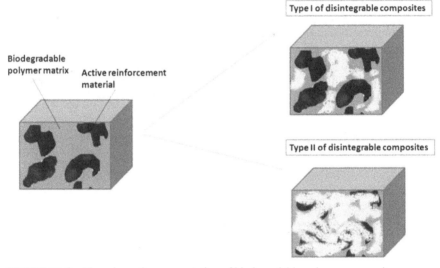

FIGURE 11.5   The schematic representation of biodegradable polymer composites.

As such, biodegradable nanocomposites with polymer matrix have large application prospects in electronics. A brief schematization of their practical use in this industry can be done as follows:

- *Substrates* can be found in many electronic devices and since they represent the prevalent weight, they are dictating the degradation behavior of the product. The selection of the materials is related to the demands fabrication processing stages, thermal stability, solvent compatibility, and mechanical performance. CMOS transistors array on several biodegradable substrates made from materials polylactic-*co*-glycolic acid (PLGA), PCL, cellulose, and its derivatives which can be doped with NFs or other types of nanoparticles.[69]

- *Dielectrics* are materials with high or low permittivity (k) depending on the pursued goal. Low-k composites are suitable for reduced dissipation of electromagnetic energy, while high-k composites are adequate for fabrication of capacitors being essential for capacitive sensing and field–effect transistors. Biodegradable dielectrics are achieved by incorporation of high-κ fillers (Al_2O_3, HfO_2, NFs extracted from plants such as cotton, banana, or jute) into a degradable/green polymer matrix (DNA, poly(glycerol sebacate), cellulose)[69,72].

- *(Semi)conductors* are used for ensuring the switching mechanisms in transistors, while the conducting biodegradable composites are used for interconnects and contacts in electronic circuits. The key feature here is the conductivity that imposes the use of a biodegradable polymer with conjugated structure and conductive fillers. Moreover, the doping of the matrix must be done above the percolation threshold to create conductive pathways in the polymer host. Among the reported green composites, one may mention[69-72]: poly(D,L-lactic acid)/polypyrrole nanoparticles, polypyrrole/PCL fumarate/anionic dopants, polypyrrole/PLGA fibers, polyaniline gelatin NFs/camphorsulfonic acid, and so on.

Figure 11.6 illustrates a general picture regarding the described applications of the green polymer composites in electronic industry.

In conclusion, future directions of the biodegradable/green composites with polymer matrix, as well as the related challenges, that should be overcome are presented.

11.5 CONCLUDING REMARKS AND FUTURE PERSPECTIVES

In the context of our planet pollution, it is easy to understand why e-waste is becoming a hot topic of discussion around sustainability. There is an increasing need for immediate solutions for these growing mountains of residues. Among the measures that must be taken, one should include improved legislation, proper education, more refined recycling technologies, and better recycling programs to overcome the crisis of the electronic wastes. There are certain companies that have already changes their policies and made real efforts toward application of advanced recycling methods. For example, consumers have the opportunity to sell old phones at the shop where they buy a new one. The company then refurbishes the components that can be brought in good conditions and guarantees that the rest of electronic pieces are recycled in a sustainable way.

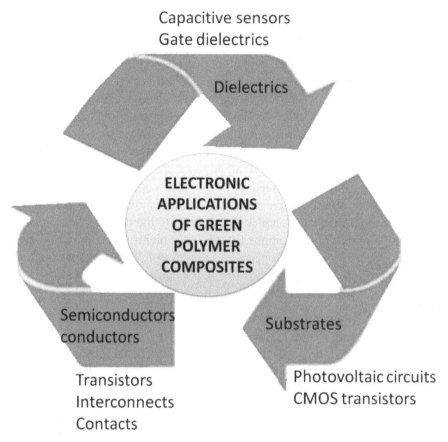

Capacitive sensors
Gate dielectrics

Dielectrics

ELECTRONIC
APPLICATIONS
OF GREEN
POLYMER
COMPOSITES

Semiconductors
conductors

Substrates

Transistors
Interconnects
Contacts

Photovoltaic circuits
CMOS transistors

FIGURE 11.6   The main practical uses of green polymer composites in electronic industry.

Another way to address e-waste on a global level is that the manufacturers collaborate with researchers in replacing as much as possible the toxic substances from the devices. In this background, green chemistry and corresponding materials have gained lots of attention. Green polymers composites made from biodegradable counterparts represent the future for green electronics. A huge number of natural and synthetic composite materials display practical electronic features as dielectrics, (semi)conductors, and flexible substrates. Electrical conductivity is determined by the characteristics of both polymer and dispersed phase. Synthetic materials enable better control regarding mechanical properties, but they are not biodegradable. Thus, a conjugation-breaking degradation approach through mechanisms like UV-light exposure, oxidation, or enzymes, without diminishing the conductivity must be developed. Polymers derived from natural sources, even if they are biodegradable they do not reach the same electrical performance as the synthetic ones so their applicability is limited. Addition of reinforcements into the polymer medium leads to enhancement of conductivity. However, there is still work to do for improving the dispersion state of fillers in such systems to achieve the desired conductivity. Future studies should also consider the interactions of electronic biodegradable devices with different physiological environments and to investigate degradation processes through the highly reactive oxygen and nitrogen radical species formed by macrophages upon meeting foreign bodies.

KEYWORDS

- **green polymers**
- **composites**
- **surface processing**
- **physical properties**
- **applications**

REFERENCES

1. Ellis, B. Environmental Issues in Electronics Manufacturing: A Review. *Circuit World* **2000,** *26*, 17–21.
2. Needhidasan, S.; Samuel, M.; Chidambaram, R. Electronic Waste—An Emerging Threat to the Environment of Urban India. *J. Environ. Health Sci. Eng.* **2014,** *12*, 36 (1–9).
3. Sivaram, N. M.; Gopal, P. M.; Barik, D. Toxic Waste from Textile Industries. In *Energy from Toxic Organic Waste for Heat and Power Generation*; Barik, D., Ed.; Woodhead Publishing Series in Energy: UK, 2019; pp 43–54.
4. Barbuta, M.; Bucur, R. D.; Cimpeanu, S. M.; Paraschiv, G.; Bucur, D. Wastes in Building Materials Industry. In *Agroecology*; Pilipavicius, V., Ed.; InTech: Croatia, 2015; pp 80–100.
5. Sitaramaiah, Y.; Kumari, M. K. Electronic Waste Leading to Environmental Pollution. *J. Chem. Pharm. Sci. Nat. Semin. Impact Toxic Metals, Miner. Solvents lead Environ. Pollut.* **2014,** *2014*, 39–42.

6. Lucier, C. A.; Gareau, B. J. Electronic Waste Recycling and Disposal: An Overview. In *Assessment and Management of Radioactive and Electronic Wastes*. InTech: Croatia, 2019; pp 1–12.

7. Industry Solutions for the World's Fastest-Growing Waste Stream. https://www.ewaste-expo.com/about/ (accessed on December 2019).

8. Electronic Devices 'Need to Use Recycled Plastic'. https://www.bbc.com/news/science-environment-50046859 (accessed on December 2019).

9. Irimia-Vladu, M. "Green" Electronics: Biodegradable and Biocompatible Materials and Devices for Sustainable Future. *Chem. Soc. Rev.* **2014**, *43*, 588–610.

10. Cheng, H. N.; Smith, P. B.; Gross, R. A. Green Polymer Chemistry: A Brief Review. In *Green Polymer Chemistry: Biocatalysis and Materials* II; Cheng, H. N., Gross, R. A., Smith, P. B., Eds.; American Chemical Society: USA, 2013.

11. Baumgartner, M.; Coppola, M. E.; Sariciftci, N. S.; Glowacki, E. D.; Bauer, S.; Irimia-Vladu, M. Emerging "Green" Materials and Technologies for Electronics. In *Green Materials for Electronics*; Irimia-Vladu, M., Glowacki, E. D., Sariciftci, N. S., Bauer, S., Eds.; Wiley: New York, 2018; pp 1–54.

12. Saha, M.; Sutradhar, P. Advances in Polymer Composites: Green and Nanotechnology. In *Green Polymer Composites Technology. Properties and Applications*; Inamuddin, Ed.; CRC Press: USA, 2017.

13. Visakh, P. M.; Bayraktar, O. *Bio-monomers for Green Polymeric Composite Materials*; Wiley: USA, 2019.

14. Rose, C. Designing for Composites: Traditional and Future Views. In *Green Composites: Polymer Composites and the Environment*; Baillie, C., Ed.; CRC Press: USA, 2004; pp 9–23.

15. Rahaman, M.; Khastgir, D.; Aldalbahi, A. K. *Carbon-Containing Polymer Composites*; Springer: Singapore, 2019.

16. Khan, W. S.; Hamadneh, N. N.; Khan, W. A. Polymer Nanocomposites—Synthesis Techniques, Classification and Properties. In *Science and Applications of Tailored Nanostructures*; Di Sia, P., Ed.; One Central Press (OCP): Italy, 2017; pp 50–67.

17. Oliveira, M.; Machado, A. V. Preparation of Polymer-Based Nanocomposites by Different Routes. In *Nanocomposites: Synthesis, Characterization and Applications*; Wang, X., Ed.; NOVA Science Publishers: China, 2013; pp 73–94.

18. Guo, Y.; Peng, F.; Wang, H.; Huang, F.; Meng, F.; Hui, D.; Zhou, Z. Intercalation Polymerization Approach for Preparing Graphene/Polymer Composites. *Polymers* **2018**, *10*, 61 (1–28).

19. Motaung, T. E.; Mochane, M. J.; Linganiso, Z. L.; Mashigo, A. P. In Situ Polymerization of Nylon-Cellulose Nano Composite. *Polym. Sci.* **2017**, *3* (1), 2 (1–8).

20. Reddy, R. J. Preparation, Characterization and Properties of Injection Molded Graphene Nanocomposites, Master's Thesis, Mechanical Engineering, Wichita State University: Wichita, Kansas, USA, 2010.

21. Pomogailo, A. D. Polymer Sol–Gel Synthesis of Hybrid Nanocomposites. *Colloid J.* **2005**, *67*, 658–677.

22. Tanahashi, M. Development of Fabrication Methods of Filler/Polymer Nanocomposites: With Focus on Simple Melt-Compounding-Based Approach without Surface Modification of Nanofillers. *Materials* **2010**, *3*, 1593–1619.

23. Volpe, V.; Lanzillo, S.; Affinita, G.; Villacci, B.; Macchiarolo, I.; Pantani, R. Lightweight High-Performance Polymer Composite for Automotive Applications. *Polymers (Basel)* **2019**, *11*, 326 (1–16).

24. Mitsoulis, E. Calendering of Polymers. *Advances in Polymer Processing: From Macro-to Nanoscales*; Woodhead Publishing Limited: Sawston, Cambridge, United Kingdom, 2009; pp 312–351.

25. Park, C. H.; Lee, W. I. Compression Molding in Polymer Matrix Composites. In *Manufacturing Techniques for Polymer Matrix Composites (PMCs)*; Woodhead Publishing Limited: Sawston, Cambridge, United Kingdom; 2012; pp 47–94.

26. Blow Molding. https://www.substech.com/dokuwiki/doku.php?id=blow_molding (accessed on December 2019).

27. Rodriguez-Castellanos, W.; Martínez-Bustos, F.; Rodrigues, D.; Trujillo-Barragán, M. Extrusion Blow Molding of a Starch-Gelatin Polymer Matrix Reinforced with Cellulose. *Eur. Polym. J.* **2015**, *73*, 335–343.

28. Rotational Molding. https://en.wikipedia.org/wiki/Rotational_molding (accessed on December 2019).

29. Hanana, F. E.; Rodrigue, D. Rotational Molding of Polymer Composites Reinforced with Natural Fibers. *Plast. Eng.* **2015**, *71*, 28–31.

30. Sohn, J. S.; Ryu, Y.; Yun, Ch.-S.; Zhu, K.; Cha, S. W. Extrusion Compounding Process for the Development of Eco-Friendly SCG/PP Composite Pellets. *Sustainability* **2019**, *11*, 17209 (1–12).

31. Thermoforming. https://en.wikipedia.org/wiki/Thermoforming (accessed on December 2019).

32. Stabik, J.; Dybowska, A. Methods of Preparing Polymeric Gradient Composites. *J. Achiev. Mater. Manuf. Eng.* **2007**, *25*, 67–70.

33. Hashmi, S. R. A.; Dwivedi, U. K. Estimation of Concentration of Particles in Polymerizing Fluid during Centrifugal Casting of Functionally Graded Polymer Composites. *J. Polym. Res.* **2007**, *14*, 75–81.

34. Li, H.; Lambros, J., Cheesemas, B. A.; Santare, M. H. Experimental Investigation of the Quasi-Static Fracture of Functionally Graded Materials. *Int. J. Solids Struct.* **2000**, *37*, 3715–3732.

35. Chung, H.; Das, S. Processing and Properties of Glass Bead Particulate-Filled Functionally Graded Nylon-11 Composites Produced by Selective Laser Sintering. *Mater. Sci. Eng. A* **2006**, *437*, 226–234.

36. Jepson, L.; Beaman, J.; Bourell, D.; Wood, K. *SLS Processing of Functionally Gradient Materials*. In Solid Freeform Fabrication Proceedings, Austin, USA, 1997; pp 67–80.

37. Lee, N. J.; Jang, J. The Effect of Fibre-Content Gradient on the Mechanical Properties of Glass-Fibre-Mat/Polypropylene Composites. *Compos. Sci. Technol.* **2000**, *60*, 209–217.

38. Sikorski, R. T. *Foundations of Chemistry and Technology of Polymers*; PWN: Warsaw, 1985 (in Polish).

39. Sierra-Romero, A.; Chen, B. Strategies for the Preparation of Polymer Composites with Complex Alignment of the Dispersed Phase. *Nanocomposites* **2018**, *4*, 137–155.

40. Duong, H. M.; Gong, F.; Liu, P.; Tran, T. Q. Advanced Fabrication and Properties of Aligned Carbon Nanotube Composites: Experiments and Modeling. In *Carbon Nanotubes—Current Progress of their Polymer Composites*; Berber, M., Hafez, I. F., Eds.; InTech: Croatia, 2016; pp 47–72.

41. Beigmoradi, R.; Samimi, A.; Mohebbi-Kalhori, D. Engineering of Oriented Carbon Nanotubes in Composite Materials. *Beilstein J. Nanotechnol.* **2018**, *9*, 415–435.

42. Cole, M. T.; Cientanni, V.; Milne, W. I. Horizontal Carbon Nanotube Alignment. *Nanoscale* **2016**, *8*, 15836–15844.

43. Iakoubovskii, K. Techniques of Aligning Carbon Nanotubes. *Cent. Eur. J. Phys.* **2009**, *7*, 645–653.

44. Zhao, H.; Zhou, Z.; Dong, H.; Zhang, L.; Chen, H.; Hou, L. A Facile Method to Align Carbon Nanotubes on Polymeric Membrane Substrate. *Sci. Rep.* **2013**, *3*, 3480 (1–5).

45. Ren, Z.; Lan, Y.; Wang, Y. Technologies to Achieve Carbon Nanotube Alignment, Aligned Carbon Nanotubes. *Physics, Concepts, Fabrication and Devices*; Springer: Berlin, 2013; pp 111–156.

46. Benedict, L. X.; Louie, S. G.; Cohen, M. L. Static Polarizabilities of Single-Wall Carbon Nanotubes. *Phys. Rev. B* **1995**, *52*, 8541–8549.

47. Zhang, Y.; Chang, A.; Cao, J.; Wang, Q.; Kim, W.; Li, Y.; Morris, N.; Yenilmez, E.; Kong, J.; Dai, H. Electric-Field-Directed Growth of Aligned Single-Walled Carbon Nanotubes. *Appl. Phys. Lett.* **2001**, *79*, 3155–3157.

48. Ural, A.; Li, Y.; Dai, H. Electric-Field-Aligned Growth of Single-Walled Carbon Nanotubes on Surfaces. *Appl. Phys. Lett.* **2002**, *81*, 3464–3466.

49. Remillard, E.M.; Zhang, Q.; Sosina, S.; Branson, Z.; Dasgupta, T.; Vecitis, C. D. Electric-Field Alignment of Aqueous Multi-Walled Carbon Nanotubes on Microporous Substrates. *Carbon* **2016**, *100*, 578–589.

50. Amani, A.; Hashemi, S. A.; Mousavi, S. M.; Pouya, H.; Vojood, A. Electric Field Induced Alignment of Carbon Nanotubes: Methodology and Outcomes. In *Carbon Nanotubes— Recent Progress*; Rahman, M., Asiri, A. M., Eds.; InTech: London, 2018; pp 71–88.

51. Kordás, K.; Mustonen, T.; Tóth, G.; Vähäkangas, J.; Jantunen, H.; Gupta, A.; Rao, K. V.; Vajtai, R.; Ajayan, P. M. Magnetic-Field Induced Efficient Alignment of Carbon Nanotubes in Aqueous Solutions. *Chem. Mater.* **2007**, *19*, 4, 787–791.

52. Correa-Duarte, M.A.; Grzelczak, M.; Salgueiriño-Maceira, V.; Giersig, M.; Liz-Marzán, L. M.; Farle, M.; Sierazdki, K.; Diaz, R. Alignment of Carbon Nanotubes under Low Magnetic Fields through Attachment of Magnetic Nanoparticles. *J. Phys. Chem. B* **2005**, *109*, 19060–19063.

53. Jang, B. K.; Sakka, Y.; Woo, S. K. Alignment of Carbon Nanotubes by Magnetic Fields and Aqueous Dispersion. *J. Phys. Conf. Ser.* **2009**, *156*, 012005.

54. Ariu, G.; Hamerton, I.; Ivanov, D. Positioning and Aligning CNTs by External Magnetic Field to Assist Localised Epoxy Cure. *Open Phys.* **2016**, *14*, 508–516.

55. Kim, P.; Lieber, C. M. Nanotube Nanotweezers. *Science* **1999**, *286*, 2148–2150.

56. Maeda, M.; Hyon, C. K.; Kamimura, T.; Kojima, A.; Sakamoto K.; Matsumoto, K. Growth Control of Carbon Nanotube Using Various Applied Electric Fields for Electronic Device Applications. *Jpn. J. Appl. Phys.* **2005**, *44*, 1585–1587.

57. Law, J. B. K.; Koo C. K.; Thong, J. T. L. Horizontally Directed Growth of Carbon Nanotubes Utilizing Self-Generated Electric Field from Plasma Induced Surface Charging. *Appl. Phys. Lett.* **2007**, *91*, 2431081–2431083.

58. Jang, Y. T.; Ahn, J. H.; Ju B. K.; Lee, Y. H. Lateral Growth of Aligned Mutilwalled Carbon Nanotubes under Electric Field. *Solid State Commun.* **2003**, *126*, 305–308.

59. Barzic, R. F. Study of Thermophysical Properties of Polymer Nanocomposites. PhD Thesis, "Gheorghe Asachi" Technical University, Faculty of Mechanics, 2013.

60. Stoica, I., Barzic, A. I., Hulubei, C. Polyimide Embedding in lyotropic Polymer Matrix for Surface-Related Applications: Rheological and Microscopy Investigations. *Rev. Roum. Chim.* **2016**, *61*, 575–581.

61. Ernst, B.; Navard, P. Band Textures in Mesomorphic (Hydroxypropyl) Cellulose Solutions. *Macromolecules* **1989**, *22*, 1419–1422.

62. Cosutchi, A. I.; Hulubei, C.; Stoica, I.; Ioan, S. A New Approach for Patterning Epiclon-Based Polyimide Precursor Films using a Lyotropic Liquid Crystal Template. *J. Polym. Res.* **2011**, *18*, 2389–2402.

63. Barzic, A. I.; Hulubei, C.; Avadanei, M. I.; Stoica, I.; Popovici, D. Polyimide Precursor Pattern Induced by Banded Liquid Crystal Matrix: Effect of Dianhydride Moieties Flexibility. *J. Mater. Sci.* **2015**, *50*, 1358–1369.

64. Marsano, E.; Carpaneto, L.; Ciferri, A. Formation of a Banded Texture in Solutions of Liquid Crystalline Polymers: 1. Hydroxypropylcellulose in H_2O. *Mol. Cryst. Liq. Cryst.* **1988**, *158*, 267–278.

65. Godinho, M. H.; Fonseca, J. G.; Ribeiro, A. C.; Melo, L. V.; Brogueira, P. Atomic Force Microscopy Study of Hydroxypropylcellulose Films Prepared from Liquid Crystalline Aqueous Solutions. *Macromolecules* **2002**, *35*, 5932–5936.

66. Almeida, A. P. C.; Canejo, J. P.; Fernandes, S. N.; Echeverria, C.; Almeida, P. L.; Godinho, M. H. Cellulose-Based Biomimetics and their Applications. *Adv. Mater.* **2018**, *30*, 1703655 (1–31).

67. De Jonge, N.; Allioux, M.; Doytcheva, M.; Kaiser, M.; Teo, K. B. K.; Lacerda, R. G.; Milne, W. I. Characterization of the Field Emission Properties of Individual Thin Carbon Nanotubes. *Appl. Phys. Lett.* **2004**, *85*, 1607–1609.

68. Cole, M. T.; Doherty, M.; Parmee, R.; Dawson P.; Milne, W. I. Ultra-Broadband Polarisers Based on Metastable Free-Standing Aligned Carbon Nanotube Membranes. *Adv. Opt. Mater.* **2014**, *2*, 929–937.

69. Feig, V. R.; Tran, H.; Bao, Z. Biodegradable Polymeric Materials in Degradable Electronic Devices. *ACS Cent. Sci.* **2018**, *4*, 337–348.

70. Li, R.; Wang, L.; Yin, L. Materials and Devices for Biodegradable and Soft Biomedical Electronics. *Materials* **2018**, *11*, 2108 (1–23).

71. Liu, H.; Jian, R.; Chen, H.; Tian, X.; Sun, C.; Zhu, J.; Yang, Z.; Sun, J.; Wang, C. Application of Biodegradable and Biocompatible Nanocomposites in Electronics: Current Status and Future Directions. *Nanomaterials* **2019**, *9*, 950 (1–31).

72. Huang, Y.; Kormakov, S.; He, X.; Gao, X.; Zheng, X.; Liu, Y.; Sun, J.; Wu, D. Conductive Polymer Composites from Renewable Resources: An Overview of Preparation, Properties, and Applications. *Polymers* **2019**, *11*, 187 (1–32).

CHAPTER 12

Sustainable Eco-Friendly Polymer-Based Membranes Used in Water Depollution for Life-Quality Improvement

ADINA MARIA DOBOS*, MIHAELA DORINA ONOFREI, and ANCA FILIMON

Department of Physical Chemistry of Polymers, Petru Poni Institute of Macromolecular Chemistry, Iassy, Romania

Corresponding author. E-mail: necula_adina@yahoo.com

ABSTRACT

As a result of the recent environmental problems, such as pollution and water contamination, in-depth studies have been carried out to develop systems for wastewater treatment and cleaning, some of these systems involving the use of membranes technology. There are three types of wastewater resulting from industry, agriculture, and those from households, these sectors generating the largest amount of wastewater. The chapter presents a series of membranes based on eco-friendly polymers designed to remove the main contaminants of the water, such as dyes, metal ions, fertilizers, feces, bacteria, or oils. The technique of membrane systems using is effective, ease of implementation, and does not involve high costs, especially when they are obtained from natural resources. The creation of the improved systems containing new and more efficient membranes with enhanced porosity, selectivity, antifouling, and antibacterial properties is and will be the subject of future research because it ensures the decontamination of the environment, improving the life quality.

12.1 INTRODUCTION TO MEMBRANE TECHNOLOGY

The hazards on the environmental and human health generated by the release of various synthetic chemicals are under constant debate. There is no doubt that this debate will continue until the science resolves the uncertainties concerning toxicological data (exposure, transport, etc.) and risk analysis. Industry and society have traditionally dealt with the reduction of these risks focusing on the exposure decrease. The risk can be reduced through pollution prevention, defined as the use of processes, practices, materials, or products that avoid or reduce the pollution from the source. Pollution prevention may include recycling, process changes, mechanism for environment safety control, efficient resources use, and materials substitution. In addition, the pollution control can be achieved through an environmentally safe treatment, providing an opportunity to devise creative strategies for the protection of the human health and environment, and to formulate innovative approaches for building a sustainable planet.

Two of the greatest global challenges facing the 21st century involve the sustainable supply of clean water and energy, two highly interrelated resources, at affordable costs. Water contaminations, resulting from extensive industrialization and population growth, are the prime cause of environmental and human health degrading because it is directly linked to chemicals with a high degree of toxicity emitted by over 80% of the places with harmful waste. There is a lot of water on the planet, but of inadequate quality (purity) for the human consumption or other beneficial (e.g., industrial/agricultural) purposes. Increasingly, the drinking water resources are showing evidence of contamination with several kinds of pollutants, such as macromolecular proteins, oils, volatile organic solvents, heavy metals, pesticides, asbestos, humic acids, dust particles, and so on.[1] It has been recognized that the water resources depletion combined with their environmental impact pose a major issue for our society and that the new technologies are the key to current and future needs to meet these challenges.[2]

Traditional water treatment methods do not effectively remove the contaminants which pollute the water sources, and consequently are not capable to meet high water quality standards.[3–5] Instead, the membrane technology is expected to continue to dominate the water purification technologies owing to its energy efficiency. However, there is a need for improved membranes that present a higher flux, are more selective, less prone to various types of fouling, and more resistant to the chemical environment of these processes. Therefore, deficiencies of current membranes are identified, and the opportunities to resolve them through innovative polymer chemistry and physics will address the fundamental issues. Additionally, new technologies for designing of the integrated processes will develop and implement strategies for achieving their objectives, having a critical role in pollution prevention.[2] Thus, advanced technologies for water purification are an essential part of meeting the current and future water needs.

12.1.1 THE KEY FOR THE POLLUTION PREVENTION AND WASTE REDUCTION

Pollution is the consequence of many complex and interrelating actions, decisions, and technologies involving multiple parties including individuals, companies, and regulatory bodies. The difference between the actual and possible performance must be realized through an integrated approach involving the regulatory process, executive-level understanding, and the application of technology to design integrated production facilities. Thus, the following three activities have the *greatest impact on* the *pollution prevention*, namely, the regulatory process that creates the method by which the companies engage to achieve products/materials; the executive understanding, including both the people and systems that they use to develop strategies for the direction targeted, and the applied technology to plan, design, construct, operate, maintain, and improve the production facilities. Pollution will be reduced or eliminated only if all three activities work together in an integrated way to solve the problem. Consequently, *integration is the key to pollution prevention.*

Pollution prevention and waste reduction will occur when these parties act in an integrated or global manner to address the subject realistically and fundamentally. Therefore, the main challenge is *to identify and implement the technology*. Thus, there are three interrelated key concepts[6,7]:

- first concept involves the processes in which the components act in a coordinated manner to transform raw materials into products and to achieve the desired objectives (e.g., polymeric membranes);
- second concept involves the union of the parts/components to form the whole, defining the integration—the key for pollution prevention;
- third concept of the process integration is one of the most important developments in technological design. In this stage, it is recognized that processes can be analyzed, understood, and designed as integrated systems.

Currently, there is sufficient technology to make significant progress in preventing pollution and waste reduction. The *effective understanding and deployment of technology* constitute other interesting points, leading to the idea that technology is not the limiting factor. Additionally, science and technology represent the physical reality governing pollution prevention. Science also represents the physics, chemistry, thermodynamics, mathematics, and other laws that govern the physical world, manages information that support the process control, collection, storage, retrieval, analysis, and use. Moreover, science includes the biological effects or impact of pollutants on the health and environment.

To solve the problem properly, any designed or integrated system that facilitates the development and implementation of the proposed technology must accomplish three things: (1) to correctly identify which technologies and technical information are relevant; (2) to achieve the proper level of understanding of this information and technologies; and (3) to correctly apply the information and technologies to the targeted problem.

12.1.2 TECHNOLOGICAL DEVELOPMENTS

The evergrowing population, environmental pollution, and ecological degradation cause suffering to human health due to chemicals and other water contaminants, such as heavy metals, pesticides, and insecticides. In this context, water is one of the survival needs of the whole community, but the clean water sources are decreasing, more and more, as a result of the high demands and contamination from various human activities which degrade the quality of river water, groundwater, ponds, and springs. The remedial measures taken to eradicate the contaminants depend on several parameters, such as its physical nature, solubility in water, chemical properties, and so on.[8,9] Hence, attempts are made to purify water by advanced technologies employing smart materials.

Advanced wastewater treatment processes must be applied to eliminate the organic or inorganic pollutants that are difficult to remove within the conventional treatment stages such as mechanical, chemical, and biological. Besides the technologies that are essential for safe reuse, the public participation is also necessary regarding the decision support and awareness of the risks and benefits of the reuse. Recently, the membranes' development for effective water decontamination has attracted a lot of attention due to their extensive features such as long-life span, low cost, and high-mechanical, chemical, and thermal stability. Statistically, about 54% of the whole processes for fresh and clean water producing is carried out with the membranes help.[8,10,11]

Since 1960, the membrane technology has been transformed from laboratory development to proven industrial applications.[12] More than 95% of applications are for liquid separations. Thus, the membranes are used for desalination of the seawater and brackish water, for potable water production and treating industrial effluents, and for water reclamation and reuse. Additionally, the membranes are used for the concentration and purification of food and pharmaceutical products, in base chemicals production, and energy conversion devices such as fuel cells. Membranes are also used in medical devices, namely hemodialysis, blood oxygenators, and controlled drug delivery products. The membrane separation processes are increasingly integrated with the conventional technologies as hybrid membrane systems to reduce the energy consumption and to minimize the environmental impact. Therefore, the utilization of membranes leads to the improvement of the industrial design. This makes the membrane technology to be more economical for the water filtration, separation, and catalysis.[13]

Four stages of developments are considered to be responsible for the membrane technology transferring from laboratory to advanced industrial applications[14]:

- development of high-efficiency membrane elements/modules with large surface areas;
- creation of advanced materials with controllable capabilities to separate molecularly similar components (e.g., gases, salts, colloids, proteins);
- tailoring the membrane morphology for controlling the microscopic transport phenomena; and
- manufacturing the membrane elements economically and reliably.

Continuing advances in development of new membranes with improved thermal, chemical, and transport properties have led to new possible applications. The advanced technology aids to control the structural and chemical functionality in the materials based on polymers/biopolymers which generates novel membrane modules for water purification by molecular level rationally designed approaches.[13]

A wide range of raw materials was available for the membranes' achievement, according to which the membranes are divided into *polymeric* and *ceramic*. The *ceramic membranes* are very popular because of their chemical and thermal stability, with a long-life span and high porosity; however, these membranes are expensive and brittle. Instead, the polymer-based membranes are flexible, cheaper, and more efficient in removing the soluble organic and inorganic species.[13] However, they present some limitations, such as high hydrophobicity, low-mechanical strength, and predisposition to biofouling.[8,13] Currently, the attention toward natural polymers has accelerated because these materials are nontoxic and biodegradable and are substitutes to petroleum-based polymers with extensive applications. Therefore, the natural polymeric membranes are widely used for water treatment from agrofood waste streams,[15] textile,[16] and petroleum industries,[17] for removal of pollutants from drinking water,[18] allowing the concentrate to be treated or removed, reducing thus the direct or indirect discharge of the contaminants in wastewater.[15,19,20]

Over the past decade, numerous trials have been devoted to the manufacture of synthetic membranes for particular applications, having appropriate features, such as permeability, selectivity, and specific chemical and physical properties. To reach this target, various techniques have been performed, namely track-etching, stretching, sintering, phase inversion, electrospinning, and interfacial polymerization.[21,22] Thus, the development of novel membrane modules and operating procedures has provided a key stimulus for the membranes industry growth. Because of their energy efficiency, the membranes will gain more importance compared to other technologies. This, in combination with the need for water purification, represents an opportunity for the membrane technology development. Better membranes are needed to meet these challenges. As the membranes of major interest are the polymeric ones, this also represents an opportunity for polymer science development.

Therefore, the purpose of this chapter is to review the current state of the polymeric membranes, to provide more information on membrane technology for water purification, and also to identify improvement areas via polymer science.

12.2   ECO-FRIENDLY POLYMER-BASED MEMBRANES FOR WASTEWATER TREATMENT

Lately, in the field of scientific research has been a continuous concern regarding the obtaining of the membrane materials based on compounds from natural sources that can be used in the treatment and water purification. Cellulose and chitosan are the most representative and widely distributed natural polymers. The main sources of the cellulose extraction are plants (stems and woody parts) but also algae, fungi, or bacteria, whereas chitosan is obtained from the shell of crustaceans (crab, shrimps), fungi, or from the insects' cuticles.[23,24] These polymers present properties as biodegradability, nontoxicity, biocompatibility, versatility, and ease of structural modification, relatively low-cost characteristics that recommend them for application in various fields of activity including the environmental remediation.[25,26] In this sense, there is a series of literature data that confirms the possibility of obtaining from these polymers of membranes with improved porosity, permeability, selectivity, and antimicrobial activity, properties that make them capable of performing certain functions necessary for wastewater cleaning.[27,28]

12.2.1 BASIC FUNCTIONS OF THE MEMBRANES DESIGNED FOR THE WASTEWATER TREATMENT

In the wastewater treatment process, the natural polymer-based membranes can be inert or catalytically active and can perform functions such as filtration, adsorption, catalysis, or antimicrobial/antibiofouling.

12.2.1.1 FILTRATION

Generally, the filtration processes involve the use of a membrane that acts as a "barrier" which separates the molecules and other particles or chemicals from the rest of the fluid, by accelerating the flow, under pressure on the polymeric membrane surface. The membranes will retain the molecules/particles that are larger than the pores' dimensions, allowing at the same time for smaller molecules to pass through them (Fig. 12.1). As was mentioned in the literature, the separation processes can be differentiated on the basis of the pressure exerted to drive the membrane transport and on the size of the membrane pores. Thus, four types of processes were identified as follows[29]:

- *Microfiltration* (MF)—the undissolved particles can be removed; the membranes' pores have dimensions > 0.01 μm;
- *Ultrafiltration* (UF)—both undissolved particles and substances with high molecular weights are removed; the pores' dimension varies between 0.005 and 0.01 μm;
- *Nanofiltration* (NF)—the microcontaminants and polyvalent ions are removed; the membrane pores sizes varies between 0.001 and 0.005 μm;
- *Reverse osmosis* (RO, hyperfiltration)—the microcontaminants, polyvalent and monovalent ions can be removed; the membrane pores sizes are > 0.001 μm.

In Figure 12.1, the four types of membranes with the characteristic porosity, pressure exerted, and materials removed are schematically represented.

MF and UF membranes are usually utilized in wastewater cleaning, for oil–water separation. The UF membranes are suitable for removing low concentrations of oils (retained oil concentration is about 10%), whereas UF membranes are used to remove oil from emulsions resulted from the metal processing industry. The MF membranes are symmetrical with open pores, of the same size, through which the molecules of smaller dimensions pass, and those of larger dimensions are retained. It can be stated that the MF is carried out according to the principle of mechanical elimination of the dimensions and that an important role in this process it has the steric structure of the molecules. Unlike the MF membranes, those of UF are asymmetrical, containing two layers: an active one, with pores of variable submicrometric dimensions, and the other one, which serves as a support, with open pores.

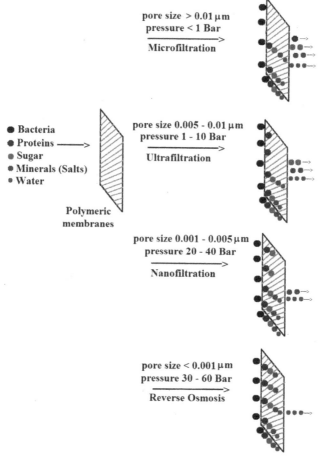

FIGURE 12.1 Types of membranes with the characteristic porosity, pressure exerted, and biomolecules/particles/chemicals which they can remove (adapted on Ref. [29]).

The retention capacity of these membranes does not depend on the size of the particles but depends on the polarity of the membrane. Thus, it was found that the hydrophilic membranes show higher permeate flows compared to the lipophilic ones. A common feature of the two types of membranes is the tangential flux of the emulsion relative to the membrane, which minimizes, through tangential gravitational forces, the layer of deposits from the membrane surface which could decrease its efficiency.

The most used MF and UF membranes are the organic ones obtained from cellulose acetate (CA). Usually, by using these membranes, the oil content from retentates varies between 30 and 45% (the oil can be thermally recycled or subjected to a rerefining process), while small molecular organic substances are retained in permeate and further treating steps are necessary.

NF and *RO membranes*, by comparison with those of *MF* and *UF*, are semipermeable or diffusion, the separation process depending on the dissolution capacity of the material from which the membranes are processed. The use of these membranes in wastewater treatment, when the water comes from waste places, requires a high-operating pressure reaching even 120 bar. The electric charge loading of the NF membranes favors a separation of the particles of same size but differently charged. Due to the charge and NF properties, these membranes are capable of separating the small molecular compounds but also polyvalent ions (heavy metals).

The RO membranes are exclusively obtained from organic compounds (cellulose and chitosan or appropriate mixtures) and are used for desalination of drinking water in regions where is no freshwater.[30,31] In this case, the desalination process does not involve the use of other chemicals or of thermal energy. Literature mentions that there are many membranes on the market, but their choice is made according to the physicochemical characteristics of the wastewater. In the RO process, the first used were CA membranes in 1965.[32] In addition, the CA membranes play an important role in removing bacteria, viruses, pesticides, oils, and heavy metals, their filtration capacity being improved especially by adding additives such as bentonite, modified coal, boehmite, and so on.[33–35]

12.2.1.2 ADSORPTION

The adsorption is a convenable and easy technique that consists in capturing of pollutants (heavy metals as Pb(II), Cd(II), and Cr(IV), phenol, or dyes) from wastewater using physical or chemical methods. The physical adsorption occurs in any system of the solid/liquid or solid/gas type and consists in binding of the adsorbate on the adsorbent surface (polymeric membrane) through Van der Waals forces. As opposed to this, the chemical adsorption, or activated adsorption, consists in two processes: transport of the pollutant from the solution mass to the outer surface of the adsorbent membrane; the internal transfer of the polluting substance is made by the diffusion pores from the outer surface to the inner one, and the second process which involves the pollutants adsorption on the active sites of the adsorbent polymeric membrane.[36]

Usually, the adsorption membranes are porous membranes of MF and UF, provided on the surface with functional groups, of carboxyl or amino type, capable of bonding with heavy metals[37] (Fig. 12.2). Heavy metal ions will be removed from the wastewater even when their dimensions are smaller than the pore size, through a process that involves a low energy consumption and a high permeation flow.

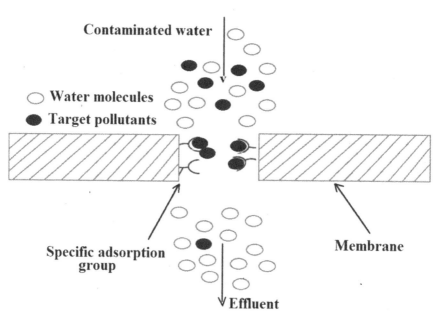

FIGURE 12.2 The working principle of the adsorbent membrane (*source*: reprinted with permission from Ref. [36], 2018 Springer).

From the multitude of the adsorption membranes those based on chitosan have attracted considerable attention, due to the large number of free amino and hydroxyl groups which have a high adsorption capacity of heavy metal ions.[38] It has been shown that the metal cations can be adsorbed by chelation to the amino groups, in neutral solutions, while the sorption of the metal anions is done by electrostatic forces to the protonated amino groups, in acidic solution. The membranes sorption properties are greatly influenced by the presence of ligands but also of the pH value and structural parameters of the polymer (crystallinity or degree of diacetylation). The identification of the sorption limits represents a challenge for the researchers in finding other forms of the chitosan to obtain new and more efficient adsorption membranes. Thus, by chemical modifications (grafting with sulfur compounds), the properties of the chitosan are changed and the selectivity of the corresponding membranes is improved. These interactions of the heavy metals with chitosan have a role not only in the decontamination of wastewater but also in the recovery of the precious metals. In addition, another way to increase the efficiency of the membranes in water treatment is the development of optical sensors and adsorption membranes as parallel technologies. Thus, new membrane that can visually detect and remove heavy metals by adsorption, especially lead ions, has been designed. This type of membrane based on a mixture of chitosan/cellulose, and that contains dithizone, a sensitive optical ligand immobilized on their surface, was tested in terms of its optical response and adsorption capacity of lead ions. As the literature mentions, this seems to be the first type of membrane that fulfills both functions.[37]

In case of the cellulose adsorbent membrane, the positively charged metal ions are adsorbed on the negatively charged membrane surface. Due to the small surface area of the membrane, its pores are covered very quickly with the metal ions causing a decrease in the ions rejection. To eliminate this disadvantage and to increase the membranes' efficiency in the removal of the heavy metals from the wastewater, in the cellulose membranes are introduced either fillers (humic acid) or adsorbent particles (iron, silica). The main characteristic of these additives is their ability to bind with metal ions. In literature[39,40] have been demonstrated that the iron oxides from cellulosic membranes form a negatively charged surface that favors the fixation of U(VI) and Pb(II) ions by complexation process. As in the case of chitosan membranes, the heavy metal removal efficiency of the cellulose composite membranes is highly dependent on the pH, dopant nature, metal ions concentration, and temperature.[41]

12.2.1.3 CATALYSIS

Low drinking water resources, together with the environmental pollution concerns, have expanded the research field of the scientists. Thus, besides the organic adsorption membranes presented previously, very often used in the wastewater treatment, the development of the photocatalytic membranes reactors (PMRs) represents another challenge for the scientists in this field. The PMRs are multifunctional reactors characterized by the combination of a catalytic reaction with a membrane separation process. This technique is very efficient and implies a photocatalytic degradation of the wastewater impurities by using a proper fotocatalyst, in the presence of luminous radiation.[42–46] Thus, as literature mention, depending on the catalyst nature these PMRs can be classified in those *with solubilized* or *suspended photocatalyst* and those with *photocatalyst immobilized in/on a membrane*.[47,48] Both types of membrane reactors have advantages and disadvantages depending on the applicability domain. In case of *PMRs with solubilized/suspended* photocatalyst, the catalyst can be retained in the reactor when a membrane with appropriate molecular weight cut-off is used. The retention ability of the catalyst can be improved by catalyst extending in the form of catalysts bound to a soluble polymer, a superbranched polymer or dendrimers.[49] Besides the fact that it retains the catalysts and pollutants, the membrane has the role of confining the degradation intermediates in the reaction environment. Usually, in a common PMR, the degradation intermediates remain in the treated effluent, which reduce the process efficiency because these intermediates are much more dangerous than the pollutants as such. For this reason, the use of a photocatalyst in combination with a separation membrane can improve the filtration property so that both the pollutants and degradation intermediate are removed and the effluent is of greater purity. Thus, the PMRs with solubilized/suspended photocatalyst can be:

- *of integrative type*—the photocatalytic reaction and membrane separation processes occur within the same equipment, in other words, the polymeric/inorganic membrane is immersed in a photocatalytic suspension;
- *of split type*—the photocatalytic reaction and membrane separation processes take place in two different equipment, which means that both the photocatalytic and membrane systems are separated but are appropriately coupled for the intended purpose.[47,50]

In addition, in the literature[51] is specified that PMRs with solubilized/suspended photocatalyst may be different, depending on the position of the light source: (a) the light source is above but inside of the membrane system, and (b) the light source is above, but inside of an additional vessel located between the power supply and membrane system.

As opposed to PMRs with solubilized/suspended photocatalyst, where the photocatalyst is solubilized or suspended into the reaction environment, in the case of PMRs with immobilized photocatalyst, the photocalyst is immobilized *in* or *on* the membrane forming a photocatalytic membrane (PM). This type of membrane works both as a selective barrier for contaminants that they retain in the reaction medium and as support photocatalytic. The whole system is intrinsically integrative, which means that the photocatalytic reaction and membrane separation processes occur at the same installation. It was found that between the two types of reactors, those with solubilized/suspended photocatalyst are more efficient because they ensure a greater contact surface with the pollutants.[47,48] Therefore, the most widely studied are the PMRs with solubilized/suspended photocatalyst to expand their fields of applicability.[52,53] However, for this type of PMRs the fouling is a process that limits its efficiency, and for this reason, the PMs have been chosen to improve the performance in the wastewater treatment processes. By creating reactive oxygen radicals by irradiation with light, these PMs have the role of preventing the formation of the layer on the membrane surface, decreasing the concentration of pollutants in the effluent, and increasing the quality of the permeate. In addition, when PMs are used, are prevented the light scattering—a phenomenon that happens in the case of PMRs with suspended photocatalysts, where the light is scattered by the photocatalyst particles. Therefore, a significant progress in the process of wastewater treatment and cleaning was found when a reactor with PMs was used.[54]

Ideal membranes that exhibit properties of solute rejection, high flux and antifouling properties are those obtained from CA. In addition to the properties above mentioned, other features that recommend CA for being used for obtaining the porous membranes for wastewater treatment are high strength, hydrophilicity, relatively low cost, but also the ability to be processed in different forms and to combine with different additives.[55,56] In literature has been shown that the incorporation of nanoparticles in polymeric membranes improves their properties. As example, TiO_2, a good inorganic additive, was used as photocatalyst because of its good physical and chemical properties.[57,58] Within a PM, this metallic nanopaths will function as a catalyst in a catalytic reaction. Thus, in case in which semiconductors as TiO_2 or ZnO are involved, the use of UV source or sunlight is also necessary to initiate the reaction. TiO_2 becomes active under the action of light radiation below 387 nm. Only under these conditions, it will generate electron–hole pairs, and subsequent charge separations will facilitate the obtaining of highly reactive oxygen that has the possibility to mineralize the persistent organic pollutants.[59] The modification of the TiO_2 nanoparticles surface with noble metal nanoparticles as Au, Ag, Pt, and so on, and combination with a semi-conductor such as CdS, Ag_2O, V_2O_5, was performed to improve their photo-catalytic activity in visible light.[60–62] Of these, Ag or Ag_2O are most often used because of their relatively low cost, efficiency, but also of high absorption capacity of electrons under UV irradiation and sensitizers for transforming the UV response in the visible region.[63] As result of these features, they were successfully used as oxidant, water purifying agent, catalyst, and in combination with TiO_2 is recommended in obtaining photocatalysts with superior properties used in visible light.[64] By combining TiO_2 with organic polymers, composite polymeric membranes with immobilization capacity can be obtained. In this respect, the membranes obtained from chitosan, the second most widespread natural polymer after cellulose, with properties of biocompatibility and biodegradability and role of transporter for TiO_2, have often been used as a result of the adsorption capacity of the heavy metals and the adsorption–photocatalysis process of the organic pollutants.[65,66]

The application of the catalytic membranes was limited to remove the synthetic dyes, and nitroaromatic compounds but also for the photocatalytic removal of the microbes.[67–69]

12.2.1.4 ANTIMICROBIAL/FOULING

Fouling represents the most inconvenient in the use of membranes in the wastewater treatment. This phenomenon consists in the deposition on the membrane surfaces or in membrane pores of the particles and microorganisms that decrease the efficiency of the polymeric membranes but also the possibility of being used for several times. The schematically representation of fouling process can be observed in Figure 12.3. Depending on the nature of the pollutants, the fouling processes can be *inorganic, colloidal, organic,* or *biofouling*. *Inorganic fouling* results from precipitation of the inorganic calcium carbonate ($CaCO_3$), calcium sulfate ($CaSO_4$) on the membrane surface, due to its supersaturation during the filtration process. *Colloidal fouling* is a consequence of the accumulation of the inorganic/organic colloids (oxides/hydroxides of iron, silica, heavy metal, proteins) and suspended materials on the membrane surface and pore clogging. *Organic fouling* is generated by the dissolved organic matter such as soluble microbial products, polysaccharides, proteins. It is assumed that the natural organic materials cause an irreversible fouling as a result of their hydrophobic nature, generating a stronger adsorption at the membrane surface and leading to a diminution of the permeation flow.[70–73]

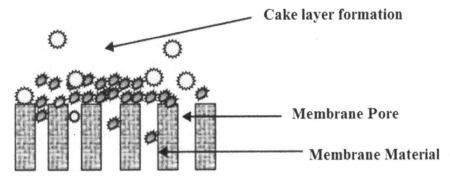

FIGURE 12.3   The schematically representation of the fouling process (*source*: reprinted with permission from Ref. [70], 2018 Elsevier).

The deposition onto polymeric membrane surface of the microorganisms such as bacteria, fungi, proteins, lipids, and lipoproteins is called *biofouling* and is the most widespread type of fouling. It is considered to be the most persistent due to the microorganisms ability to develop and multiply in the presence of the nutrients from the feed water.[74] Microorganisms attach to the polymeric surface through interactions as hydrogen bonding, hydrophobic or weak Van der Waals forces, initially forming sessile colonies, and then releasing the extracellular polymeric secretions that generate the biofilm due to their accumulation on the membrane surface.[75]

Taking into account this classification, the development of the polymeric membranes that prevent pollutants to cling to the membrane, coagulation, and the emergence of the so-called clogging phenomenon was followed.[76,77] In addition to designing the most suitable polymeric membranes, there are also several possibilities to prevent biofouling, namely, by suppressing the adhesion of the bacteria to the membrane surface, toxic release, or killing by contact. The first of the three ways of prevention implies the modification of the membrane surfaces characteristics, such as hydrophilicity, roughness, while the last two involve the incorporation into the polymeric matrix of particles with antimicrobial activity. The most used are silver particles, ammonium salts, carbon nanotubes but also the organic antimicrobial agents. The latter are less often utilized due to the inability to be easily processed under adverse conditions and are represented by the inorganic antimicrobial agents. In the literature are mentioned studies performed on the organic polymeric membranes of cellulose, chitosan, or their combinations with silver nanoparticles (AgNPs).[60,61] CA membranes, as was mentioned above, were used in wastewater treatment for their ability to remove bacteria, pesticides, dyes, oil, heavy metals, and even dissolved salts since the 1960s. However, enhancing their properties was, and still is, a basic concern for the researchers. In this sense, in many of their studies, the scientists have incorporated in the cellulose matrix AgNPs, which besides the antimicrobial action have the role of improving the features of the cellulose membrane surface. The resistance of this kind of membrane to biofouling is given by the amount of silver and also by the way in which the nanoparticles are distributed in the matrix. The same as silver, the copper oxide, and semiconductor oxides (TiO_2 and ZnO) in cellulose membranes were evaluated concerning their antimicrobial activity. The photocatalytic antibacterial activity is sustained by reactive oxygen species which were obtained by irradiation and that can penetrate the cell wall of bacteria inhibiting their development.[78,79] Moreover, as previous was mentioned, besides the antimicrobial activity, the presence of semiconductor oxides in the cellulose membranes it confers photocatalytic properties, and so these can be used for wastewater pollutants degradation.[80–82]

Apart from the three techniques presented above, there are other methods of fouling reducing which implies *physical cleaning* via backwash, relaxation before the filtration processes, *chemical cleaning*—using acids, caustic soda or hypochlorite *or ultrasound connection before installation*.[83] Since, these methods of pretreatment with physical or chemical cleaning can cause the polymeric membrane damaging, the most approached are the techniques of membrane surface modifying through addition of additive materials, such as chitosan, starch, $FeCl_3$.[84,85] However, to avoid fouling, most often are controlled the flow and flow velocity, as can be seen from Figure 12.4.

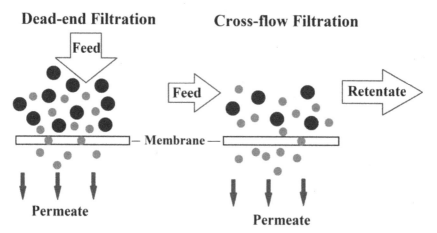

FIGURE 12.4 Schematically representation of dead end (left) and cross-flow filtration (right) (adapted on Ref. [86]).

Two types of flow have been mentioned in the literature: *dead end flow* and *cross-flow*. In the first case, the feeding direction coincides with that of the membrane, so that the possibility of forming a layer of microorganisms on the surface that block the passage through the membrane is very high. In cross-flow, the feeding direction is opposite or tangential to the membrane, so that the deposition of pollutants on its surface is diminished.[70,86,87]

12.2.2 CHARACTERISTICS OF THE MEMBRANES REQUIRED BY THE WATER DEPOLLUTION TECHNOLOGY

The main considerations underlying the technology of water depollution using membranes are the membrane properties and implicitly their processing technique. In this sense, to perform the abovementioned functions (filtration, adsorption, catalysis, antifouling/antimicrobial), the membranes must have certain porosity and pore distribution, sensitivity to moisture, flexibility, mechanical resistance, chemical stability (resistance to pH, oxidants, or chlorides), antimicrobial activity, and fouling resistance.[88] The membrane selectivity is influenced, in turn, by the surface properties that involve the membrane surface charge. Small roughness, high adhesion of particles and microorganisms, and also low resistance to cleaning agents are factors that negatively influence the performance of membranes, reducing their lifetime.[89]

12.2.2.1 POROSITY

In literature, many studies have been performed on improving the porosity properties of the membranes. In this regard, a while ago, polymeric membranes have been obtained with controllable pore sizes using a template with calcium carbonate nanoparticles. This technique was developed on industrial scale, culminating in the formation of a commercially available drinking water filter. The manufacturing of such filter was easy but it based on the use of sophisticated equipment. Voicu et al. have obtained CA membranes with pores of different shapes and sizes, adding to the casting CA solutions of 10 wt% concentration, in *N*,*N*-dimethylformamide, tree different cationic surfactants (dimethyl-dioctodecyl ammonium bromide, alkyl-benzyl-dimethyl ammonium chloride, *N*-dodecyl-pyridinium chloride). First of all, they have obtained the solutions by adding the CA in solvent, gradually, to avoid the formation of the insoluble blocks. After the tree solutions have been obtained, the mentioned surfactants were added in concentration of 10^{-5} M and dispersed in mass solution by ultrasonic for 30 min. To observe the surfactant impact on the CA membranes porosity they performed scanning electron microscopy studies. These analyses have shown that membranes from pure CA (Fig. 12.5a) present pores with dimension between 10 and 100 μm, those from CA/alkyl-benzyl-dimethyl ammonium chloride and CA/N-dodecyl-pyridinium chloride have pore size of about 20 μm (Fig. 12.5c,d). The membrane with the largest dimensions of the pores, of about 180 μm, with uniform surface distribution is that obtained from CA/dimethyl dioctodecyl ammonium bromide (Fig. 12.5b). By comparison with the others, this seems to be the most stable membrane because does not fold after drying.

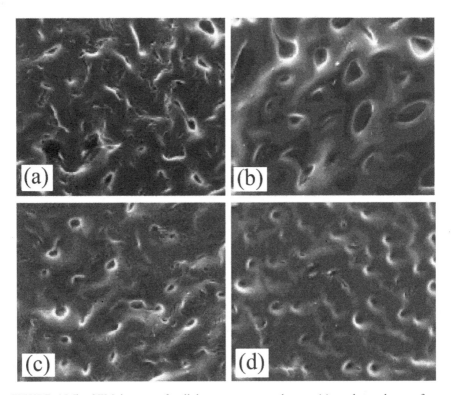

FIGURE 12.5 SEM images of cellulose acetate membranes (a), and membranes from cellulose acetate/dimethyl-dioctodecil ammonium bromide (b), cellulose acetate/alkyl-benzyl-dimethyl ammonium chloride (c), and cellulose acetate/N-dodecyl-pyridinium chloride (d) (adapted from Ref. [90]).

In addition, the literature data show that the efficiency of Loeb–Sourirajan CA membranes, with a heterogeneous microporous surface can be improved by a process of water pumping over their backside, for a sufficient period of time. The membranes were used in measurements of RO with the surface layer oriented toward the feed solution. It was found that by exposing the normal and non-contracted membranes to a pressure of 400 psig, for 85 h, the product rate increased over 20%, without a decrease in solute separation rate in RO assays, at a pressure of 600 psig when a 0.5-wt% NaCl–H$_2$O feed solution was used. For RO measurements performed at a pressure of 1500 psig and membranes subjected to another process of backpressure treatment, the results were similar. Moreover, it has been observed that during the RO the compaction effect of a normal membrane or a membrane subjected to backpressure is the same, as well as the effects of backpressure treatment on a compacted or normal membrane are the same. In addition, from the pure water permeability measurements was concluded that the small pores from the cellulose membrane surface are more open than the large ones, indicating that the pressure treatment of the posterior part of the membranes implies morphological changes.

Other organic porous membranes, as those from chitosan, were obtained based on the principle of cryogenically induced phase separation, by expanding the thermally induced phase separation for the polymer/solvent system to the polymer solution, at room temperature. It was noticed that in the process of separation in layers, corresponding to thermally induced solid–liquid phase separation, a specific lacy pore structure and porosity that can exceed 70% can appear. By studying the processes of separation in layers and that of the membrane obtaining was observed that the membrane consists in stacked porous lamella; the porosity increases and the pore diameter decreases as the solidifying temperature is increased, while with increasing of chitosan concentration, the two parameters decrease.[91]

12.2.2.2 SELECTIVITY AND PERMEABILITY

The first CA membranes that were obtained could not be used in practical application, even if they have a very good selectivity, the water permeability was very small. For this reason, the cellulose diacetate membranes were created, but, although they were more effective were predisposed to the microbial attacks. Thus, the production of new physically, chemically and biologically stable cellulose membranes (thermostable, resistant to the action of some chemical and microbial agents) was required. Cellulose triacetate membranes appear to be much more efficient, taking into account these considerations.[92,93] Thus, has been demonstrated that the membrane filtration performance is highly dependent on the acetylation degree of the cellulose. According to literature studies, a CA membrane with 40 wt% acetate and 2.7° of acetylation shows a salt rejection of approximately 99%. However, the increase of the selectivity, progressively with the degree of cellulose acetylation results in a decrease of the water permeability.[94] In addition, it was found that, that the CA membranes are stable when are used for feed solution with pH varying between 4 and 6, while in acidic/basic media the hydrolysis reaction occurs, decreasing the membrane selectivity. Although the efficient CA membranes that can be used in wastewater treatment were synthesized, studies regarding the improvement of their performance have been made. The dispersion of AgNPs on the CA membranes surface not only gives them biological stability against certain bacterial species but also maintains their permeability and salt removal capacity.[95] A high flow of water and fouling resistance is provided when a phospholipid polymer is deposited on the CA membrane. According to literature studies,[96] the hollow fiber CA membrane modified with the water-soluble amphiphilic 2-methacryloyloxyethyl phosphorylcholine (MPC) copolymer (poly(MPC-co-n-butyl methacrylate) present a good permeability and low membrane fouling property comparatively with unmodified CA membrane.

Although in recent years, the development of the composite membranes used in the wastewater treatment has been laid, those from CA remain on the market due to the high chlorine resistance. The water treatment and its purification involve the use of disinfectants, and the CA membrane appears to be the one with the highest chemical stability.

As was mentioned, besides cellulose, the chitosan was also chosen as a suitable biopolymer for obtaining membranes used for wastewater treatment. In the separation processes, the chitosan membranes will have the same disadvantage; the particles or microorganisms will be attached to the polymeric membrane surface. To avoid this, the chitosan can be cross-linked with polyethylene glycol (PEG), a polymer that prevents fouling, increases the water permeability, and is more chemically stable. In this combination, the chitosan/PEG membranes were used to remove the oils from the wastewater, the results being interpreted depending on the amount of PEG used.

12.2.2.3 ANTIMICROBIAL ACTIVITY

The antimicrobial activity of chitosan and its derivatives has been demonstrated early,[97] but recent studies have been done in the sense of understanding as accurately as possible of its bactericidal or bacteriostatic effect. According to literature data, its activity is bacteriostatic rather than bactericidal and is influenced by its type, molecular weight, and degree of acetylation. However, it was proved that the molecular mass has a greater impact on chitosan antimicrobial activity than the degree of acetylation.[98,99]

The antimicrobial activity of chitosan was justified in three ways:

- *By electrostatic interactions* between the positively charged of chitosan (provided by NH_3^+ or Ca^{2+}), and negatively charged of the microbial cell membranes.[100] These electrostatic interactions inhibit the development of microorganisms either by altering the permeability of the cell membrane, thus causing an internal osmotic imbalance that prevents the growth and multiplication of bacteria,[101] or by hydrolysis reaction that occurs in bacterial cell wall. In the latter case, the potassium ions, proteins, glucose, or others low molecular weight protein constituents leak inside the cell, inhibiting the microorganisms development.[102]

    Raafat and collaborators[98] have investigated the antimicrobial activity of chitosan against *Staphylococcuss simulans 22*. They have observed, through transmission electron microscopy assays, that the chitosan molecules have attached on the bacteria cell surface, and that, in contact points, the cell membrane detaches from the cell wall forming vacuoles. By this detachment is formed ions and also an efflux of water which causes a decrease of the internal pressure of the bacteria.[103] It was shown that, as the chitosan concentration increases, the content of polycationic chitosan capable of binding to the charged bacterial surface is apparently diminished.[104] Moreover, in high concentration, regardless of the type of analyzed bacteria, the chitosan has a tendency to settle over them and not to attach to their surface.[105]

    Besides these, in the antimicrobial activity explaining, the bacterial surfaces polarity plays an important role. The cell membrane of the gram-negative species consists in lipopolysaccharides, with phosphates and pyrophosphates groups, which confer a negative charge, superior to that of the gram-positive species, whose basic component is peptitoglycan. Therefore, in contact with chitosan, the intracellular material leakage, in the case of gram-negative bacteria, is higher than in the case of gram-positive bacteria.[106,107] Some of the literature data[108] show that the chitosan exhibits antimicrobial activity more specifically for gram-positive bacterial species (*Listeria monocytogenes, Bacillus megaterium, Bacillus cereus, Staphylococcus aureus*) than for the gram-negative bacteria (*Escherichia coli, Pseudomonas fluorescens, Salmonella typhymurium, Vibrio parahaemolyticus*) are in contradiction with other data, by which the antimicrobial activity of chitosan is related to the hydrophilicity of bacteria.[109] According to these, the gram-negative bacteria are more hydrophilic than the gram-positive ones and more sensitive in the presence of chitosan. However, the charge density on the cell surface represents the main factor that influences the amount of the adsorbed chitosan. Modifications in the structure and cell membrane permeability are more evident as the amount of adsorbed chitosan is higher. Consequently, the antimicrobial effect of chitosan depends very much by the nature of host bacteria.[110]

- *By binding chitosan to microbial DNA*—Chitosan penetrates the microorganism nucleus inhibiting the synthesis of the messenger ribonucleic acid (mRNA) and proteins.[111] Although the confocal laser scanning microscopy images have revealed the presence of chitosan oligomer in *E. coli*, it is assumed that this inhibition mechanism occurs very rarely, most often considering that the chitosan appears as an agent that alters the bacterial membrane, than one that penetrates its nucleus.[98]

- *By chelating with metals*—Chitosan, through the amino groups that have the ability to absorb the metal cations, binds very well to metals.[112,113] According to literature data,[98] this way of the microbial development inhibiting has better results under high pH conditions, when the pair of electrons belonging to the amine nitrogen is available for metal ion donation. Thus, in the case of the chitosan–Cu system, for a pH < 6, the complexation involves one $-NH_2$ group and three $-OH$ groups or H_2O molecules, while for a pH ranging between 7 and 9 the complexation is provided by two $-NH_2$ and two $-OH$ groups dissociated. It is very obvious that the chitosan molecules that surround the bacteria can complex the metal, preventing the bacterial species development.[114] Nevertheless, this evidently, is not a determinant antimicrobial action since the sites available for interaction are limited and the complexation reach saturation in function of the metal concentration.[115]

The UF membranes from chitosan or mixtures of chitosan and cellulose were chosen as an alternative to the present membranes based on nonbiodegradable petrochemical compounds that not only require a high amount of energy but greatly damage the ecosystem. In this context, Weng and collaborators[116] have obtained NF membranes from bamboo cellulose and chitosan (cell/CS), testing besides the antimicrobial activity, mechanical resistance, and salt and dyes filtration performance. As the results of the antimicrobial activity studies show the cellulose membranes do not present an inhibition zone, while for cellulose/chitosan mixture-based membranes the inhibition zone diameter against *E. coli* increases as the amount of chitosan from the mixture increases (Fig. 12.6). The mechanism by which the bacteria growth and development are stopped is based on the interaction between the positively charged of chitosan molecules and negatively charged of the bacterial cell membranes. These types of interactions determine a decrease in membrane permeability, a leakage of intracellular components, factors that cause the *E. coli* destruction.

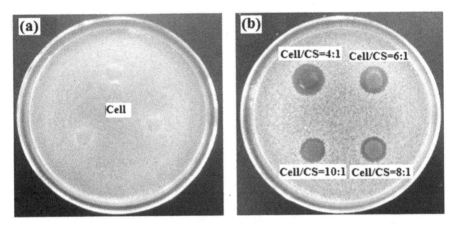

FIGURE 12.6 The diameter of the inhibition zone of cell membranes (a), and cell/CS based membranes against *E. coli* (b). *Source*: reprinted with permission from Ref. [116]. 2017 MDPI.

In the same sense, Isik et al. have obtained bacterial cellulose membranes[117] and have tested their antimicrobial activity against *E. coli*, for the purpose of textile wastewater treatment. To improve the antimicrobial properties of the bacterial cellulose membranes, these were subjected to some special treatments which include the immersion in $AgNO_3$ solution, to obtain, at membrane surface, AgNPs of different concentration (2, 4, and 8 mM), these being known as having antibacterial properties. The results of the antimicrobial activity assays, performed by the agar diffusion method,[118] have shown that the diameter of the inhibition zone of the bacterial species analyzed increases with the increasing of the AgNPs content at the cellulose membrane surface. Thus, ecological membranes of bacterial cellulose that can be used to remove the color and chemical oxygen demand from textile wastewater have been created.

12.3 PROVEN LABORATORY APPLICATIONS OF ECO-FRIENDLY POLYMER-BASED MEMBRANES

Various pollutants detected in water may have unfavorable repercussions on the human health and aquatic ecosystems, and for this reason, the need for clean water, free of toxic chemical substances, becomes bigger and bigger. Different kinds of waste from industry, agriculture, or from the households, produce a large volume of wastewaters which comprise organic and inorganic compounds (dyes, heavy metal ions, antibiotics, feces, pesticides, herbicides, and other contaminants). Principal sources of water pollution are presented in Figure 12.7.[119]

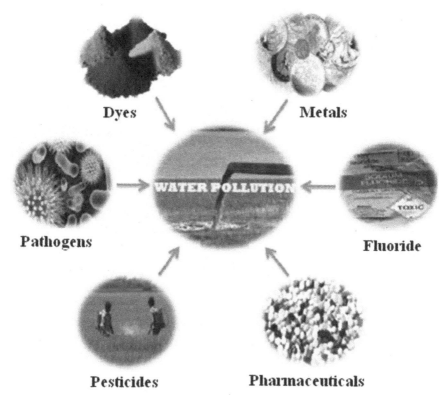

FIGURE 12.7   Sources of water pollution. *Source*: reprinted with permission from Ref. [119], 2012 OMICS Group.

The elimination, by membrane technology, of these pollutants from wastewater represents a great importance for the water conservation.[120]

Currently, different kinds of eco-friendly polymer-based membranes were used, as a result of the increased researchers' interest in developing of the materials of low costs, biodegradable, biocompatible, and bioregenerable that can be used in applications of wastewater treatment and purification.[121] In literature are mentioned methods of wastewater treatment, some of these methods being exemplified below.[122,123]

12.3.1 DYES REMOVAL

Due to their acute toxicity, carcinogenicity, visible color, complex molecular structure, and synthetic origin, the toxic dyes are considered the most primitive pollutants from wastewater. Their elimination represents a very important issue, because they cause serious harm to humans. Toxic dyes are delivered from different industries as cosmetics, paper coloring, dyes fabrication, textiles, mining, electroplating, pharmaceuticals, and skin tanning products, and by the serious effects to aquatic environment that they produce, damage the ecological stability.[123,124] In this context, an effect of dyes is represented by the inhibition of light to penetrate the water, disturbing thus, the aquatic plants' photosynthesis. In addition, the carcinogenic effects of a certain dyes are very obvious, their degradation products being observed on the water surface. From this reason, the cleaning or pretreating of these dye effluents, before their release into water resources, is very important.[124] In this context, the use of low-cost membranes, with high salt/dye selectivity and high flux, is ideal for an economic and eco-friendly treatment of the dye wastewater. The literature data exhibit advances in development of the polymeric membrane materials, for water purification using biorenewable polymers such as cellulose and chitosan.[125] These polymers display ideal adsorption specifications, useful for elimination of the pollutants, including the dyes from colored wastewater.[122] The highly colored wastewaters, discharged from the industry and agriculture, have serious repercussions for the environment, not only due to toxicity, carcinogenicity and non-biodegradability but also due to the visibility problems.[126] UF and RO are the two major membrane processes for treatment of dyes from wastewater.[127] Chitosan was usually used in decolorization research because of its specific feature of flocculation, adsorption, and chelation.[128–132] In addition, the compound has been recognized as membrane forming, with an increasing role in color removing, having affinity especially for the anionic dyes.[133] Many authors have studied the efficiency of chitosan, due to its capability to interact with various dyes. In literature[134–136] have been demonstrated the usefulness of chitosan for the reactive dyes' removal. This polysaccharide shows a higher capacity for dyes adsorption than the commercial activated carbon and other low-cost adsorbents.[137] The interactions between the chitosan and anionic dyes have also been intensively investigated by several authors, indicating that the chitosan has a natural selectivity for dye molecules, being very useful in the wastewater treating.[138-141] For example, an complex UF membrane of chitosan/activated carbon was utilized and was demonstrated the effectiveness for decolorizing of Acid Red B water solutions.[142] Usually, many of the industrial chemical processes employ salts in aqueous systems for different purposes. Sodium chloride is one of the most used inorganic materials and its separation of small molecular weight components is frequently necessary. Literature data show that the regenerated cellulose membranes, prepared from trimethylsilyl cellulose, are used for artificial dye effluents' treatment. The scientists have used a concentrated solution of Congo Red containing a high concentration of sodium chloride, and have observed that for this solution, the membranes show an impressive removal of the dyes, but not of the salt, along with a high flow. Furthermore, the membrane exhibited a good antifouling behavior with nearly 100% flux recovery in prolonged experiments (up to 75 h). In conclusion, the study may provide an alternative for dye effluent treating, where high values of salts are present.[143,144] Additionally, the phenols discharged in aqueous systems can produce an uncomfortable odor that is very harmful to human health. As a result of that, various techniques such as conventional methods of distillation and extraction, chemical oxidation, biodegradation, and sorption, used to remove the phenol and its derivatives from wastewater have been applied. In literature has been investigated the phenol removal from aqueous solution by using a polymer inclusion membrane, based on mixture of cellulose triacetate and CA as support, calyx[4]resorcinarene derivative as a carrier and 2-nitrophenyl octyl ether as plasticizer. The results have indicated that the polymer inclusion membrane present efficiency in phenols removing from wastewaters.[145,146] Other studies conducted in the sense of the water purification have used the cellulose nanocrystals, as functional entities in chitosan matrix, via freeze–drying process, followed by compacting. The nanoporous membrane structure was provided by chitosan that linked the cellulose nanocrystals to each other, and then stabilized by crosslinking with gluteraldehyde vapors (25%), at room temperature, for 48 h.[147] For this study, the scientists have selected three positively charged dyes (Victoria Blue 2B, Methyl Violet 2B, and Rhodamine 6G), of a real problem in industrial wastewater. The results have shown that the adsorption capacity of composite membrane decreased with the increasing of dyes concentration as can be seen from Table 12.1.

TABLE 12.1 Percentage of Different Types of Positively Charged Dyes Removed by Composite Membranes, for Various Dyes Concentrations.

Dyes concentration (mg/L)	Dyes removed (%)		
	Victoria Blue 2B	Methyl Violet 2B	Rhodamine 6G
1	98	91	70
5	95	60	31
10	88	48	13

Source: Adapted from Ref. [147].

The percentages of dye removed from all solutions, along the entire day, when a cross-linked and a noncross-linked membrane are used and also the process of dyes removal using a cross-linked composite membrane can be observed in Table 12.2 and Figure 12.8, respectively.

TABLE 12.2 Percentage of Different Types of Positively Charged Dyes Removed by Cross-linked and Noncross-linked Composite Membrane.

Dyes	Dyes removed (%)	
	Cross-linked membrane	Noncross-linked membrane
Victoria Blue 2B	98	98
Methyl Violet 2B	84	90
Rhodamine 6G	69	71

Source: Adapted from Ref. [147].

The obtained results have indicated that the membranes showed a good absorption for all dyes, especially for Victoria Blue, followed by Methyl Violet and Rhodamine 6G, whether or not the membrane is cross-linked.[147] Yu and collaborators have obtained the chitosan/activated carbon complex UF membrane and have proved its efficiency for decolorization of the Acid Red B water solutions.[142] In the same context, the chitosan UF membrane was used for colored wastewater with methylene blue and Acid Red B and was found that the decolorization was more pronounced than for the other commercial membranes.[148] However, challenges to improve the eco-friendly polymeric membrane used for elimination of dyes from wastewater still remain, making continuous efforts to meet the growing expectations.

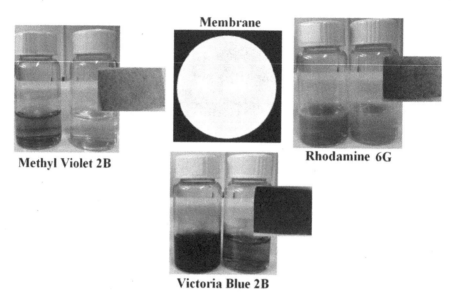

FIGURE 12.8   Process of dyes removal using a cross-linked composite membrane *Source*: reprinted with permission from Ref. [147], 2014 Elsevier.

12.3.2   HEAVY METALS REMOVAL

The increase in energy production and large utilization of heavy metals in different industrial processes has led to broad quantities of industrial waste that include heavy metals. Hence, heavy metals are delivered in environment (air, water, soil, and biosphere) in excessive quantity, on daily, and their removal from the wastewater before being evacuated is of vital importance because, as the clean water supplies are decreasing there will be a huge deficiency of water in the world.[149] The most common heavy metals which are associated with the pollution and toxicity problems are cadmium, copper, chromium, mercury, nickel, lead, and zinc. These metals are very quickly absorbed by the plants and aquatic species, entering in the food chains, and consequently in the human body, leading to serious health effects, such as stomach cramps, vomiting, skin irritations, multiple organ damage, and nerve system damages.[149] Therefore, finding a solution for the heavy metals removal represents an important task to protect the environment.[150,151] In this sense, the health authorities have adopted the legal limits for discharge of effluent containing heavy metal, while the researchers have focused on development of treatment techniques. As literature mentioned the most used techniques for the heavy metals removal comprise the membrane filtration, adsorption, coagulation, chemical precipitation, ionexchange, electrochemical, biological treatments, and advanced oxidation processes.[152] Recently, the considerable research work, on a series of biopolymers, has been done in the sense of their transformation and uses in the elimination of heavy metal ions. The most effective adsorbents of heavy metals from wastewater were found to be the polysaccharides such as chitosan and cellulose. The literature data show that the chitosan is a suitable bioabsorbent, due to interesting properties as hydrophilicity, biodegradability, biocompatibility, nontoxicity, and presence of very reactive amino ($-NH_2$) and hydroxyl ($-OH$) groups in its backbone, mainly used for household purposes.[153] The scientists have studied and compared the adsorption capacities both for pure chitosan and chitosan composites, chitosan nanoparticles, cross-linked chitosan, chitosan nanofibers, porous chitosan.[154] In literature, a novel chitosan/hydroxyapatite composite membrane, for Ni(II), Pb(II), and Co(II) ions removal, from aqueous solutions, was presented. The results obtained show that this composite membrane presents a maximum adsorption capacity for Pb(II) (296.7 mg/g) as compared to Ni(II) (213.8 mg/g) and Co(II) (180.2 mg/g).[155] In addition, the chitosan/clinoptilolite composite membrane shows an increased adsorption for Co(II) (467.9 mg/g), while chitosan/silica composite membrane has a high ability to adsorb Ni(II) (254.3 mg/g).[156] For the latter, it was found that, the adsorption has increased with increasing of the polymer concentration. In addition, has been demonstrated that the chitosan is an efficient material for the removal of the heavy metals from gold mining effluent. The adsorption studies on zinc, lead, mercury, and copper have shown that a maximum removal takes place for higher pH values (4 and 6).[157] Ghaee and coworkers have obtained the macroporous chitosan membranes, and have tested and compared their adsorption properties for copper and nickel ions. From the measurements they made they found that the copper ions adsorption was higher than that of the nickel.[158] Membranes from blends of cellulose diacetate/chitosan were used for heavy metals removal, by chelation of Cu(II) ions, from aqueous solutions.[159] The chitosan membranes can be used as individual supports or as modifying layer for inorganic supports. In this sense, a chitosan porous film was coated on inner surface of tubular ceramic membrane, to enhance the removal capacity of the copper ions from water. The results suggest that by using such ceramic membranes coated with chitosan, almost complete removal of cooper was achieved.[160] In addition, another study investigated the potential of the arsenic removal using a membrane based on blends of quaternized hydroxyethyl cellulose ethoxylate/regenerated cellulose. The process of arsenic removal was analyzed by polymer-enhanced UF using quaternized hydroxyethyl cellulose ethoxylate as a sorbent. The total removal of the arsenic from solution took place after the electrochemical oxidation of the arsenic (III and IV), in presence of quaternized hydroxyethyl cellulose ethoxylate, was performed. The obtained data have shown that this kind of cationic polysaccharides may represent a promising material for new possible applications.[161] In the same sense, the chitosan impregnated ceramic membranes, obtained from local clay and kaolin support followed by immersion in chitosan solution, were tested regarding the removal property of arsenic and mercury ions from water by polymer enhanced UF technique, using as chelating agent the polyvinyl alcohol. It was found that the concentration of heavy metals in permeate is higher as their initial concentration increases, and decreases as the initial polyvinyl alcohol (PVA) concentration increase.[162]

Along with chitosan, cellulose, a biopolymer used for many years as energy source, by chemical transformations, suffers modifications of the properties as hydrophobicity or hydrophilicity, adsorptivity, elasticity, microbial and mechanical resistance, and modifications that lead to an increase of its adsorption ability for heavy metals in aqueous and nonaqueous media.[153] Liu and Bai[163] have prepared porous adsorptive hollow fiber membranes from chitosan and CA and have used them for water copper ion removal. The results revealed that a higher content of chitosan in the mixtures from which the membranes were obtained, resulted in membranes with more affinity for the copper ions. In addition, the possibility of chitosan/ CA membranes to adsorb the heavy metal, through adsorption-assisted filtration mechanism, from water or wastewater, was explained. Other studies, made on chitosan/poly(vinyl alcohol) thin adsorptive membranes embedded with amino-functionalized multiwalled carbon nanotubes, have shown that these membranes are suitable for copper ion (Cu(II)) adsorption. The copper ion adsorption was more obvious for high content of chitosan, at high temperatures[164] The adsorption potential, from wastewater, of the Pb(II) and Cd(II) ions, was also investigated by using the poly(vinyl alcohol)/ chitosan nanofibers membranes, fabricated by electrospun method, proving that these membranes are effective adsorbent for the specified ions[165] In a literature study, the carboxylmethyl cellulose was investigated as chelating agent, to eliminate different heavy metals ions from synthetic wastewater solutions. Surprising results were obtained, in the sense that approximately 99.1–99.9% of Cu(II), Pb(II), Ni(II), and Cr(VI) ions were removed. The membrane selectivity was confirmed by the simultaneous separation of the three metal ions in one boot.[166,167]

12.3.3 BACTERIA REMOVAL

Actually, the water purification by harmful bacteria, using inexpensive and eco-friendly filters or membranes, exhibits a great interest. Because the contaminated drinking water causes various diseases such as cholera and other waterborne diseases, its decontamination could save a lot of lives.[168–170] Wood-based cellulose pulp fibers are today used as disposable everyday filters, some of the first membranes for MF being produced from cellulose nitrate.[171] The MF membranes, with pore size less than 0.5 μm, remove from water bacteria such as *E. coli* and *Vibrio cholera*, which are approximately 1–2μm.[172,173] Recently, literature data have been concentrate on manufacturing of the biobased membranes for MF and UF from cellulose nanofibrils.[174,175] A simple method for the water filtration is the use of paper filters made from cellulose fibers. Some authors debated this issue by incorporating the antibacterial metal nanoparticles (AgNPs and copper nanoparticles) into cellulose-based water filters.[95,176,177] Literature data show that cellulose membranes functionalized with amines were developed to evaluate their capability to remove the microbial contamination. In this sense, samples of water with several levels of *E. coli* inoculum were filtered through membranes, and for evaluation of the systems' efficiency the tests have been done for several times. It was observed that the membranes were capable to filter the water samples in a few seconds, and partially or completely remove the microorganism. Thus, it was concluded that the amine-functionalized membranes present a considerable retention capacity in samples of water with high bacterial concentrations, being useful in decontamination of drinking water.[178] In this context, the cellulosic nanomaterials are considered to be potential candidates for wastewater treatment. This finding is also supported by other studies where palm fruit stalks cellulose nanofibers (CNF), TEMPO-oxidized CNF (OCNF), and activated carbons (AC) were used to prepare membranes for *E. coli* removal from water. The researchers have obtained two types of layered membranes: a single layer setup of cross-linked CNF and a two-layer setup of AC/OCNF (bottom) and cross-linked CNF (up) on hardened filter paper. The results indicated that both types of membranes have ability for *E. coli* removing, with slightly higher efficiency for the AC/OCNF/CNF membrane than for the CNF ones. Moreover, *E. coli* and *S. aureus* are sensitive on the upper CNF surface of the AC/OCNF/CNF membrane, while the single layer of CNF membrane did not show resistance against growth of the bacteria.[179] Another study was focused on the removal of the bacteria from drinking water sample by using nanocomposites membrane coated with chitosan loaded with bioactive compounds (from *Streptomyces* species). Membrane filtration method and multiple tube fermentation techniques were the basis for quality water obtaining. The results evidenced a complete removal of bacteria from the water.[180] Wang and collaborators[181] have investigated a novel class of MF membrane, based on a fibrous scaffold of two-layered polyacrylonitrile/polyethylene terephthalate including infused functional CNF. The results have shown that the composite nanofibrous membrane presents a high surface area, high porosity, and high charge density, with superior properties of bacteria (*E. coli*) and viruses retention. According to obtained data these membranes can be considered suitable for application in drinking water purification.[181] In addition, due to the properties as biocompatibility and nontoxicity, chitosan represents a good candidate for the water disinfection. The antibacterial activity of chitosan-based membranes against gram-positive and gram-negative species has been studied by many researchers. Therefore, the chitosan (positively charged) can be used to obtain membranes for filters to adsorb and even kill microorganisms (negatively charged) in a natural way, its antimicrobial effects being attributed to its flocculation and bactericidal activities. In the literature, a mechanism of bacterial coagulation has been reported and illustrated in the present chapter in Figure 12.9. As can be observed, the chitosan molecules connect on the microbial cell surface forming an impermeable layer around it. By this clamping mechanism, the chitosan will alter the cell permeability causing a leakage of intracellular constituents and finally leading to the bacteria destruction.[182,183]

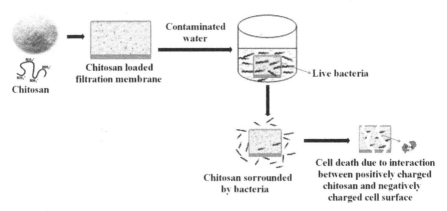

FIGURE 12.9 The chitosan bactericidal action against of wastewater bacteria.

As can be seen from the ones presented above, the membranes based on natural polymers have an essential role in water treating and purifying, studies in the direction of the development of other new types of such membranes with special properties, continuing to be a challenge for the scientists.

12.4 CONCLUSION

With the increase of the environmental pollution problems, considerable efforts have been made to restore the drinking water reserves and the ecosystems "health." Membrane technology is a powerful candidate with rapid development that plays an important role in the environmental issues. The membranes can cover the entire range of applications concerning the water treatment from the water desalination, through their treatment and recovery, to surface water treatment.

In this context, the membrane technology continues to advance by improved membranes and processes, the current studies focusing especially on the development of membranes based on natural polymers in detriment of the synthetic ones. The research directions are aimed on obtaining of low-cost membrane materials and improving the filtering, selectivity, and permeability properties, increasing of the thermal and mechanical stability, antimicrobial activity, and resistance to fouling. The chapter briefly presents the current state of the eco-friendly polymer-based membranes, some of the characteristic properties and functions underlying the wastewater treatment and purification, and along with some of proven laboratory, applications can constitute the basis for future studies on the membrane technology development in sense of life-quality improvement.

KEYWORDS

- **eco-friendly polymers**
- **membranes**
- **pollution prevention**
- **wastewater treatment**

REFERENCES

1. Aschengrau, A.; Rogers, S.; Ozonoff, D. Perchloroethylene-Contaminated Drinking Water and the Risk of Breast Cancer: Additional Results from Cape Cod. *Environ. Health Perspect.* **2003**, *111*, 167–173.
2. Geise, G. M.; Lee, H.-S.; Miller, D. J.; Freeman, B. D.; McGrath, J. E.; Paul, D. R. Water Purification by Membranes: The Role of Polymer Science. *J. Polym. Sci. B: Polym. Phys.* **2010**, *48*, 1685–1718.
3. Romanos, G. E.; Athanasekou, C. P.; Likodimos, V.; Aloupogiannis, P.; Falaras, P. Hybrid Ultrafiltration/Photocatalytic Membranes for Efficient Water Treatment. *Ind. Eng. Chem. Res.* **2013**, *52*, 13938–13947.
4. Qu, X.; Brame, J.; Li, Q.; Alvarez, P. J. J. Nanotechnology for a Safe and Sustainable Water Supply: Enabling Integrated Water Treatment and Reuse. *Acc. Chem. Res.* **2012**, *46*, 834–843.
5. Cheng, X. Q.; Wang, Z. X.; Jiang, X.; Li, T.; Lau, C. H.; Guo, Z.; Ma, J.; Shao, L. Towards Sustainable Ultrafast Molecular-Separation Membranes: From Conventional Polymers to Emerging Materials. *Prog. Mater. Sci.* **2018**, *92*, 258–283.
6. Shanley, A. Pollution Prevention: Reinventing Compliance. *Chem. Eng.* **1993**, *100*, 30–43.
7. Rossiter, A. P.; Spriggs, H. D.; Klee, H., Jr. Apply Process Integration to Waste Minimization. *Chem. Eng. Prog.* **1993**, *89*, 30–36.
8. Veglio, F.; Beolchini, F. Removal of Metals by Biosorption: A Review. *Hydrometallurgy* **1997**, *44*, 301–316.
9. DeMarco, M. J.; Sen Gupta, A. K.; Greenleaf, J. E. Arsenic Removal Using a Polymeric/ Inorganic Hybrid Sorbent. *Water Res.* **2003**, *37*, 164–176.
10. Parhi, P. K. Supported Liquid Membrane Principle and Its Practices: A Short Review. *J. Chem.* **2012**, *2013*, 618236.
11. Richards, L. A.; Richards, B. S.; Schafer, A. I. Renewable Energy Powered Membrane Technology: Salt and Inorganic Contaminant Removal by Nanofiltration/Reverse Osmosis. *J. Membr. Sci.* **2011**, *369*, 188–195.
12. Koros, W. J. Evolving Beyond the Thermal Age of Separation Processes: Membranes can Lead the Way. *AIChE J.* **2004**, *50*, 2326–2334.
13. Dongre, R. S.; Sadasivuni, K. K.; Deshmukh, K.; Mehta, A.; Basu, S.; Meshram, J. S.; Al-Maadeed, M. A. A.; Karim, A. Natural Polymer Based Composite Membranes for Water Purification: A Review. *Polym. Plast. Technol. Mat.* **2019**, *58*, 1295–1310.
14. Strathmann, H. Membrane Separation Processes: Current Relevance and Future Opportunities. *AIChE J.* **2001**, *47*, 1077–1087.
15. Castro-Muñoz, R.; Yáñez-Fernández, J.; Fíla, V. Phenolic Compounds Recovered from Agro-Food By-products Using Membrane Technologies: An Overview. *Food Chem.* **2016**, *213*, 753–762.
16. Van der Bruggen, B.; Curcio, E.; Drioli, E. Process Intensification in the Textile Industry: The Role of Membrane Technology. *J. Environ. Manage.* **2004**, *73*, 267–274.
17. Alzahrani, S.; Wahab, A. Challenges and Trends in Membrane Technology Implementation for Produced Water Treatment: A Review. *J. Water Process Eng.* **2014**, *4*, 107–133.
18. Kim, J.; van der Bruggen, B. The Use of Nanoparticles in Polymeric and Ceramic Membrane Structures: Review of Manufacturing Procedures and Performance Improvement for Water Treatment. *Environ. Pollut.* **2010**, *158*, 2335–2349.

19. Castro-Muñoz, R.; Barragán-Huerta, B. E.; Fíla, V.; Denis, P. C.; Ruby-Figueroa, R. Current Role of Membrane Technology: From the Treatment of Agro-Industrial By-products Up to the Valorization of Valuable Compounds. *Waste Biomass Valori.* **2018**, *9*, 513–529.

20. Van der Bruggen, B.; Lejon, L.; Vandecasteele, C. Reuse, Treatment, and Discharge of the Concentrate of Pressure-Driven Membrane Processes. *Environ. Sci. Technol.* **2003**, *37*, 3733–3738.

21. Lalia, B. S.; Kochkodan, V.; Hashaikeh, R.; Hilal, N. A Review on Membrane Fabrication: Structure, Properties and Performance Relationship. *Desalination* **2013**, *326*, 77–95.

22. Ursino, C.; Castro-Muñoz, R.; Drioli, E.; Gzara, L.; Albeirutty M. H.; Figoli, A. Progress of Nanocomposite Membranes for Water Treatment. *Membranes* **2018**, *8*, 18.

23. Azizi Samir, M. A. S.; Alloin, F.; Dufresne, A. Review of Recent Research Intocellulosic Whiskers, Their Properties and Their Application in Nanocomposite Field. *Biomacromolecules* **2005**, *6*, 612–626.

24. Jayakumar, R.; Prabaharan, M.; Sudheesh Kumar, P. T.; Nair, S. V.; Furuike, T.; Tamura, H. Novel Chitin and Chitosan Materials in Wound Dressing. In *Biomedical Engineering, Trends in Materials Science*; Laskovski, A., Ed.; Intech Publisher: Croatia, 2011; pp 1–24.

25. Muzzarelli, R. Nanochitins and Nanochitosans: Paving the Way to Eco-friendly and Energy-Saving Exploitation of Marine Resources. *Polym. Sci.* **2012**, *10*, 153–164.

26. Rajawat, D. S.; Kardam, A.; Srivastava, S.; Satsangee, S. P.; Nanocellulosic Fiber-Modified Carbon Paste Electrode for Ultra-trace Determination of Cd(II)and Pb(II) in Aqueous Solution. *Environ. Sci. Pollut. Res.* **2013**, *20*, 3068–3076.

27. Wang, X.; Ba, X.; Cui, N.; Ma, Z.; Wang, L.; Wang, Z.; Gao, X. Preparation, Characterisation, and Desalination Performance Study of Cellulose Acetate Membranes with MIL-53(Fe) additive. *J. Membr. Sci.* **2019**, *590*, 117057.

28. Du, Y.; Li, Y.; Wu, T. A Superhydrophilic and Underwater Superoleophobic Chitosan–TiO_2 Composite Membrane for Fast Oil-in-Water Emulsion Separation. *RSC Adv.* **2017**, *7*, 41838–41846.

29. https://www.smartmembranesolutions.co.nz/membrane-classifications/

30. Yangr, T.; Zall, R. Chitosan Membranes for Reverse Osmosis Application. *J. Food Sci.* **2006**, *49*, 91–93.

31. El-Gendi, A.; Abdallah, H.; Amin, A.; Amin, S. K. Investigation of Polyvinylchloride and Cellulose Acetate Blend Membranes for Desalination. *J. Mol. Struct.* **2017**, *1146*, 14–22.

32. Zhang, L.; Hsieh, Y. L. Ultrafine Cellulose Acetate Fibres with Nanoscale Structural Features. *J. Nanosci. Nanotechnol.* **2008**, *8*, 4461–4469.

33. Pagidi, A.; Thuyavan, Y. L.; Arthanareeswaran, G.; Ismail, A. F.; Jaafar, J.; Diby Paul, D. Polymeric Membrane Modification Using SPEEK and Bentonite for Ultrafiltration of Dairy Wastewater. *J. Appl. Polym. Sci.* **2015**, *132*, 41651.

34. Kiran, S. A.; Arthanareeswaran, G.; Thuyavan, Y. L.; Ismail A. F. Influence of Bentonite in Polymer Membranes for Effective Treatment of Car Wash Effluent to Protect the Ecosystem. *Ecotoxicol. Environ. Safety* **2015**, *121*, 186–192.

35. Zirehpour, A.; Rahimpour, A.; Seyedpour, F.; Seyedpour, F.; Jahanshahi, M. Developing New CTA/CA-Based Membrane Containing Hydrophilic Nanoparticles to Enhance the Forward Osmosis Desalination. *Desalination* **2015**, *371*, 46–57.

36. Khulbe, K. C.; Matsuura, T. Removal of Heavy Metals and Pollutants by Membrane Adsorption Techniques. *Appl. Water Sci.* **2018**, *8*, 19.

37. Zhang, L.; Bai, R. Novel Multi-Functional Membrane Technology for Visual Detection and Enhanced Adsorptive Removal of Lead Ions in Water and Wastewater. *Water Sci. Technol.* **2011,** *11,* 113–120.

38. Guibal, E. Interactions of Metal Ions with Chitosan-Based Sorbents: A Review. *Sep. Purif. Technol.* **2004,** *38,* 43–74.

39. Gholami, A.; Moghadassi, A. R.; Hosseini, S. M.; Shabani, S.; Gholami F. Preparation and Characterization of Polyvinyl Chloride-Based Nanocomposite Nanofiltration-Membrane Modified by Iron Oxide Nanoparticles for Lead Removal from Water. *J. Ind. Eng. Chem.* **2014,** *20,* 1517–1522.

40. Rodriguez, R. V.; Montero-Caberera, M. E.; Esparza-Ponce, H. E.; Herrera-Peraza, F.; Ballinas-Casarrubias, M. L. Uranium Removal from Water Using Cellulose Triacetate Membranes Added with Activated Carbon. *Appl. Rad. Isotopes* **2012,** *70,* 872–881.

41. Shenvi, S. S.; Isloor, A. M.; Ismail, A. F.; Shilton, S. J.; Ahmed, A. A. Humic Acid Based Biopolymeric Membrane for Effective Removal of Methylene Blue and Rhodamine B. *Ind. Eng. Chem. Res.* **2015,** *54,* 4965–4975.

42. Saravanan, R.; Gupta, V. K.; Mosquera, E.; Gracia, F. Preparation and Characterization of V_2O_5/ZnO Nanocomposite System for Photocatalytic Application. *J. Mol. Liq.* **2014,** *198,* 409–412.

43. Rajendran, S.; Khan, M. M.; Gracia, F.; Qin, J.; Gupta, V. K.; Arumainathan, S. Ce(3+)-Ion-Induced Visible-Light Photocatalytic Degradation and Electrochemical Activity of ZnO/CeO_2 Nanocomposite. *Sci. Rep.* **2016,** *6,* 3164.

44. Saravanan, R.; Gupta, V. K.; Mosquera, E.; Gracia, F.; Narayanan, V.; Stephen, A. Visible Light Induced Degradation of Methyl Orange Using β-$Ag_{0.333}V_2O_5$ Nanorod Catalysts by Facile Thermal Decomposition Method. *J. Saudi Chem. Soc.* **2015,** *19,* 521–527.

45. Saravanan, R.; Sacari, E.; Gracia, F.; Khan, M. M.; Mosquera, E.; Gupta, V. K. Conducting PANI Stimulated ZnO System for Visible Light Photocatalytic Degradation of Coloured Dyes. *J. Mol. Liq.* **2016,** *221,* 1029–1033.

46. Galiano, F.; Castro-Muñoz, R.; Mancuso, R.; Gabriele, B.; Figol, A. Membrane Technology in Catalytic Carbonylation Reactions. *Catalysts* **2019,** *9,* 614.

47. Zheng, X.; Shen, Z.-P.; Shi, L.; Cheng, R.; Yuan, D.-Y. Photocatalytic Membrane Reactors (PMRs) in Water Treatment: Configurations and Influencing Factors. *Catalysts* **2017,** *7,* 224.

48. Molinari, R.; Lavorato, C.; Argurio, P. Recent Progress of Photocatalytic Membrane Reactors in Water Treatment and in Synthesis of Organic Compounds. A Review. *Catal. Today* **2017,** *281,* 144–164.

49. Dioos, B. M. L.; Vankelecom, I. F. J.; Jacobs, P. A. Aspects of Immobilisation of Catalysts on Polymeric Supports. *Adv. Synth. Catal.* **2006,** *348,* 1413–1446.

50. Molinari, R.; Lavorato, C.; Argurio, P. Photocatalytic Reduction of Acetophenone in Membrane Reactors under UV and Visible Light Using TiO_2 and Pd/TiO_2 Catalysts. *Chem. Eng. J.* **2015,** *274,* 307–316.

51. Mozia, S. Photocatalytic Membrane Reactors (PMRs) in Water and Wastewater Treatment. A Review. *Sep. Purif. Technol.* **2010,** *73,* 71–91.

52. Lavorato, C.; Argurio, P.; Mastropietro, T. F.; Pirri, G.; Poerio, T.; Molinari, R. Pd/TiO_2 Doped Faujasite Photocatalysts for Acetophenone Transfer Hydrogenation in a Photocatalytic Membrane Reactor. *J. Catal.* **2017,** *353,* 152–161.

53. Hairom, N. H. H.; Mohammad, A. W.; Kadhum, A. A. H. Effect of Various Zinc Oxide Nanoparticles in Membrane Photocatalytic Reactor for Congo Red Dye Treatment. *Sep. Purif. Technol.* **2014,** *137,* 74–81.

54. Argurio, P.; Fontananova, E.; Molinari, R.; Drioli, E. Review Photocatalytic Membranes in Photocatalytic Membrane Reactors. *Processes* **2018,** *6,* 162.

55. Waheed, S. S.; Ahmad, A. W.; Khan, S. M.; Hussain, M.; Jamil, T.; Zuber, M. Synthesis, Characterization and Permeation Performance of Cellulose Acetate/Polyethylene glycol-600 Membranes Loaded with Silver Particles for Ultra Low Pressure Reverse Osmosis. *J. Taiwan Inst. Chem. Eng.* **2015,** *57,* 129–138.

56. Wang, X.; Yang, J.; Zhu, M.; Li, F. Characterization and Regeneration of Pd/Fe Nanoparticles Immobilized in Modified PVDF Membrane. *J. Taiwan Inst. Chem. Eng.* **2013,** *44,* 386–392.

57. Gupta, V. K.; Jain, R.; Mittal, A.; Saleh, T. A.; Nayak, A.; Agarwal, S.; Sikarwar, S. Photo-Catalytic Degradation of Toxic Dye Amaranth on TiO_2/UV in Aqueous Suspensions. *Mater. Sci. Eng. C Mater.* **2012,** *32,* 12–17.

58. Saleh, T. A.; Gupta, V. K. Photo-Catalyzed Degradation of Hazardous Dye Methyl Orange by Use of a Composite Catalyst Consisting of Multi-walled Carbon Nanotubes and Titanium Dioxide. *J Colloid Interface Sci.* **2012,** *371,* 101–106.

59. Schüler, E.; Gustavsson, A. K.; Hertenberger, S.; Sattler, K. Solar Photocatalytic and Electrokinetic Studies of TiO_2/Ag Nanoparticle Suspensions. *Sol. Energy* **2013,** *96,* 220–226.

60. Kim, M.; Kim, Y. K.; Lim, S. K.; Kim, S.; Ina, S. I. Efficient Visible Light-Induced H_2 Production by Au@CdS/TiO_2 Nanofibers: Synergistic Effect of Core–Shell Structured Au@CdS and Densely Packed TiO_2 Nanoparticles. *Appl. Catal. B* **2015,** *166,* 423–431.

61. Zhou, W.; Liu, H.; Wang, J.; Liu, D.; Du, G.; Han, S.; Lin, J.; Wang, R. Interface Dominated High Photocatalytic Properties of Electrostatic Self-Assembled Ag_2O/TiO_2 Heterostructure. *Phys. Chem. Chem. Phys.* **2010,** *12,* 15119–15123.

62. Lichtenberger, J.; Amiridis, M. D. Catalytic Oxidation of Chlorinated Benzenes Over V_2O_5/TiO_2 catalysts. *J. Catal.* **2004,** *223,* 296–308.

63. Xu, M.; Han, L.; Dong, S. Facile Fabrication of Highly Efficient g-C_3N_4/Ag_2O Heterostructured Photocatalysts with Enhanced Visible-Light Photocatalytic Activity. *ACS Appl. Mater. Interfaces* **2013,** *5,* 12533–12540.

64. Liu, R.; Wang, P.; Wang, X.; Yu, H.; Yu, J. UV- and Visible-Light Photocatalytic Activity of Simultaneously Deposited and Doped Ag/Ag(I)–TiO_2 Photocatalyst. *J. Phys. Chem. C* **2012,** *116,* 17721–17728.

65. Tran, C. D.; Duri, S.; Delneri, A.; Franko, M. Chitosan-Cellulose Composite Materials: Preparation, Characterization and Application for Removal of Microcystin. *J. Hazard. Mater.* **2013,** *252,* 355–366.

66. DelaiáSun, D. Facile Fabrication of Porous Chitosan/TiO_2/Fe_3O_4 Microspheres with Multifunction for Water Purifications. *New J. Chem.* **2011,** *35,* 137–140.

67. Bai, H.; Liu, Z.; Sun, D. D. A Hierarchically Structured and Multifunctional Membrane for Water Treatment. *Appl. Catal. B* **2012,** *111–112,* 571–577.

68. Taha, A. A.; Wu, Y. N.; Wang, H.; Li, F. Preparation and Application of Functionalized Cellulose Acetate/Silica Composite Nanofibrous Membrane via Electrospinning for Cr(VI) Ion Removal from Aqueous Solution. *J. Environ. Manage.* **2012,** *112,* 10–16.

69. Meng, J.; Zhang, X.; Ni, L.; Tang, Z.; Zhang, Y.; Zhang, Y.; Zhang, W. Antibacterial Cellulose Membrane via One-Step Covalent Immobilization of Ammonium/Amine Groups. *Desalination* **2015,** *359,* 156–166.

70. Zahid, M.; Rashid, A.; Akram, S.; Rehan, Z. A.; Razzaq, W. A Comprehensive Review on Polymeric Nano-Composite Membranes for Water Treatment. *J. Membr. Sci. Technol.* **2018,** *8,* 1000179.

71. Al-Amoudi, A. S.; Farooqe, A. M. Performance Restoration and Autopsy of NF Membranes Used in Seawater Pretreatment. *J. Desal.* **2005,** *178,* 261–271.

72. Meng, F.; Chae, S. R.; Drews, A.; Kraume, M.; Shin, H. S.; Yang, F. Recent Advances in Membrane Bioreactors (MBRs): Membrane Fouling and Membrane Material. *Water Res.* **2009,** *43,* 1489–1512.

73. Nilson, J. A.; DiGiano, F. A. Influence of NOM Composition on Nanofiltration. *Am. Water Works Assoc. J.* **1996,** *88,* 53–66.

74. Vrijenhoek, E. M.; Hong, S.; Elimelech, M. Influence of Membrane Surface Properties on Initial Rate of Colloidal Fouling of Reverse Osmosis and Nanofiltration Membranes. *J. Membr. Sci.* **2001,** *188,* 115–128.

75. Misdan, N.; Ismail, A. F.; Hilal, N. Recent Advances in the Development of (bio) Fouling Resistant Thin Film Composite Membranes for Desalination. *J. Desal.* **2016,** *380,* 105–111.

76. Kochkodan, V.; Hilal, N. A Comprehensive Review on Surface Modified Polymer Membranes for Biofouling Mitigation. *J. Desal.* **2015,** *356,* 187–207.

77. Rezakazemi, M.; Dashti, A.; Harami, H. R.; Hajilari, N. Inamuddin, Fouling-Resistant Membranes for Water Reuse. *Environ. Chem. Lett.* **2018,** *16,* 715–763.

78. Ben-Sasson, M.; Zodrow, K. R.; Genggeng, Q.; Kang, Y.; Giannelis, E. P.; Elimelech, M. Surface Functionalization of Thin-Film Composite Membranes with Copper Nanoparticles for Antimicrobial Surface Properties. *Environ. Sci. Technol.* **2013,** *48,* 384–393.

79. Khan, S. B., Alamry, K. A., Bifari, E. N., Asiri, A. M.; Yasir, M.; Gzara, L.; Ahmad, R. Z. Assessment of Antibacterial Cellulose Nanocomposites for Water Permeability and Salt Rejection. *J. Ind. Eng.* **2015,** *24,* 266–275.

80. Abedini, R.; Mousavi, S. M.; Aminzadeh, R. A Novel Cellulose Acetate (CA) Membrane Using TiO$_2$ Nanoparticles: Preparation, Characterization and Permeation Study. *Desalination* **2011,** *277,* 40–45.

81. Zeng, J.; Liu, S.; Zhang, L. TiO$_2$ immobilized in cellulose matrix for photocatalytic degradation of phenol under weak UV light irradiation. *J. Phys. Chem. C* **2010,** *114,* 7806–7811.

82. Hiremath, L.; Gupta, P. K.; Kumar, S. N.; Kumar Srivastava, A.; Narayan A. V.; Rajendran, S.; Ravi, K. M.; Narasanagi, M.; Sukanya, P. Functionalized Cellulose Nanofiber Composite Membranes for Waste Water Treatment—A Review. *J. Nanotechnol. Mat. Sci.* **2018,** *5,* 35–43.

83. Demirkol, G. T.; Dizge, N.; Acar, T. O.; Salmanli, O. M.; Tufekci, N. Influence of Nanoparticles on Filterability of Fruit-Juice Industry Wastewater Using Submerged Membrane Bioreactor. *Water Sci. Technol.* **2017,** *76,* 705–711.

84. Lee, J.; Chae, H. R.; Won, Y. J.; Lee, K.; Lee, C. H.; Lee, H. H.; Kim, I. C.; Lee, J. Graphene Oxide Nanoplatelets Composite Membrane with Hydrophilic and Antifouling Properties for Wastewater Treatment. *J. Membr. Sci.* **2013,** *448,* 223–230.

85. Xu, Z.; Zhang, J.; Shan, M.; Li, Y.; Li, B.; Niu, J.; Zhou, J.; Qian, X. Organosilane-Functionalized Graphene Oxide for Enhanced Antifouling and Mechanical Properties of Polyvinylidene Fluoride Ultrafiltration Membranes. *J. Membr. Sci.* **2014,** *458,* 1–13.

86. Ketola, A. Determination of Surfactants in Industrial Waters of Paper and Board Mills, Thesis, Huss Group, February 2016.

87. Yu, Z.; Liu, X.; Zhao, F.; Liang, X.; Tian, Y. Fabrication of a Low-Cost Nano-SiO$_2$/PVC Composite Ultrafiltration Membrane and Its Antifouling Performance. *J. Appl. Polym. Sci.* **2015,** *132,* 41267.

88. Van der Bruggen, B.; Vandecasteele, C.; Van Gestel, T.; Doyen, W.; Leysen, R. A Review of Pressure-Driven Membrane Processes in Wastewater Treatment and Drinking Water Production. *Environ. Prog.* **2003,** *22,* 46–56.

89. Wilf, M.; Tech, T. Membrane Types and Factors Affecting Membrane Performance. *Advance Membrane Technology*; Stanford University: Stanford, 2008; pp. 92.

90. Voicu, Ş. I.; Miculescu, M.; Ninciuleanu, C. M. Cellulose Acetate Membranes with Controlled Porosity and Their Use for the Separation of Amino Acids and Proteins. *J. Optoelectron. Adv. Mater.* **2014,** *16,* 903–908.

91. Gu, Z. Y.; Xue, P. H.; Li, W. J. Preparation of the Porous Chitosan-Membrane by Cryogenic Induced Phase Separation. *Polym. Adv. Technol.* **2001,** *12,* 665–669.

92. Holloway, R. W.; Achilli, A.; Cath, T. Y. The Osmotic Membrane Bioreactor: A Critical Review. *Environ. Sci. Water Res. Technol.* **2015,** *1,* 581–605.

93. Shaulsky, E.; Karanikola, V.; Straub, A. P.; Deshmukh, A.; Zucker, I.; Elimelech, M. Asymmetric Membranes for Membrane Distillation and Thermo-osmotic Energy Conversion. *Desalination* **2019,** *452,* 141–148.

94. Baker, R. W. *Membrane Technology and Applications*; Wiley-Blackwell: Hoboken, NJ, USA, 2012.

95. Chou, W.-L.; Yu, D.-G.; Yang, M.-C. The Preparation and Characterization of Silver-Loading Cellulose Acetate Hollow Fiber Membrane for Water Treatment. *Polym. Adv. Technol.* **2005,** *16,* 600–607.

96. Ye, S. H.; Watanabe, J.; Iwasaki, Y.; Ishihara, K. In Situ Modification on Cellulose Acetate Hollow Fiber Membrane Modified with Phospholipid Polymer for Biomedical Application. *J. Membr. Sci.* **2005,** *249,* 133–141.

97. Sudarshan, N. R.; Hoover, D. G.; Knorr, D. Antibacterial Action of Chitosan. *Food Biotechnol.* **1992,** *6,* 257–272.

98. Raafat, D.; von Bargen, K.; Haas, A.; Sahl, H. G. Insights into the Mode of Action of Chitosan as an Antibacterial Compound. *Appl. Environ. Microbiol.* **2008,** *74,* 3764–3773.

99. Sekiguchi, S.; Miura, Y.; Kaneko, H.; Nishimura, S. I.; Nishi, N.; Iwase, M.; Tokura, S. Molecular Weight Dependency of Antimicrobial Activity by Chitosan Oligomers. In *Food Hydrocolloids: Structures, Properties and Functions*; Nishinari, K., Doi, E., Eds.; Plenum Press: New York, 1994.

100. Tsai, G. J.; Su, W. H. Antibacterial Activity of Shrimp Chitosan Against *Escherichia coli. J. Food. Prot.* **1999,** *62,* 239–243.

101. Shahidi, F.; Arachchi, J.; Jeon, Y. J. Food Applications of Chitin and Chitosans. *Trends Food Sci. Technol.* **1999,** *10,* 37–51.

102. Devlieghere, F.; Vermeulen, A.; Debevere, J. Chitosan: Antimicrobial Activity, Interactions with Food Components and Applicability as a Coating on Fruit and Vegetables. *Food Microbiol.* **2004,** *21,* 703–714.

103. Fisher, J. F.; Meroueh, S. O.; Mobashery, S. Nanomolecular and Supramolecular Paths Toward Peptidoglycan Structure. *Microbe* **2006,** *1,* 420–427.

104. Lührmann, A.; Haas, A. A Method to Purify Bacteria-Containing Phagosomes from Infected Macrophages. *Methods Cell Sci.* **2000,** *22,* 329–341.

105. Dodane, V.; Vilivalam, V. D. Pharmaceutical Applications of Chitosan. *Pharm. Sci. Technol. Today* **1998,** *1,* 246–253.

106. Chung, Y. C.; Chen, C. Y. Antibacterial Characteristics and Activity of Acid-Soluble Chitosan. *Bioresour. Technol.* **2008,** *99,* 2806–2814.

107. Prescott, L. M.; Harley, J. P.; Klein, D. A. *Microbiology*; McGraw-Hill Co.: New York, 2002.

108. Dutta, P. K.; Tripath, S.; Mehrotra, G. K.; Dutta, J. Perspective for Chitosan Based Antimicrobial Films in Food Applications. *Food Chem.* **2009**, *114*, 1173–1182.

109. Simunek, J.; Tishchenko, G.; Hodrová, B.; Bartonová, H. Chitinolytic Activities of *Clostridium* sp. JM₂ Isolated from Stool of Human Administered Per Orally by Chitosan. *Folia Mocrobiol.* **2006**, *51*, 306–308.

110. Másson, M.; Holappa, J.; Hjálmarsdóttir, M.; Rúnarsson, Ö. V.; Nevalainen, T.; Järvinen, T. Antimicrobial Activity of Piperazine Derivatives of Chitosan. *Carbohydr. Polym.* **2008**, *74*, 566–571.

111. Sebti, I.; Martial-Gros, A.; Carnet-Pantiez, A.; Grelier, S.; Coma, V. Chitosan Polymer as Bioactive Coating and Film Against *Aspergillus niger* Contamination. *J. Food Sci.* **2005**, *70*, M100–M104.

112. Helander, I. M.; Nurmiaho-Lassila, E. L.; Ahvenainen, R.; Rhoades, J.; Roller, S. Chitosan Disrupts the Barrier Properties of the Outer Membrane of Gram-Negative Bacteria. *Int. J. Food Microbiol.* **2001**, *30*, 235–244.

113. Roller, S.; Covill, N. The Antifungal Properties of Chitosan in Laboratory Media and Apple Juice. *Int. J. Food Microbiol.* **1999**, *47*, 67–77.

114. Kumar, A. B. V.; Varadaraj, M. C.; Gowda, L. R.; Tharanathan, R. N. Characterization of Chito-Oligosaccharides Prepared by Chitosanolysis with the Aid of Papain and Pronase, and Their Bactericidal Action against *Bacillus cereus* and *Escherichia coli. Biochem. J.* **2005**, *391*, 167–175.

115. Goy, R. C.; Britto, D.; Assis, O. B. G. A Review of the Antimicrobial Activity of Chitosan. *Polímeros* **2009**, *19*, 241–247.

116. Weng, R.; Chen, L.; Lin, S.; Zhang, H.; Wu, H.; Liu, K.; Cao, S.; Huang, L. Preparation and Characterization of Antibacterial Cellulose/Chitosan Nanofiltration Membranes. *Polymers* **2017**, *9*, 116.

117. Isik, Z.; Unyayar, A.; Dizge, N. Filtration and Antibacterial Properties of Bacterial Cellulose Membranes for Textile Wastewater Treatment. *Avicenna J. Environ. Health Eng.* **2018**, *5*, 106–114.

118. Dizge, N.; Ozay, Y.; Bulut Simsek, U.; Elif Gulsen, H.; Akarsu, C.; Turabik, M.; Unyayar, A.; Ocakoglu, K. Preparation, Characterization and Comparison of Antibacterial Property of Polyethersulfone Composite Membrane Containing Zerovalent Iron or Magnetite Nanoparticles. *Membr. Water Treat.* **2017**, *8*, 51–71.

119. Reddy, D. H. K.; Lee, S. M. Water Pollution and Treatment Technologies. *J. Environ. Anal. Toxicol.* **2012**, *2*, 1000e103.

120. Hahn, T.; Zibek, S. Sewage Polluted Water Treatment via Chitosan: A Review. In *Chitin-Chitosan: Myriad Functionalities in Science and Technology*; Dongre, R., Ed.; Intech Publisher: Croatia, 2018; pp 119–142.

121. Mohammed, N.; Grishkewich, N.; Tam, K. C. Cellulose Nanomaterials: Promising Sustainable Nanomaterials for Application in Water/Wastewater Treatment Processes. *Environ. Sci. Nano* **2018**, *5*, 623–658.

122. Ramazani, A.; Oveisi, M.; Sheikhi, M.; Gouranlou, F. Natural Polymers as Environmental Friendly Adsorbents for Organic Pollutants Such as Dyes Removal from Colored Wastewater. *Curr. Org. Chem.* **2018**, *22*, 1297–1306.

123. Ali, I.; Peng, C.; Naz, I.; Lin, D.; Saroj, D. P.; Alif, M. Development and Application of Novel Biomagnetic Membrane Capsules for the Removal of the Cationic Dye Malachite Green in Wastewater Treatment. *RSC Adv.* **2019**, *9*, 3625–3646.

124. Nasreen, S. A. A. N.; Sundarrajan, S.; Nizar, S. A. S.; Ramakrishna, S. Nanomaterials: Solutions to Water-Concomitant Challenges. *Membranes* **2019**, *9*, 40.

125. Thakur, V. K.; Voicu, S. I. Recent Advances in Cellulose and Chitosan Based Membranes for Water Purification: A Concise Review. *Carbohydr. Polym.* **2016**, *146*, 148–165.

126. Gorgieva, S.; Vogrinčič, R.; Koko, V. The Effect of Membrane Structure Prepared from Carboxymethyl Cellulose and Cellulose Nanofibrils for Cationic Dye Removal. *J. Polym. Environ.* **2019**, *27*, 318–332.

127. Peng, X. W.; Yang. Y. X. Application of Membrane Separation Technology in the Treatment of Priting and Dyeing Wastewater. *Jiangxi Chem. Ind.* **2003**, *1*, 21–23.

128. Chen, L.; Chen, D. H. Application of Chitosan Coagulant to Water Treatment. *Ind. Water Treat.* **2000**, *20*, 4–7.

129. Huang, C. D.; Yang, X. K.; Wang, K. M. *Dyeing Wastewater Treating*; Textile Industry Press: Beijing, 1987.

130. Long, J. J; Lu, T. Q. Application Progress of Chitin/Chitosan and Their Derivatives for Processing Dye Wastewater. *J. Donghua Univ. Nat Sci.* **2003**, *29*, 117–120.

131. Zhang, Q. H.; Tang, X. Q. Treatment of Dyeing Wastewater with Carboxymethyl Chitosan. *Environ. Pollut. Control* **1995**, *17*, 7–9.

132. Chen, X.; Sun, H.; Pan, J. Decolorization of Dyeing Wastewater with Use of Chitosan Materials Ocean. *Sci. J.* **2006**, *41*, 221–226.

133. Barbara, K. Membrane-Based Processes Performed with Use of Chitin/Chitosan Materials. *Sep. Purif. Technol.* **2005**, *31*, 305–312.

134. Wu, F. C.; Tseng, R. L.; Juang, R. S. Kinetic Modeling of Liquid-Phase Adsorption of Reactive Dyes and Metal Ions on Chitosan. *Water Res.* **2001**, *35*, 613–618.

135. Wu, F. C.; Tseng, R. L.; Juang, R. S. Enhanced Abilities of Highly Swollen Chitosan Beads for Color Removal and Tyrosinase Immobilization. *J. Hazard. Mater.* **2001**, *81*, 167–177.

136. Wu, F. C.; Tseng, R. L.; Juang, R. S. Comparative Adsorption of Metal and Dye on Flake-and Bead-Types of Chitosans Prepared from Fishery Wastes. *J. Hazard. Mater.* **2000**, *73*, 63–75.

137. Crini, G. Recent Developments in Polysaccharide-Based Materials used as Adsorbents in Wastewater Treatment. *Prog. Polym. Sci.* **2005**, *30*, 38–70.

138. Gibbs, G.; Tobin, J. M.; Guibal, E. Sorption of Acid Green 25 on Chitosan: Influence of Experimental Parameters on Uptake Kinetics and Sorption Isotherms. *J. Appl. Polym. Sci.* **2003**, *90*, 1073–1080.

139. Gibbs, G.; Tobin, J. M.; Guibal, E. Influence of Chitosan Preprotonation on Reactive Black 5 Sorption Isotherms and Kinetics. *Ind. Eng. Chem. Res.* **2004**, *43*, 1–11.

140. Guibal, E.; McCarrick, P.; Tobin, J. M. Comparison of the Sorption of Anionic Dyes on Activated Carbon and Chitosan Derivatives from Dilute Solutions. *Sep. Sci. Technol.* **2003**, *38*, 3049–3073.

141. Guibal, E.; Touraud, E.; Roussy, J. Chitosan Interactions with Metal Ions and Dyes: Dissolved-State *vs.* Solid-State Application. *World J. Microbiol. Biotechnol.* **2005**, *21*, 913–920.

142. Yu, S. F.; Ye, J. Z.; Lang, X. M. Preparation of Chitosan Activated Carbon Ultrafiltrating Compound Membrane. *Technol. Water Treat.* **1999**, *25*, 255–258.

143. Puspasari, T.; Pradeep, N.; Peinemann, K. V. Crosslinked Cellulose Thin Film Composite Nanofiltration Membranes with Zero Salt Rejection. *J. Membr. Sci.* **2015**, *491*, 132–137.

144. Puspasari, T.; Peinemann, K. V. Application of Thin Film Cellulose Composite Membrane for Dye Wastewater Reuse. *J. Water Proc. Eng.* **2016**, *13*, 176–182.

145. Benosmane, N.; Boutemeur, B.; Hamdi, S. M.; Hamdi, M. The Removal of Phenol from Synthetic Wastewater using Calix[4]resorcinarene Derivative Based Polymer Inclusion Membrane. *Algerian J. Environ. Sci. Technol.* **2016,** *2,* 26–33.
146. Benosmane, N.; Boutemeur, B.; Hamdi, S. M.; Hamdi, M. Removal of Phenol from Aqueous Solution Using Polymer Inclusion Membrane Based on Mixture of CTA and CA. *Appl. Water Sci.* **2018,** *8,* 17.
147. Karim, Z.; Mathew, A. P.; Grahn, M.; Mouzon, J.; Oksman, K. Nanoporous Membranes with Cellulose Nanocrystals as Functional Entity in Chitosan: Removal of Dyes from Water. *Carbohydr. Polym.* **2014,** *112,* 668–676.
148. Feng, B. L.; Ye, J. Z.; Lang, X. M. Preparation of Chitosan Ultrafiltration Membrane and its Application to Treat Printing and Dyeing Wastewater. *Ind. Water Treat.* **1998,** *18,* 16–18.
149. Dassanayake, R. S.; Acharya, S.; Abidi, N. Biopolymer-Based Materials from polysaccharides: Properties, Processing, Characterization and Sorption Applications. In *Advanced Sorption Process Applications*; Edebali, S., Ed.; Intech Publisher: Croatia, 2018; pp 1–24.
150. Gebru, K. A.; Das, C. Removal of Pb(II) and Cu(II) Ions From Wastewater Using Composite Electrospun Cellulose Acetate/Titanium Oxide (TiO$_2$) Adsorbent. *J. Water Process. Eng.* **2017,** *16,* 1–13.
151. Gebru, K. A.; Das, C. Removal of Chromium(VI) Ions from Aqueous Solutions Using Amine-Impregnated TiO$_2$ Nanoparticles Modified Cellulose Acetate Membranes. *Chemosphere* **2018,** *191,* 673–684.
152. Carolin, C. F.; Kumar, P. S.; Saravanan, A.; Joshiba, G. J.; Naushad, Mu.; et al. Efficient Techniques for the Removal of Toxic Heavy Metals from Aquatic Environment: A Review. *J. Environ. Chem. Eng.* **2017,** *5,* 2782–2799.
153. Ahmad, M.; Ahmad, S.; Swami, B. L.; Ikram, S. Adsorption of Heavy Metal Ions: Role of Chitosan and Cellulose for Water Treatment. *Int. J. Pharmacogn.* **2015,** *2,* 280–289.
154. Zia, Q.; Tabassum, M.; Gong, H.; Li, J. A Review on Chitosan for the Removal of Heavy Metals Ions. *J. Fiber Bioeng. Informat.* **2019,** *12,* 103–128.
155. Aliabadi, M.; Irani, M.; Ismaeili, J.; Najafzadeh, S. Design and Evaluation of Chitosan/Hydroxyapatite Composite Nanofiber Membrane for the Removal of Heavy Metal Ions from Aqueous Solution. *J. Taiwan Inst. Chem. E* **2014,** *45,* 518–526.
156. Vijaya, Y.; Popuri, S. R.; Boddu, V. M.; Krishnaiah, A. Modified Chitosan and Calcium Alginate Biopolymer Sorbents for Removal of Nickel(II) through Adsorption. *Carbohydr. Polym.* **2008,** *72,* 261–271.
157. Benavente, M.; Moreno, L.; Martinez, J. Sorption of Heavy Metals from Gold Mining Wastewater Using Chitosan. *J. Taiwan Inst. Chem. E* **2011,** *42,* 976–988.
158. Ghaee, A.; Shariaty-Niassar, M.; Barzin, J.; Zarghan, A. Adsorption of Copper and Nickel Ions on Macroporous Chitosan Membrane: Equilibrium Study. *Appl. Sur. Sci.* **2012,** *258,* 7732–7743.
159. Naim, M. M.; Abdel Razek, H. E. M. Chelation and Permeation of Heavy Metals Using Affinity Membranes from Cellulose Acetate–Chitosan Blends. *Desalin. Water Treat.* **2013,** *51,* 644–657.
160. Steenkamp, G.; Neomagus, H.; Krieg, H.; Keizer, K. Centrifugal Casting of Ceramic Membrane Tubes and the Coating with Chitosan. *Sep. Purif. Technol.* **2001,** *25,* 407–413.
161. Sánchez, J.; Butter, B.; Chavez, S.; Riffo, L.; Basáez, L.; Rivas, B. L. Quaternized Hydroxyethyl Cellulose Ethoxylate and Membrane Separation Techniques for Arsenic Removal. *Desalin. Water Treat.* **2016,** *57,* 25161–25169.

162. Jana, S.; Saikia, A.; Purkait, M.; Mohanty, K. Chitosan Based Ceramic Ultrafiltration Membrane: Preparation, Characterization and Application to Remove Hg(II) and As(III) Using Polymer Enhanced Ultrafiltration. *Chem. Eng. J.* **2011**, *170*, 209–219.

163. Liu, C.; Bai, R. Adsorptive Removal of Copper Ions with Highly Porouschitosan/ Cellulose Acetate Blend Hollow Fiber Membranes. *J. Membr. Sci.* **2006**, *284*, 313–322.

164. Salehi, E.; Madaeni, S. Influence of Poly(ethylene glycol) Aspore-Generator on Morphology and Performance of Chitosan/Poly(vinylalcohol) Membrane Adsorbents. *Appl. Surf. Sci.* **2014**, *288*, 537–541.

165. Salehia, E.; Daraeib, P.; Shamsabadi, A. A. A Review on Chitosan-Based Adsorptive Membranes. *Carbohydr. Polym.* **2016**, *152*, 419–432.

166. Barakat, M. A.; Schmidt, E. Polymer-Enhanced Ultrafiltration Process for Heavy Metals Removal from Industrial Wastewater. *Desalination* **2010**, *256*, 90–93.

167. Petrov, S.; Nenov, V. Removal and Recovery of Copper from Wastewater by a Complexation-Ultrafiltration Process. *Desalination* **2004**, *162*, 201–209.

168. Holmgren, J.; Bourgeois, L.; Carlin, N.; Clements, J.; Gustafsson, B.; Lundgren, A., Nygren, E.; Tobias, J., Walker, R.; Svennerholm, A. M. Development and Preclinical Evaluation of Safety and Immunogenicity of an Oral ETEC Vaccine Containing Inactivated *E. coli* Bacteria Overexpressing Colonization Factors CFA/I, CS3, CS5 and CS6 Combined with a Hybrid LT/CT B Subunit Antigen, Administered Alone and Together with dmLT Adjuvant. *Vaccine* **2013**, *31*, 2457–2464.

169. UNICEF; World Health Organization. Progress on Sanitation and Drinking Water–2015 Update and MDG Assessment, United States of America, 2015.

170. World Health Organization. *Guidelines for Drinking-Water Quality*, 4th ed.; *Incorporating the First Addendum*; World Health Organization: Switzerland, 2017.

171. Lonsdale, H. K. The Growth of Membrane Technology. *J. Membr. Sci.* **1982**, *10*, 81–181.

172. Howe, K. J.; Clark, M. M. Fouling of Microfiltration and Ultrafiltration Membranes by Natural Waters. *Environ. Sci. Technol.* **2002**, *36*, 3571–3576.

173. Tripathi, B. P., Dubey, N. C.; Stamm, M. Functional Polyelectrolyte Multilayer Membranes for Water Purification Applications. *J. Hazard. Mater.* **2013**, *252–253*, 401–412.

174. Carpenter, A. W., de Lannoy, C. F. O.; Wiesner, M. R. Cellulose Nanomaterials in Water Treatment Technologies. *Environ. Sci. Technol.* **2015**, *49*, 5277–5287.

175. Voisin, H.; Bergström, L.; Liu, P.; Mathew, A. P. Nanocellulose-Based Materials for Water Purification. *Nanomaterials* **2017**, *7*, 57.

176. Dankovich, T. A.; Gray, D. G. Bactericidal Paper Impregnated with Silver Nanoparticles for Point-of-Use Water Treatment, *Environ. Sci. Technol.* **2011**, *45*, 1992–1998.

177. Thomas, S. F.; Rooks, P.; Rudin, F.; Atkinson, S.; Goddard, P.; Bransgrove, R.; Mason P. T.; Allen, M. J. The Bactericidal Effect of Dendritic Copper Microparticles, Contained in an Alginate Matrix, on *Escherichia coli*. *PLoS One* **2014**, *9*, e96225.

178. Peña-Gómez, N.; Ruiz-Rico, M.; Fernández-Segovia, I.; Barat, J. M. Development of Amino-Functionalized Membranes for Removal of Microorganism. *Innovative Food Sci. Emerging Technol.* **2018**, *48*, 75–82.

179. Hassan, M.; Abou-Zeid, R.; Hassan, E.; Berglund, L.; Aitomäki, Y.; Oksman, K. Membranes Based on Cellulose Nanofibers and Activated Carbon for Removal of *Escherichia coli* Bacteria from Water. *Polymers* **2017**, *9*, 335.

180. Rajendran, R.; Abirami, M.; Prabhavathi, P.; Premasudha, P.; Kanimozhi, B.; Manikandan, A. Biological Treatment of Drinking Water by Chitosan Based Nanocomposites. *Afr. J. Biotechnol.* **2015**, *14*, 930–936.

181. Wang, R.; Guan, S.; Sato, A.; Wang, X.; Wang, Z.; Yang, R.; Hsiao, B. S.; Chu, B. Nanofibrous Microfiltration Membranes Capable of Removing Bacteria, Viruses and Heavy Metal Ions. *J. Membr. Sci.* **2013,** *446,* 376–382.

182. Qin, C.; Li, H.; Xiao, Q.; Liu, Y.; Zhu, J.; Du, Y. Water Solubility of Chitosan and Its Antimicrobial Activity. *Carbohydr. Polym.* **2006,** *63,* 367–374.

183. Chopra, H.; Ruhi, G. Eco-friendly Chitosan: An Efficient Material for Water Purification. *J. Pharm. Innov.* **2016,** *5,* 92–95.

Index